European Handbook of Crowdsourced Geographic Information

Edited by
Cristina Capineri, Muki Haklay,
Haosheng Huang, Vyron Antoniou,
Juhani Kettunen, Frank Ostermann
and Ross Purves

]u[

ubiquity press
London

Published by
Ubiquity Press Ltd.
6 Windmill Street
London W1T 2JB
www.ubiquitypress.com

Cover design by Amber MacKay
Cover illustration by Rob Lemmens with Tagul.com
Word cloud based on keywords extracted from this book.
For more information, see: Lemmens, R, Falquet, G, Métral, C 2016 Towards
Linked Data and ontology development for the semantic enrichment of
volunteered geo-information. Link-VGI workshop at 19th AGILE conference,
Helsinki, Finland. http://www.cs.nuim.ie/~pmooney/LinkVGI2016/

Printed in the UK by Lightning Source Ltd.
Print and digital versions typeset by Siliconchips Services Ltd.

ISBN (Paperback): 978-1-909188-79-2
ISBN (PDF): 978-1-909188-80-8
ISBN (EPUB): 978-1-909188-81-5
ISBN (Kindle): 978-1-909188-82-2
DOI: http://dx.doi.org/10.5334/bax

The full text of this book has been peer-reviewed to ensure high academic
standards. For full review policies, see http://www.ubiquitypress.com/

Suggested citation: Capineri, C, Haklay, M, Huang, H, Antoniou, V, Kettunen, J,
Ostermann, F and Purves, R 2016 *European Handbook of Crowdsourced
Geographic Information*. London: Ubiquity Press. DOI: http://dx.doi.
org/10.5334/bax. License: CC-BY 4.0

To read the free, open access version of this
book online, visit http://dx.doi.org/10.5334/bax
or scan this QR code with your mobile device:

Table of Contents

Part V: VGI in mobility

Part VI: VGI in spatial planning

Acknowledgements

We would like to thank the below individuals for reviewing this book and providing valuable feedback to the authors.

Gennedy Andrienko, DE
Natalya Andrienko, DE
Vyron Antoniou, UK
Maria Attard, MT
Andrea Ballatore, UK
Sofia Basiouka, GR
Maria Antonia Brovelli, IT
Bénédicte Bucher, FR
Michele Campagna, IT
Cristina Capineri, IT
Ivan Chorbev, MK
Eleonora Ciceri, IT
Cheli Cresswell, UK
Stefano De Sabbata, UK
Gilles Falquet, CH
Hongchao Fan, DE
Karoly Farkas, HU
Alexandrea Fonseca, PT
Gianfranco Gliozzo, UK

Cristina Gouveia, PT
Mark Graham, UK
Muki Haklay, UK
Haosheng Huang, AT
Clemens Jacobs, DE
Bin Jiang, SE
Jamal Jokar, DE
Pinar Karagoz, TR
Juhani Kettunen, FI
Jakub Krukar, DE
Rob Lemmens, NL
Imre Lendak, RS
Claudine Metral, CH
Amin Mobasheri, DE
Peter Mooney, IE
Hristo Nikolov, BG
Frank Ostermann, NL
Ross Purves, CH
Femke Reitsma, NZ
Bernd Resch, AT
Sven Schade, IT
Christoph Schlieder, DE
Dragon Stojanovic, RS
Guillaume Touya, FR
Nico Van de Weghe, BE
Stephan Winter, AU
Jianhong Xia, AU

Supporting Institutions

COST (European Cooperation in Science and Technology) is a pan-European intergovernmental framework. Its mission is to enable break-through scientific and technological developments leading to new concepts and products and thereby contribute to strengthening Europe's research and innovation capacities.

It allows researchers, engineers and scholars to jointly develop their own ideas and take new initiatives across all fields of science and technology, while promoting multi- and interdisciplinary approaches. COST aims at fostering a better integration of less research intensive countries to the knowledge hubs of the European Research Area. The COST Association, an International not-for-profit Association under Belgian Law, integrates all management, governing and administrative functions necessary for the operation of the framework. The COST Association has currently 36 Member Countries. www.cost.eu

This book is based upon work from COST Action, supported by COST (European Cooperation in Science and Technology)"

EUROPEAN COOPERATION
IN SCIENCE AND TECHNOLOGY

COST is supported by the EU Framework Programme Horizon 2020

CHAPTER I

Introduction

Cristina Capineri*, Muki Haklay, Haosheng Huang,
Vyron Antoniou, Juhani Kettunen,
Frank Ostermann and Ross Purves
*cristina.capineri@unisi.it

This book features contributions stemming from the activities of the ENERGIC (European Network Exploring Research into Geospatial Information Crowdsourcing: software and methodologies for harnessing geographic information from the crowd) scientific network. Researchers from 23 European countries participate in ENERGIC. It is funded as action IC1203 by the COST (Cooperation in Science and Technology) programme, which is a European framework supporting trans-national cooperation among scientists, engineers, and scholars across Europe.

The ENERGIC network was born out of scientific connections in the area of Geographic Information Science and friendships that can be traced back over 20 years ago. Indeed, the first important event was the specialist meeting on Volunteered Geographic Information (VGI) organised in 2007 in Santa Barbara (California) under the auspices of NCGIA, Los Alamos National Laboratory, the Army Research Office and The Vespucci Initiative (www.vespucci.org). A number of fundamental questions were examined at this meeting and the results showed that VGI was a field of great potential, but lacking methodological and functional developments (NCGIA 2007).

How to cite this book chapter:
Capineri, C, Haklay, M, Huang, H, Antoniou, V, Kettunen, J, Ostermann, F and Purves, R. 2016. Introduction. In: Capineri, C, Haklay, M, Huang, H, Antoniou, V, Kettunen, J, Ostermann, F and Purves, R. (eds.) *European Handbook of Crowdsourced Geographic Information*, Pp. 1–11. London: Ubiquity Press. DOI: http://dx.doi.org/10.5334/bax.a. License: CC-BY 4.0.

Five years later, in 2012, the ENERGIC action started exploring new VGI sources, sharing and developing data retrieval software, assessing VGI quality, defining standardization criteria for interoperability with other datasets, identifying applications and transferring them for business implementation (market analysis, risk management, advertising, etc.[1]).

The action is based on the study of the remarkable new source of geographic information that has become available in the form of user-generated content accessible over the Internet. People now consume and produce geographic information on the go via platforms such as Facebook, Twitter, Flickr, Instagram and others. The availability of cheap GPS allows everyone to survey and map and contribute to projects like Wikimapia and OpenStreetMap. The exploitation, integration and application of these sources, termed crowd-sourced or user generated information, offer to multidisciplinary scientists an unprecedented opportunity to conduct research on a variety of topics at multiple scales.

The most popular definition of such content that possesses a geographic reference data is Volunteered Geographic Information (VGI), first coined by Michael Goodchild in 2007: its success is revealed by the growing number of articles published since 2007. By simply searching Google scholar for references which match the term 'volunteered geographic information' from 2007 to 2014 an interesting trend emerges: from 83 articles in 2007 to 1,720 in 2014 (Fig 1)!

The growing volume of scientific production on the topic cover multiple domains but some major threads may be identified, although intertwined and often coexisting, to build a narrative on the development of VGI. After an initial phase concerned mostly with conceptualizing and defining the new phenomenon (Coleman 2010; Elwood,2008; Capineri & Rondinone 2011; See et al. 2016; Sui et al. 2012) and types of participation (Bonney et al. 2009; Coleman, Georgiadou & Labonte 2009; Haklay 2010; Haklay 2013; Goodchild & Li 2012), a first relevant thread in the literature is dedicated to the critical aspects of quality (Ali & Schmidt 2014; Antoniou 2016; Foody et al. 2013), among which accuracy and precision of geo-location and of observations, completeness and intelligibility of contents, as well as the reliability of information and the trustworthiness of the data source (Bishr& Kuhn. 2007; Bishr & Janowicz 2010) and at the same time, the first applications and experimentations appear and show the use of VGI in natural disaster management (Zook et al. 2010; Goodchild & Glennon 2010; Ostermann & Spinsanti. 2012, Spinsanti & Ostermann 2013), land use (Antoniou et al. 2016; Perger et al. 2012), tourism (Girardin et al. 2008; Sun et al. 2013;Teobaldi & Capineri 2014), environmental monitoring

---

[1] www.vgibox.eu

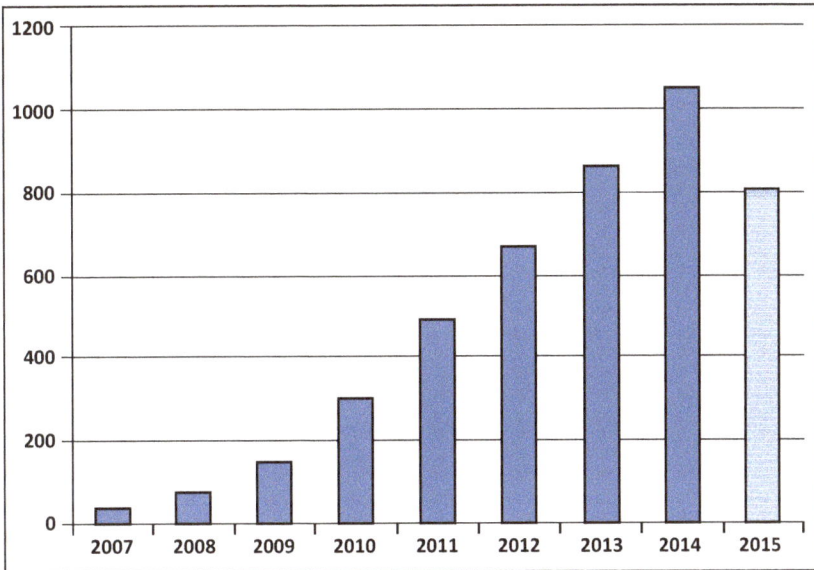

Figure 1: Articles on Google Scholar on 'volunteered geographic information' (2007–2015 January).
Source: Google Scholar (January 2015).

(Gouveia & Fonseca 2008; Connors et al. 2012), the integration of VGI and spatial data infrastructures (Craglia 2007; McDougall 2009), the relationship with the GIS world (Kuhn 2007; Goochild 2016) and mapping (in particular with reference to the OpenStreetMap project) (Neis & Zipf 2012; Dodge & Kitchin 2013) and finally ontology (Tardy et al. 2016). Another relevant thread addresses the role of VGI in the inner worlds of geography such as place (Corbett 2013; Purves & Edwardes 2008; Hardy et al. 2012; Hecht & Gergle 2010; Ostermann et al. 2015; Purves & Derungs 2015), the definition of 'vague places' like downtown, livelihoods and vernacular geography (Hollenstein & Purves 2010); the dynamics of urban cores (Aubrecht 2011; Jiang & Jia 2011; Sagl et al. 2012) and space-time relationships (Li, Goodchild, Xu, 2013). More recently analysis on the societal implications of ICT have employed VGI to discuss the digital divide metaphor (Elwood 2010; Elwood et al.2012; Graham et al. 2014).[2]

---

[2] The references mentioned above are only a small selection and certainly do not pay justice to all the valuable scholars involved in the debate and research on VGI. Apologies for all the authors that have not been quoted despite being relevant. The contents of this essay also draw from the activities of the COST Action IC1203 ENERGIC, *European Network Exploring research into Geospatial Information Crowdsourcing*, which the authors belong to.

In conclusion, a very wide panorama which proves the broad penetration of VGI in the scientific community and the diverse research paths.

The collection of the papers in this book touch on many of these threads.

The Book Structure

The book includes peer-reviewed chapters, organised in six parts, which try to address some fundamental questions: what motivates citizens to provide such information in the public domain, and what factors govern/predict its validity? What methods might be used to validate such information? Can VGI be framed within the larger domain of sensor networks, in which inert and static sensors are replaced by, or combined with, intelligent and mobile humans? What limitations are imposed on VGI by differential access to broadband Internet, mobile phones and other communication technologies, and by concerns over privacy? How do VGI and crowdsourcing enable innovation applications to benefit human society?

PART I: Theoretical and social aspects

Part I deals with the nature and features of crowdsourced geographic information: the different sources and typologies of VGI, the fundamental aspect of participation. Capineri (Chapter 2) discusses the main features of crowdsourced geographic information by focusing on the components of such data: the geographical reference, the contents and the producers' profile in order to show the potentialities and critical aspects posed by these sources. Haklay (Chapter 3) looks at participation inequality and explains how it emerges in VGI and citizen science projects at both temporal and spatial scales, and also evaluates its implication on the use of VGI and citizen science data. Campagna (Chapter 4) introduces the concept of Social Media Geographic Information (SMGI) as a specific type of VGI for expressing pluralism in such domains as spatial planning and governance.

PART II: Quality: Criteria and methodologies

This part covers issues relevant to the quality evaluation of VGI datasets. The evaluation of spatial data quality elements and the development and adoption of new quality criteria and methodologies for VGI is presented here. Criscuolo et al. (Chapter 5) offer an analysis of strategies for quality control and describe a simple representation of the components of quality in crowdsourced geographic information. Jacobs (Chapter 6) explores methods of (semi)automatic validation of observation data especially in field of citizen science projects. Ballatore

(Chapter 7) provides an overview of the semantic issues experienced in VGI and what potential solutions are emerging from research in geo-semantics and in the Semantic Web. Antoniou et al. (Chapter 8) present how named features in Open Street Map behave and change in terms of type, name and location by analysing the volatility of places and points-of-interest (POIs). Ali (Chapter 9) presents an approach for rule-guided classification for VGI projects which consists in a learning and a guiding phase. Finally, Bucher et al. (Chapter 10) present findings from the literature and from the experience of the French National Mapping Agency, IGN, about quality management of geographical data and focus on the potential of context and tasks modelling to address quality issues

PART III: Data analytics

Part III focuses on data analytics and visualization methods for deriving knowledge from varying VGI sources such as Twitter, Flickr and OpenStreet-Map. Purves and Mackaness (Chapter 11) provide an overview of methodologies used to extract meaning from the analysis of geotagged images based on research in natural language processing and statistical and exploratory techniques. Gennady and Natalia Andrienko (Chhapter 12) offer an analysis of geographically referenced posts published in social media, such as Twitter, Flickr and YouTube and an overview of visual analytics approaches to extracting various kinds of information and knowledge. Jiang (Chapter 13) proposes the head/tail breaks is a powerful tool for visualizing city structures and dynamics, and uses social media location data to illustrate its effectiveness in visualization. Lemmens et al. (Chapter 14) present ways to enrich the unstructured nature of VGI through semantic enrichment and explain how folksonomies and ontologies play a fundamental role. Karagoz et al. (Chapter 15) focus on detecting real-world events by following posts in microblogs by employing a process for toponym recognition and location estimation. Song and Xia (Chapter 16) concentrate on spatio-temporal variation of sentiment polarity patterns of georeferenced Tweets, with a view to understanding how opinions evolve on Twitter over space and time and across communities of users. Stojanocski et al. (Chapter 17) present a methodology for detecting and identifying social hotspots from Twitter stream data and applying sentiment analysis on the data in New York. Finally, Steiger et al. (Chapter 18) provide a state of the art survey on social media data analysis and mining from a GIScience perspective.

PART IV: VGI and crowdsourcing in environmental monitoring

Part IV presents several case studies where crowdsourced geographic information has been applied especially in the field of environmental monitoring. Kettunen et al. (Chapter 19) describe several examples to illustrate the changing role of citizens in environmental monitoring in Finland. Jokar Arsanjani and Fonte (Chapter 20) investigate the completeness, thematic accuracy and fitness for use of OpenStreetMap features for land mapping purposes in Europe. Lupia and Estima (Chapter 21) evaluate the adequacy of the geotagged photos available at Panoramio for monitoring Land Use/Cover (LULC) in urban areas, taking Rome (Italy) as the study area. Oltra et al. (Chapter 22) introduce AtrapaelTigre.com, a citizen science project focusing on the Asian tiger mosquito in Spain, and describe lessons they have learned. De Albuquerque et al. (Chapter 23) close the section by reviewing the use of crowdsourced geographic information for disaster risk management and improving urban resilience and suggesting future research directions.

PART V: VGI in mobility applications

Part V includes several case studies and research activities of VGI in smart cities and mobility applications. Zipf et al. (Chapter 24) investigate the use of OpenStreetMap for routing and navigation for mobility-impaired persons and describe existing challenges in this aspect. Farkas (Chapter 25) introduces a crowdsourcing based smart timetable service for public transportation. Lendák (Chapter 26) reviews existing mobile crowdsourcing projects in smart city, focusing on environmental monitoring, citizen collaboration, urban mobility, health/fitness and social networking. Finally, Stojanovic et al. (Chapter 27) present a framework and some applications to illustrate how mobile crowdsourcing can be used for enabling smart urban mobility.

PART VI: VGI in spatial planning

Part V includes a broad variety of case studies and research activities of VGI in spatial planning. Huang and Gartner (Chapter 28) illustrate how mobile crowdsourcing and social media data can be used to study people's affective responses to different environments, as well as the potential applications of these affective data. Massa and Campagna (Chapter 29) introduce Spatext, a tool that allows integration of VGI and authoritative data, and present a case study to illustrate its application in urban planning. Basiouka and Potsiou (Chapter 30) investigate the use of VGI and crowdsourcing in Cadastre design, and propose a crowdsourcing cadastral model for official cadastral surveys. And in the last chapter, Fan and Zipf (Chapter 31) investigate the generation of 3D city models by using OpenStreetMap data and introduce the OSM-3D project.

In addition, the book consolidates the references and information from all the chapters in a rich bibliography and a glossary, thereby providing a state-of-the-art of research on crowdsourced and volunteered geographic information, as well as valuable reference for PhD candidates and senior researchers from many disciplines who aim to tap into the potential of new geographic data sources.

Concluding Remarks

As this collection demonstrates, research into crowdsourced geographic information or VGI (depending on the definition that the researchers prefer), has made significant inroads in a relatively short period of 7 or 8 years. Yet the book opens questions and points to new research directions, in addition to the findings that each of the researchers demonstrate.

As can be seen from this book, the crowdsourcing techniques and methods and the VGI phenomenon have motivated a multidisciplinary research community to identify both fields of applications and quality criteria depending on the use of VGI. Besides harvesting tools and storage of these data, many research attentions have been paid to these information resources, in an age when information is one of the most important drivers of development. The participation component is a fundamental aspect of crowdsourced information and it reveals both a new way of doing science with a problem-solving approach.

Despite rapid progress in VGI research, this book also shows that there are technical, social, political and methodological challenges that require further studies and research. We hope that the book will spark new research questions and development—and hopefully foster new research collaborations, and friendships.

References

Ali, A.L., & Schmid, F. 2014. Data Quality Assurance for Volunteered Geographic Information. In: *Geographic Information Science*, Springer International Publishing, pp. 126–141.

Antoniou, V. 2016. Volunteered geographic Information measuring quality, understanding the value. *GEOmedia*, *1*(1). Retrieved at: http://ojs.mediageo.it/index.php/GEOmedia/article/viewFile/1298/1183.

Antoniou, V., Fonte, C., See, L., Estima, J., Arsanjani, J. J., Lupia, F., Minghini, M., Foody, G., & Fritz, S. 2016. "Investigating the Feasibility of Geo-Tagged Photographs as Sources of Land Cover Input Data." *ISPRS Int. J. Geo-Inf.*, 5(5): 64.

Aubrecht, C., Ungar. J., & Freire, S. 2011. Exploring the potential of volunteered geo-graphic information for modeling spatio-temporal characteristics of urban population. In: *Proceedings of 7VCT*, pp. 11–13.

Bishr, M., & Janowicz, K. 2010 (September). Can we trust information?-the case of volunteered geographic information. In: *Towards Digital Earth Search Discover and Share Geospatial Data Workshop at Future Internet Symposium,* vol. 640. Retrieved at: http://www.opl.ucsb.edu/~jano/DE2010qp.pdf.

Bishr, M., & Kuhn, W. 2007. Geospatial Information Bottom-Up: A Matter of Trust and Semantics. In: Fabrikant, S. I., & Wachowitz, M. (Eds.) The European Information Society, Springer Verlag, pp. 365–387.

Bonney, R., Cooper, C. B., Dickinson, J., Kellin, S., Phillip, T., Rosenberg, K. V., & Shirk, J. 2009. Citizen science: a developing tool for expanding science knowledge and scientific literacy. *BioScience*, *59*(11): 977–984.

Capineri, C., & Rondinone, A. 2011. Geografie (in) volontarie. *Rivista geografica italiana*, *118*(3): 555–573.

Coleman, D. J. 2010. Volunteered geographic information in spatial data infrastructure: an early look at opportunities and constraints. In: *GSDI 12 World Conference*. Retrieved at: https://www.researchgate.net/profile/David_Coleman2/publication/228863877_Volunteered_Geographic_Information_in_Spatial_Data_Infrastructure_An_Early_Look_At_Opportnities_And_Constraints/links/5405c2c80cf23d9765a734cb.pdf.

Coleman, D. J., Georgiadou, Y., & Labonte, J. 2009. Volunteered geographic information: The nature and motivation of produsers. *International Journal of Spatial Data Infrastructures Research*, *4*(1): 332–358.

Connors, J. P., Lei, S., & Kelly, M. 2012. Citizen science in the age of neogeography: Utilizing volunteered geographic information for environmental monitoring. *Annals of the Association of American Geographers*, *102*(6): 1267–1289.

Corbett, J. 2013. "I Don't Come from Anywhere": Exploring the Role of the Geoweb and Volunteered Geographic Information in Rediscovering a Sense of Place in a Dispersed Aboriginal Community. In: *Crowdsourcing Geographic Knowledge*, Springer Netherlands, 223–241.

Craglia, M. 2007. Volunteered Geographic Information and Spatial Data Infrastructures: when do parallel lines converge. In: *Position paper for the Specialist Meeting on Volunteered Geographic Information, December 13–14, 2007, Santa Barbara, CA*. Retrieved at: http://www.ncgia.ucsb.edu/projects/vgi/docs/position/Craglia_paper.pdf.

De Longueville, B., Ostländer, N., & Keskitalo, C. 2010. Addressing vagueness in Volunteered Geographic Information (VGI)–A case study. *International Journal of Spatial Data Infrastructures Research*, *5*: 1725–0463.

Dodge, M., & Kitchin, R. 2013. Crowdsourced cartography: mapping experience and knowledge. *Environment and Planning A*, *45*(1): 19–36.

Elwood, S. 2008. Volunteered geographic information: key questions, concepts and methods to guide emerging research and practice. *GeoJournal*, *72*(3): 133–135.

Elwood, S. 2010. Geographic information science: emerging research on the societal implications of the geospatial web. *Progress in Human Geography*, 34 (3): pp. 349–357.

Elwood, S., Goodchild, M. F., & Sui, D. Z. 2012. Researching volunteered geographic information: Spatial data, geographic research, and new social practice. *Annals of the Association of American Geographers*, *102*(3): 571–590.

Foody, G. M., See, L., Fritz, S., Van der Velde, M., Perger, C., Schill, C., & Boyd, D. S. 2013. Assessing the accuracy of volunteered geographic information arising from multiple contributors to an internet based collaborative project. *Transactions in GIS*, *17*(6): 847–860.

Girardin, F., Calabrese, F., Fiore, F. D., Ratti, C., & Blat, J. 2008. Digital footprinting: Uncovering tourists with user-generated content. *Pervasive Computing, IEEE*, *7*(4): 36–43.

Goodchild, M. F. 2016. GIS in the Era of Big Data. *Cybergeo: European Journal of Geography*. Retrieved at: http://cybergeo.revues.org/27647; DOI: http://dx.doi.org/10.4000/cybergeo.27647.

Goodchild, M. F., & Glennon, J. A. 2010. Crowdsourcing geographic information for disaster response: a research frontier. *International Journal of Digital Earth*, *3*(3): 231–241.

Goodchild, M. F., & Li, L. 2012. Assuring the quality of volunteered geographic information. *Spatial statistics*, *1*: 110–120.

Gouveia, C., & Fonseca, A. 2008. New approaches to environmental monitoring: the use of ICT to explore volunteered geographic information. *GeoJournal*, *72*(3–4): 185–197.

Graham, M., Hogan, B., Straumann, R. K., & Medhat, A. 2014. Uneven geographies of user-generated information: patterns of increasing informational poverty. *Annals of the Association of American Geographers*, *104*(4): 746–764.

Haklay, M. 2010. How Good is volunteered geographical information? a comparative study of OpenStreetMap and ordnance survey datasets. *Environment and Planning B: Planning and Design*, *37*(4): 682–703.

Haklay, M. 2013. Citizen science and volunteered geographic information: Overview and typology of participation. In: *Crowdsourcing Geographic Knowledge*. Springer Netherlands, pp. 105–122.

Hardy, D., Frew, J., & Goodchild, M. F. 2012. Volunteered geographic information production as a spatial process. *International Journal of Geographical Information Science iFirst*: 1–22

Hecht, B., & Gergle, D. On the "Localness" of User-Generated Content. In: *Proceedings of the 2010 ACM conference on Computer supported cooperative work*, New York, NY, pp. 229–232.

Hollenstein, L., & Purves, R. 2010. Exploring place through user-generated content: Using Flickr tags to describe city cores. *Journal of Spatial Information Science*, *2010*(1): 21–48.

Jiang, B., & Jia, T. 2011. Zipf's law for all the natural cities in the United States: a geospatial perspective. *International Journal of Geographical Information Science*, *25*(8): 1269–1281.

Kuhn, W. 2007. Volunteered geographic information and GIScience. *NCGIA, UC Santa Barbara*. Retrieved at: https://www.researchgate.net/profile/Werner_Kuhn/publication/228563727_Volunteered_geographic_information_and_GIScience/links/0deec52cded8fcd5b3000000.pdf.

Li, L., Goodchild, M. F., & Xu, B. 2013. Spatial, temporal, and socioeconomic patterns in the use of Twitter and Flickr. *Cartography and Geographic Information Science*, 40(2): 61–77.

McDougall, K. 2009. Volunteered geographic information for building SDI. In: *Proceedings of the Surveying and Spatial Sciences Institute Biennial International Conference (SSC 2009)*, pp. 645–653.

NCGIA, Workshop on Volunteered geographic Information, 13-14 December 2007, Los Alamos, 2007 [accessible at http://ncgia.ucsb.edu/projects/vgi/]

Neis, P., & Zipf, A. (2012). Analyzing the contributor activity of a volunteered geographic information project—The case of Open Street Map. *ISPRS International Journal of Geo-Information*, 1(2): 146–165.

Ostermann, F., & Spinsanti, L. 2012. Context analysis of volunteered geographic information from social media networks to support disaster management: A case study on forest fires. *International Journal of Information Systems for Crisis Response and Management (IJISCRAM)*, 4(4): 16–37.

Ostermann, F. O., Huang, H., Andrienko, G., Andrienko, N., Capineri, C., Farkas, K., & Purves, R. S. 2015. Extracting and Comparing Places Using Geo-social Media. Retrieved at: http://real.mtak.hu/26267/1/GEOSPATIAL_WEEK_2015_submission_158.pdf.

Perger, C., Fritz, S., See, L., Schill, C., Van der Velde, M., McCallum, I., & Obersteiner, M. 2012. A campaign to collect volunteered geographic Information on land cover and human impact. *GI_Forum*: 83–91.

Purves, R. S., & Derungs, C. 2015. From space to place: place-based explorations of texts, *International Journal of Humanities and Arts Computing*, 9(1): 74–94.

Purves, R. S., & Edwardes, A. J. 2008. Exploiting Volunteered Geographic Information to describe Place. In: *Proceedings of the GIS Research UK 16th Annual Conference*, pp. 252–255.

Sagl, G., Resch, B., Hawelka, B., & Beinat, E. 2012. From social sensor data to collective human behaviour patterns: Analysing and visualising spatio-temporal dynamics in urban environments. In: *Proceedings of the GI-Forum 2012. Geovisualization, Society and Learning*, pp. 54–63.

See, L., Mooney, P., Foody, G., Bastin, L., Comber, A., Estima, J., ... & Liu, H. Y. (2016). Crowdsourcing, Citizen Science or Volunteered Geographic Information? The Current State of Crowdsourced Geographic Information. *ISPRS International Journal of Geo-Information*, 5(5): 55–65.

Spinsanti, L., & Ostermann, F. 2013. Automated geographic context analysis for volunteered information. *Applied Geography*, 43: 36–44.

Sui, D., Elwood, S., & Goodchild, M. F. (Eds.). 2012. *Crowdsourcing geographic knowledge: volunteered geographic information (VGI) in theory and practice*. Springer Science & Business Media.

Sun, Y., Fan, H., Helbich, M., & Zipf, A. 2013. Analyzing human activities through volunteered geographic information: Using Flickr to analyze spatial and temporal pattern of tourist accommodation. *Progress in Location-Based Services*: 57–69.

Tardy, C., Moccozet, L., & Falquet, G. 2016. A simple tags categorization framework using spatial coverage to discover geospatial semantics. In: *Proceedings of the 25th International Conference Companion on World Wide Web,* pp. 657–660.

Teobaldi, M., & Capineri, C. 2014. Experiential tourism and city attractivness in Tuscany, *Rivista Geografica Italiana,* 121: 259–274.

Zook, M., Graham, M., Shelton, T., & Gorman, S. 2010. Volunteered geographic information and crowdsourcing disaster relief: a case study of the Haitian earthquake. *World Medical & Health Policy,* 2(2): 7–33.

PART I

Theoretical and social aspects

CHAPTER 2

The Nature of Volunteered Geographic Information

Cristina Capineri

Dept. of Social, Political and Cognitive Sciences, University of Siena,
cristina.capineri@unisi.it

Abstract

This contribution starts from the assumption that volunteered geographic information is a technological, cultural and scientific innovation. It therefore offers first some general background on the context that has fuelled the development of VGI and the lively scientific debates that have accompanied its success. The paper then focuses on the nature of this data by describing the main elements of VGI: the geographical reference (coordinates, geotag, etc.), the contents (texts, images, etc.) and the producers' profiles. The opportunities and the criticalities offered by this data are described with examples drawn from recent literature and applications to highlight both the research challenges and the current state of the subject. The chapter aims to provide a guide to and a reference picture of this rapidly evolving subject.

Keywords

volunteered geographic information, crowdsourced information, geography

Introduction: a technological and cultural innovation.

The most cited and debated definition of volunteered geographic information was coined in 2007 by Michael Goodchild (2007) as a subset of user-generated content which carries specific spatial and temporal components:

How to cite this book chapter:
Capineri, C. 2016. The Nature of Volunteered Geographic Information. In: Capineri, C, Haklay, M, Huang, H, Antoniou, V, Kettunen, J, Ostermann, F and Purves, R. (eds.) *European Handbook of Crowdsourced Geographic Information*, Pp. 15–33. London: Ubiquity Press. DOI: http://dx.doi.org/10.5334/bax.b. License: CC-BY 4.0.

'the widespread engagement of large numbers of private citizens, often with little in the way of formal qualifications, in the creation of geographic information, a function that for centuries has been reserved to official agencies. [..] I term this *volunteered geographic information* (VGI), a special case of the more general Web phenomenon of *user-generated content'(p.2).*

Since 2007, the appeal of VGI has grown steadily and created a wide scientific community involved in the harnessing of these new sources of geographical information and in satisfying the spatial shift fuelled by the neogeography revolution which has put mapping within the grasp of almost any user who is not only in possession of suitable technology (e.g. a smartphone), but is capable of configuring it in order to capture location and skilled enough to view the resulting information and share it in space or on a map (Turner 2006; Batty 2010; Wilson & Graham 2013).

All historical transformations in means of communication have led to a redefinition of lifestyles, of time and space which are deeply connected in society: their meaning, perceptions and manifestations are linked to social practices and evolve throughout history and across cultures. The transformations addressed here refer both to the development of Web 2.0 technologies and the diffusion of sensors of different types which have profoundly modified the ways of accessing, producing, diffusing and representing geographic information: '*This is an unprecedented moment in human history: we can now know where nearly everything, from genetic to global levels, is at all times'*" (Sui & Delyser 2013: p.13).

In this sense VGI may be considered a significant innovation and as with any other innovation it combines technology, social practices and power relationships. First, the technology relies on the many location-based devices used potentially by ordinary citizens who become sensors and on Web 2.0 applications which enable information co-creation (social media, photo-sharing platforms, wiki projects, etc.) in huge quantities. Secondly, the phenomenon of user-generated content is part of a cultural change which very recently has led to the adoption of open access and a collaborative and sharing approach to information resources. This cultural turn has been defined as *collective intelligence* by the French philosopher Pierre Levy (1994) who explains that 'the collective intelligence tries to articulate in a new way the individual and the collective domains in a new space of knowledge'. This concept has also been discussed by Manuel Castells (1996, 2008) who has explained that in the information age there is a growing juxtaposition of individualism and communalism: networks of individuals which provide the basis for increasing our sociability as individuals. In sociological studies many argue that the 'bond of community' has been lost (Putman 1995) and that there is a need to improve and rebuild social capital; in this sense the debate on the contribution of social media to social capital building is questionable. It is hard to imagine how large-scale social movements and community building activities can be organized effectively on the basis of qualitative practices alone, as the logistical challenge requires an ability to plan schedules, develop strategies and master communications technologies: certainly the burgeoning relationships through social media betray new communication and social practices. Furthermore, the accelerated deterritorialization process implies the removal of barriers and limits but also the restoration of stronger social ties.

Indeed the participative and collaborative approach is emerging in contemporary society as Jeremy Rifkin (2011) explains: 'people are biologically predisposed to be empathic—that our core nature is not rational, detached, acquisitive, aggressive, and narcissistic, but affectionate, highly social, cooperative, and interdependent. *Homo sapiens* is giving way to *Homo empathicus*'. Historians tell us that empathy is the social glue that allows increasingly individualized and diverse populations to forge bonds of solidarity across broader domains so that society can cohere as a whole.

In this context the crowdsourcing process has emerged and has been defined by the journalist Jeff Howe (2008) as *'the process by which the power of the many can be leveraged to accomplish feats that were once the province of a specialized few'*. This phenomenon has been supported by the so-called 'sharing economy' as a *socio-economic syste*m built on the sharing of skills, goods and services driven by the increasing sense of urgency of resource depletion; by the open source movement which – at least in theory – enables any user to participate in the information society by sharing know-how and skills mediated by Web 2.0 tools and applications. Famous initiatives like Wikipedia, founded in 2001 or, more specifically in the realm of geography, OpenStreetMap, launched in 2004, do not need any further explanation here.

VGI is thus closely related to the concept of crowdsourcing as it is an assertive method of collecting geospatial information from people who are mainly participating in Web-based social networking sites, in citizen science initiatives or in the context of collaborative commons-based peer production networks (Benkler & Nissenbaum 2006).

The volunteering element is the subject of debate in VGI literature since contributors may produce information either consciously or unconsciously: crowdsourcing implies a process of consensus production whereby many people will provide and augment information about the same thing which will become more and more accurate thanks to a convergence of information. In the case of geographic information involuntarily produced by individuals, quality might be debated since data are often collected publicly without strict standardization and every user inserts data according to his/her personal background and point of view (Coleman 2009). In fact Harvey (2013) has suggested the definition *crowdsourced geographic information* which refers to data collected via 'opt-out' agreements which are more open-ended and offer fewer opportunities to control the data collection, and subsequently its quality and assessment. In contrast, when volunteered information production – or *geographic volunteer work* (Priedhorsky et al. 2010) – is regulated by shared rules concerning the geocoding, tagging and annotation of the data, VGI becomes part of citizen science. Citizen science has emerged from the fields of ecology, biology and nature conservation, whose projects are based on volunteering and contribution of information on areas such as biodiversity, environmental quality or endangered species both for the benefit of human knowledge and science (Haklay 2013a). This is demonstrated by the countless applications in the field of biodiversity monitoring and environmental assessment (Bonney et al 2009; Gouveia et al. 2008) as emerging problematic issues in the context of sustainable development.

Citizen science re-evaluates the separation between scientists and public, and scientists need to adjust to their new role as mediators of knowledge rather than as the sole repository of scientific truth: 'This might end up being the most important outcome of citizen science as a whole as it might eventually catalyse the education of scientists to engage more fully with society' (Haklay 2013:14).

In this context VGI embodies either the implicit or explicit relationship of the individual to the world and represents the sense of belonging, in some intrinsic way, to a larger body, whether a nation or a neighbourhood; this relationship has long been a critical part both of the individual's motivation to act in some larger interest and of the group's ability to exhort the individual to take action and participate (Curry 1997).

Before closing this introductory section, it is worth mentioning that literature related to VGI has increased enormously in the recent past (see the Introduction of this Handbook) which may lead us to ask why VGI is so appealing and why it has created so much scientific interest in geography. The main drivers of its success certainly relate to:

a) the features of this information (the non-expert producers, the participatory approach, the huge quantity, the real time accessibility, the finer-grained resolution and the scalability);

b) the extremely diversified fields of potential applications (disaster and crisis management, environmental monitoring, planning, land use, mobility, people's behaviour etc.) which are more and more employed in governance and in the management of public services;

d) the 'wow' component due to unexpected, creative and sometimes amusing topics which can be tackled spatially with these sources. This is well demonstrated by the Floatingsheep collective which started in 2009 by producing witty and entertaining explorations such as Santa Claus's homeland, the 'beer belly' of America, zombies *vs* vampires and so on: 'At FloatingSheep, we're willing to search for and analyse almost anything that falls within the realm of human experience. Sometimes this is mundane (pizza) and sometimes it is contentious (abortion) but most of the time it falls somewhere in between. Such as, where can I get a drink?';[1]

e) the experiential and perceptional nature of the content embedded in VGI which can be distilled both to achieve a better understanding of beliefs, practices and habits and potentially challenge the dominant narratives, because VGI is built on the understanding of the social world mediated by people's conversations and contributions, thus it consists of social practices (Elwood 2008).

The components of VGI

Generally speaking VGI consists of a 'big' and ongoing flow of data deriving from different tools and media (mobile phones, cameras, records of smartcard transactions, social platforms, check-ins, location-based devices, etc.); they are digital footprints, or data shadows, or by-products of human/machine interactions (Graham 2013). Such digital footprints are produced by anyone who may potentially act as a sensor and provide, more or less consciously, valuable information (Capineri & Rondinone 2011; Haklay et al. 2008; Sui 2008) by applying local and sectorial knowledge since producers are '[...] equipped with some working subset of the five senses and with the intelligence to compile and interpret what they sense, and each free to rove the surface of the planet' (Goodchild 2007).

More precisely the essential components of VGI are:

1) the geographical references (i.e. geotag, coordinates, geographic name) which enable the information to be represented on a map and thus satisfy the eternal human desire to know 'where we are' or 'where things are';
2) the stock of content which makes it possible to transform this data into information and possibly knowledge. The content may take different forms: images, texts, symbols, maps, check-ins, photos, videos, drawings, etc.
3) attributes, of various degrees of accuracy, of content users and content producers (*produsers*, Coleman et al. 2009) (such as nationality, language and possibly age and gender) and the time of the digital footprint's creation.

The combination of three components provides a powerful way to aggregate, synthesize and compare information at different scales on specific issues or events which are occurring either in real time or in longer time spans.

Thus VGI components may be represented as the three corners in a standard ternary diagram (Figure 1) to show that employment of this data may vary according to the emphasis given to one or more of the components or to the

---

[1] www.floatingsheep.org.

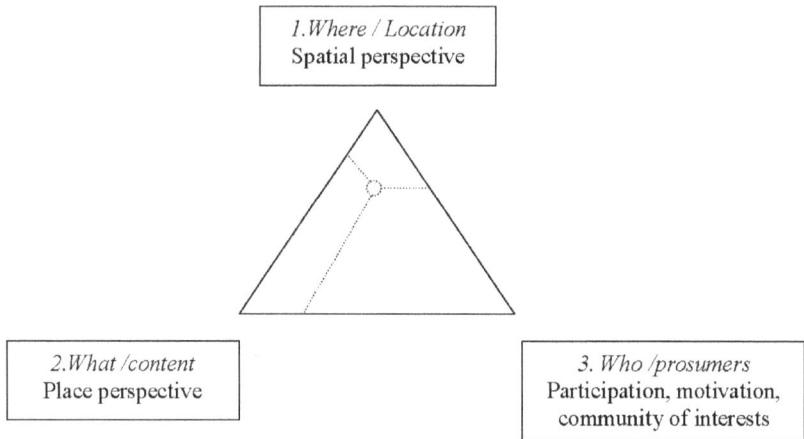

Figure 1: VGI, a diagram representation of the components.

changing balance between, for example, ecological, spatial or regional synthesis. The corners represent the components and the related perspectives: (1) the *where* of the information given by the geographical reference which serves a spatial perspective, (2) the *what* given by the stock of content which suits place-oriented and qualitative analysis, (3) the *who* given by the *produsers* which fits well the analysis of the participation and production process. The barycentre's position represents the different composition of the three VGI components and it may vary according to the chosen perspective.

Of course the representation is oversimplified since at least the time dimension needs to be added to give a fuller picture of the potentialities of the relationship between the components of VGI sources. Moreover the complexity of VGI data lies in the fact that analysis needs to draw on cognitive, psychological and anthropological inputs to fully appreciate and exploit this data, thus going beyond simple representation on a map. The fragmented individual-level contents from the crowd provide qualitative information which was unreachable in the past through traditional direct investigations (i.e. surveys, interviews, etc.) or official data (i.e. census); the employment of qualitative information is not new in geography, as it was the pillar of the perception and behavioural approach (Claval 1974), but the innovative aspects are, in addition to the quantity and the scale (from global to local and vice versa), the granularity of the topic and the timeliness that VGI allows. As Table 1 shows, very specific events or unexpected topics, like beverage-consumption habits, can be now quite easily addressed and explored.

In the following sections each of the components will be briefly discussed drawing examples both from existing literature and my own research.

Event-dependent	Activity time	User generated contents	Scale
Habemus Papam (Source:Ladest)	13–14 March 2013	10,000 Geocoded Tweets	*Global*
Palio di Siena (Source:Ladest)	01–02 July 2013	375 Geocoded Tweets	*Urban*
Hurricane Sandy in NYC (Source: Shelton et al. 2014)	24–30 October	16,000 Geocoded Tweets	*Urban*
Topic-oriented			
Church or Beer (Source: FloatingSheep)	22–28 June 2012	17,686 (church) + 14,405 (beer)	*National*

Table 1: Some examples of event-dependent or topic-oriented applications in VGI.

The geographical reference: living space and local knowledge

The geographical component represents the raw digital footprint that can be represented in space as the manifestation of the producers' activity – or inactivity – on the Web. Although the act of producing information is to a greater or lesser degree voluntary, locating the origin of this data on the Earth's surface highlights the constellations of participation on the Web thanks to geocoding attributes (geotags; geographic names, coordinates). The footprints offer a preliminary source of information which reveal the producers' appropriation of place by naming, tagging or annotating it.

The geographical component has a phenomenological value in itself since it is often a response to a stimulus, either an event or a simple desire to get in touch with friends and show where you are or where you have been. Many case studies show strong uneven geographical patterns of participation in the production of VGI which may be described through the digital divide metaphor (Graham 2014). The divide may be caused by different reasons such as the uneven diffusion of the technology but also the ability to work with or benefit from it. The distribution of geotagged Wikipedia articles (Figure 2) supports this assumption. The articles on Wikipedia are an example of pure volunteered geographic information because users have added content in Wikipedia deliberately, and despite the pervasiveness of the internet, recent analysis shows that the practice of producing content is mainly concentrated in the United States, Western Europe, Japan and in some emerging countries in South America and Asia. The uneven distribution is mainly explained by the traditional variables of wealth such as population, GDP per capita and broadband internet connections (Graham et al. 2014).

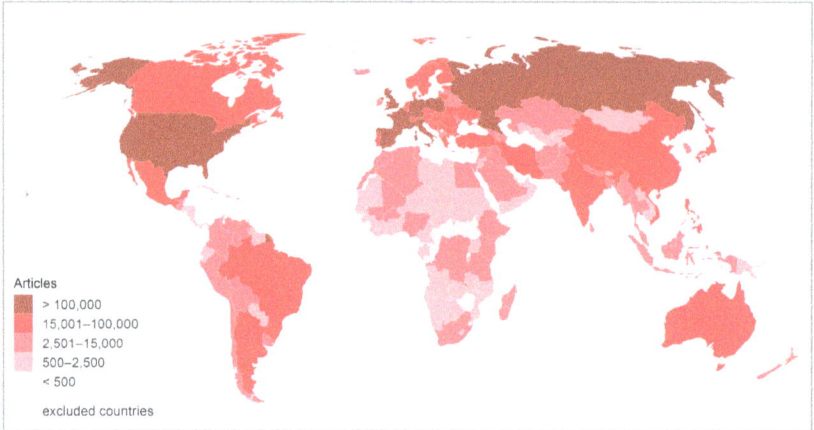

Figure 2: Geotagged Wikipedia articles per country in all languages (Source: Graham et al. 2014).

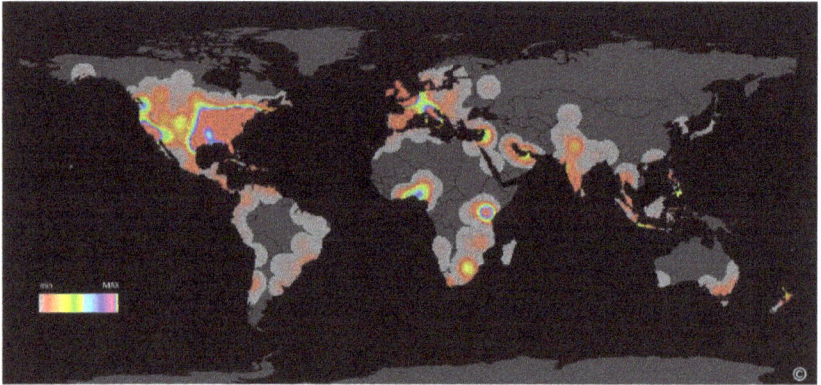

Figure 3: Habemus Papam: Tweeting activity on the Pope's election day (Source: Ladest & Unisi 2013).

When dealing with crowdsourced information derived from social media, other types of uneven distribution may emerge. The map (Figure 3) showing the Tweets sent about the election of Pope Francis in March 2013 (10,000 Tweets collected on 13-14 March 2013) is an expression of a heterogeneous community of interest which has spontaneously responded to a particular event, unaware of the fact that their Tweets might be collected and analysed: here the uneven distribution appears smoother than Wikipedia's divide due to the religious or political appeal that the Pope's election may have (note the intense activity in the Arab Gulf and in Western Africa).

This uneven participation pattern on the Pope's election day requires further investigation in order to discover the reasons for such disparities, which could include critical aspects of religious beliefs and cultural proximity: the Pope's election map is a manifestation of an event-related social activity which has converged on the election of Pope Francis and will soon fade away. The uneven participation in Wikipedia, which is a long-standing project, is more concerning than that described by the Pope's election map since it highlights a consolidated process of information production which reproduces well-known dichotomies (North–South; developed–less developed): so the apparent democratisation of information and knowledge remains debatable (Haklay 2013b).

Nevertheless, the great innovation is that this data enables us to locate spatially practices and topics (especially cultural and political ones) pertaining to people's everyday lives which reflect specific time–space configurations at different scales and with a degree of fine resolution which was impossible in the past.

From a methodological point of view, the location of the information may be questionable since georeferencing creates many ambiguities: (a) the same name may be used for more than one location, (b) the same location can have more than one name and (c) place names can be used in non-geographic contexts such as organizations, events or personal names. Furthermore, the greater or lesser awareness of the produsers' in the information generation process may affect accuracy and quality: while prosumers participating in citizen science activities adopt and share recording rules, unaware citizens just use the technologies for their own purposes and do not generally pay attention to the georeferencing of the information, despite adding relevant content. For example, if we search for 'Chianti', the famous wine-producing area in Tuscany, on Flickr, among the photos there is one of the bronze sculpture of Perseus with the head of Medusa by Benvenuto Cellini which is located near the Uffizi in Florence:[2] in this case the tag 'Chianti' does not enable us to place the photo in the correct location but nevertheless it offers other hints relevant to a place-based analysis: indeed 'Chianti' is one of the many tags of this picture (Florence, wine, holiday, Stendhal Syndrome, lovely city, Arno river) which are employed by the user to describe the spirit of his/her Tuscan holiday experience.

The stock of contents: place and qualitative information

The content reveals the 'sticky places' in the fluid information flows: a world of places of knowledge which not only tell stories of VGI 'birthplaces' but collect the added value generated by the produsers. Contents may be either 'neutral/locational' if carrying simply positional information (i.e. an address) or descriptive if they take the form of texts, comments, images, drawings or video clips. The stock of contents records points of view, values, feelings, expressions of appreciation or contempt, of happiness and unhappiness; in short they

---

[2] https://www.flickr.com/photos/75992994@N05/15607974697/in/photolist-pMdX72-pkvq8o-6JxVtE-4WVhxs-5W7cNo-78kME8-sjwVKi [Accessed April 2016].

represent the 'sense of place' engineered by the Web because VGI contributors are engaged in knowledge production processes which are grounded in social structures and sets of values, and in turn, physical place (Hardy et al. 2012: 3; Lussault 2007).

From this point of view, VGI content capitalizes the informal knowledge of the producers and becomes a collector of multiple identities and perceptions which highlight the variegated relationships with a certain place such as inclusion and exclusion, sharing and reacting and so on.

It is true in fact that VGI content incorporates the situatedness of individuals and the invisible knowledge of the producer, the location and the fluidity of perceptions (Zook & Graham 2007).

The number of content contributors combined with the ability to annotate place in the geoweb may result in dense layers of information augmenting some parts of the world which describe 'the indeterminate, unstable, context dependent and multiple realities brought into being through the subjective coming-togethers in time and space of material and virtual experience'" (Graham et. al. 2012: 465; Graham, Zook & Boulton 2013). Several scholars (Graham 2010; Crang 1996) use the metaphor of palimpsests, with reference to medieval writing blocks that could be reused while maintaining traces of earlier inscriptions: 'the countless layers of any place come together in specific times and spaces and have bearing on the cultural, economic, and political characteristics, interpretation and meaning of a place'" (Graham 2010: 422).

For example, when reading the English and Farsi versions of the description of the town Esfahan in Iran on Wikipedia, the information is slightly different both in terms of the images shown and of content: the English version has more stereotyped pictures of the town's blue and white ceramic decorations than the Farsi version, which also contains descriptions of local artefacts. In addition, the nuclear activity which takes place close to the city is omitted in the Farsi version since the topic is clearly regarded as a sensitive one.

In this way crowdsourced information becomes particularly relevant in the production and acquisition of local knowledge either through place names or practices and values. People's contributions in VGI tend to be more accurate in places the contributor knows best and is nearer to, in accordance with Tobler's law which states that 'everything is related to everything else but near things are more related than distant things'" (Tobler 1970). As such, some literature hypothesizes that (a) contributors write about nearby places more often than distant ones and that (b) this likelihood follows an exponential distance decay function (Hardy et al. 2012). Indeed, according to recent research, about 50 percent of Flickr users contribute local information on average, and over 45 percent of Flickr photos are local to the photographer (Hecht & Gergle 2010). Local knowledge deriving from VGI has remarkably been applied to vernacular geography, which had been eroded by the quantitative approach, which 'encapsulates the spatial knowledge that we use to conceptualize and communicate about space on a day-to-day basis. Importantly, it deals with areas which are typically not represented in formal administrative gazetteers and which are often considered to be vague'" (Hollenstein & Purves, 2005: 22) such as 'down-town'" (Hollestein & Purves 2010), 'neighbourhood'[3] or regions like the 'Alps'" (Purves & Derungs 2015) from the tags associated with georeferenced images and place marks.

The urban character

The settings of VGI are mainly urban: most of the crowdsourced information – especially if derived from social media platforms – is produced in urban areas which combine connection facilities (internet, free WiFi, hotspots etc.) and the concentrated critical mass of city users (residents, tourists, business people, commuters, students, visitors, etc.).

A recent study (Hecht & Stephens 2014) reveals that in the US there are 3.5 times more Twitter users *per capita* in core urban counties than rural counties; the same authors have discovered that urban users Tweet more than their rural counterparts.

The following image (Figure 4) shows the relationship between Twitter activity at night and large urban areas in Italy (2013): 52% of the georeferenced Tweets (12,000 collected in one night) fall within the boundaries of the Italian large urban zones (LUZ), as defined by EUROSTAT; the percentage would be higher if the peri-urban areas were included.

VGI has been used in interesting ways in spatial analysis of cities' urban structures. In fact, recent research has identified 'natural cities' as human settlements, or human activities in general on the Earth's surface, that are delineated from massive geographic information derived from geocoded social media data (Jiang & Liu 2012; Jiang in this book). The interesting aspect of this approach is that the employment of crowdsourced information allows us to see how cities evolve and change over time, even if the changes may simply refer to the *espace vécu* (living space) by city users. Here is an example of Jiang's methodology applied to Florence (Italy) which shows the concentration changes of geocoded Tweets for six months (May–October 2013).

VGI has contributed to the discovery of other urban issues. For example, analysis of geolocated Flickr photos has identified the most attractive spots or intra-urban tourist routes (Girardin et al. 2008) and discovered that foreigners privilege stereotyped places mainly concentrated in the city centre or at transport nodes (airports, stations) while local people's gaze seems to fall on less central locations (Crandall et al. 2009; Straumann et al. 2014).

If the annotations (comments, texts) of crowdsourced data are taken into account, narratives about urban settings may be constructed and may vary according to whether they are produced by insiders or outsiders. The following

---

[3] See http://livehoods.org/.

table (Table 2) attempts to show the different qualitative narratives about London's attractions which emerge from different VGI sources. The table has been created by selecting the 10 most cited attractions on TripAdvisor and the 10 most frequently used hashtags on the Twitter account @Londonist; it highlights the sites mentioned, the related attributes and the meaning which can be ascribed. TripAdvisor's posts are generally created by outsiders (or tourists) and highlight the persistence of global imagery and the *grandeur* ideal of certain monuments which are at the base of a model of collective knowledge which endlessly reproduces itself over time (Raffestin 1988). In contrast, the most used hashtags from the profile @Londonist may be considered the

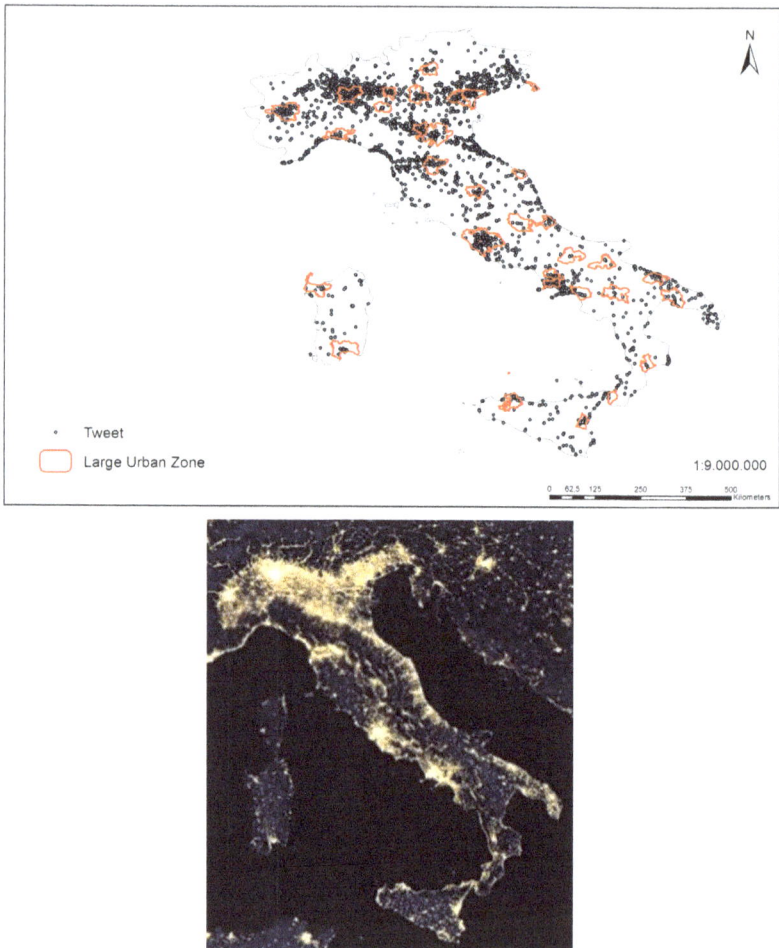

Figure 4: A comparison between Tweets sent at night and large urban zones in Italy (Source: Ladest 2016).

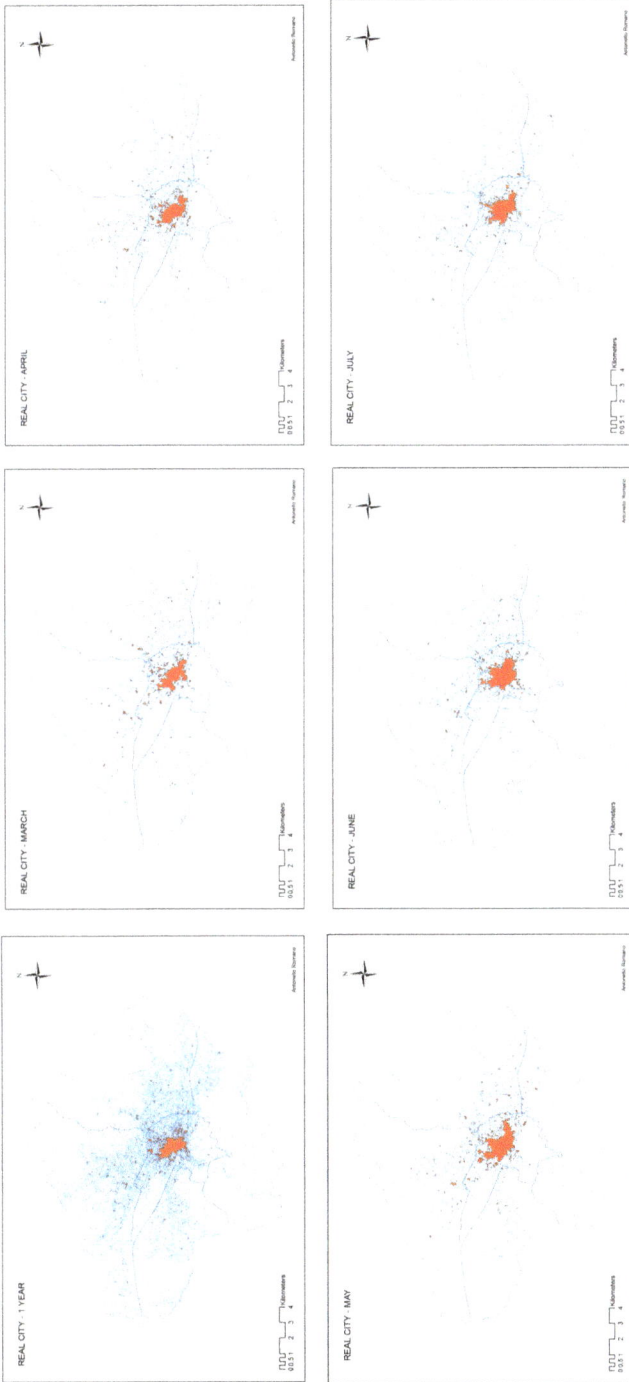

Figure 5: The urban *espace vécu* in Florence (May–October 2013). Source: Ladest Lab 2013.

Source	Sites	Comments / Attributes	Meaning /process
TripAdvisor	Tower of London, Big Ben, Tower Bridge, St. Paul's Cathedral, London Eye, Buckingham Palace, Trafalgar Square, Piccadilly Circus	Wonder, uniqueness, grandeur, beauty, excitement, surprise	Tourist place commodification Global imagery
	Covent Garden, Camden market	Curious, unusual	
@Londonist	Artistic and folk events Suburban neighborhoods (i.e. Walthamstow in East London) Wimbledon River Thames Parks	Silence Peaceful Relaxing Entertaining Cheap	Environmental quality Urban free time Local community life

Table 2: Qualitative crowdsourced information about the London urban area.

manifestation of insiders' preferences: they reveal Londoners' desire for more intimate and less crowded places and the quest for sites where they can find better environmental quality, social ties and green areas.

Conclusions

Crowdsourced information, namely volunteered geographic information (VGI), is a revolutionary source of information for increasing spatial and behavioural knowledge on different topics or phenomena in contemporary and everyday life from politics to the environment, from cultural events to natural disasters and more. The advances in geospatial technologies in the past twenty years have enabled ordinary citizens with little formal training to participate in the production of geographic data and knowledge through diverse forms of user-generated content and VGI: everyday activities may be transformed into creative expressions that can be uploaded, modified and shared in the digital world (Sui & Delyser 2012; Parks 2001). The multiple identities of VGI have been described as hybrid geographies which consider creative connections within geographies – physical and human, critical and analytical, qualitative and quantitative – aiming to integrate perspectives on place, revealing interactions and society at large (Sui & DeLyser 2012: 112). From a spatial perspective, VGI is an attempt to break free from institutional boundaries (municipalities, regions, provinces, counties, etc.) as shown by the application of the long tail model in urban development based on VGI data (Jiang 2014) or the employment of clustering methodologies based on geolocated crowdsourced data which produce areas of diffusion of interests, emotions and conflicts. From a place perspective, places – or more traditionally regions – identified are based on people who are at or experiencing a certain place and deliver different type of information which capture social practices and ongoing processes, either peaceful or conflictual.

As with any other scientific advance, VGI provides new food for thought and has raised many epistemological questions in the field of geography (Kitchin 2013, 2014) which have led to arguable definitions stating that data can speak for itself and that theory is no longer needed (Anderson 2008). But crowdsourced information is not just facts devoid of context: it may provide large quantities of geographic information which need to be distilled and exploited within clear theoretical frameworks (Kitchin 2013; Sui & Delyser 2012). Similarly to what happened in the 19th-century when geographer-explorers needed precise measuring instruments and binoculars to record their observations of the new lands and the resultant collaboration with naturalists, surveyors and biologists, nowadays, the geographer working with VGI data needs both the computing expertise to scrape and organise data from the Web and the cognitive tools to reach the inner meaning of this information. In this scenario new alliances have emerged between geography and computing and the cognitive sciences: the tools for making good use of VGI lie in the methodologies both for geolocating the information and for qualitatively analysing the contents around which discourses and narratives can be built on different scales.

Finally, we should praise VGI and its capacity to deal both with both every day and more fundamental topics that were unreachable in the past in a timely manner and with the heterogeneity of social phenomena by locating them on the Earth's surface. There is still a great deal to do to make sense of these distributions but undoubtedly the pulse of life with all its contradictions and inequalities can be grasped through a kind of information that, although it is currently concentrated in certain countries and areas, is likely to grow rather than shrink and hopefully to become ever more inclusive.

References

Anderson, C. 2008. The end of theory: the data deluge makes the scientific method obsolete. *Wired* (June 23, 2008). Available at: http://www.wired.com/science/discoveries/magazine/16-07/pb_theory (Last accessed 20 October 2014).

Aubrecht, C., Ungar, J., & Freire, S. 2011. Exploring the potential of volunteered geo-graphic information for modeling spatio-temporal characteristics of urban population. *Proceedings of 7VCT, 11*: 13.

Batty, M., Hudson-Smitha, A., Milton, R., & Crooks, A. 2010. Map mashups, web 2.0 and the GIS revolution. *Annals of GIS*, (16): 1–13.

Benkler, Y., & Nissenbaum, H. 2006. Commons-Based Peer Production and Virtue. *Journal of Political Philosophy, 14*(4): 394–419.

Bonney, R., Cooper, C. B., Dickinson, J., Kellin, S., Phillip, T., Rosenberg, K.V., & Shirk, J. 2009. Citizen science: a developing tool for expanding science knowledge and scientific literacy. *BioScience, 59*(11): 977–984.

Boulding, K. 1956. The image: knowledge and life in societies. Michigan: Michigan University Press.

Capineri, C., & Rondinone, A. 2011. Geografie. In volontarie. *Rivista Geografica Italiana, 118*: 559–577.

Castells, M. 1996. The Rise of the Network Society, The Information Age: Economy, Society and Culture. Oxford: Blackwell.

Castells, M. 2008. Communication power. Oxford: Oxford University Press.

Claval, P. 1974. La géographie et la perception de l'espace. *L'espace gèographique, 39*: 174–189.

Coleman, D. J., Georgiadou, Y., & Labonte, J. 2009. Volunteered Geographic Information: the nature and motivation of produsers. *International Journal of Spatial Data Infrastructures Research, 4*(1): 332–358.

Cooper, A. K., Coetzee, S., Kaczmarek, I., Kourie, D.G., Iwaniak, A., & Kubik, T. 2011. Challenges for quality in volunteered geographical information. Retrieved at: http://researchspace.csir.co.za/dspace/bitstream/10204/5057/1/Cooper1_2011.pdf.

Crandall, D. J., Backstrom, L., Huttenlocher, D., & Kleinberg, J. 2009. Mapping the world's photos. *Proceedings of the 18th international conference on World wide web*. ACM: pp. 761–770.

Crang, M. 1996. Envisioning Urban Histories: Bristol as Palimpsest, Postcards, and Snapshots. *Environment and Planning A, 28*: 429–452.

Curry, M. 1997. The digital individual and the private realm. *Annals of the Association of American Geographers, 87*(4): 681–699.

Curry, M. 1998. Digital Places: Living with Geographic Information Technologies. London: Routledge.

De Longueville, B., Ostländer, N., & Keskitalo, C. 2010. Addressing vagueness in Volunteered Geographic Information (VGI)–A case study. *International Journal of Spatial Data Infrastructures Research, 5*: 1725–0463.

Dodge, M., & Kitchin, R. (2013). Crowdsourced cartography: mapping experience and knowledge. *Environment and Planning A, 45*(1): 19–36.

Dykes, J., Purves, R., Edwardes, A., & Wood, J. 2008. Exploring volunteered geographic information to describe place: visualization of the 'Geograph British Isles' collection. *Proceedings of the GIS Research UK 16th Annual Conference* GISRUK: pp. 256–267. Retrieved at: https://www.researchgate.net/profile/Alistair_Edwardes/publication/242083251_Exploring_Volunteered_Geographic_Information_to_Describe_Place_Visualization_of_the_'Geograph_British_Isles'_Collection/links/02e7e52c2862197174000000.pdf.

Elwood, S. 2008. Volunteered geographic information: key questions, concepts and methods to guide emerging research and practice. *GeoJournal, 72*(3): 133–135.

Elwood, S. 2010. Geographic information science: emerging research on the societal implications of the geospatial web. *Progress in Human Geography, 34*(3): 349–357.

Elwood, S., Goodchild, M. F., & Sui, D. Z. 2012. Researching volunteered geographic information: Spatial data, geographic research, and new social practice. *Annals of the Association of American Geographers, 102*(3): 571–590.

Girardin, F., Calabrese, F., Fiore, F. D., Ratti, C., & Blat, J. 2008. Digital footprinting: Uncovering tourists with user-generated content. *Pervasive Computing, IEEE, 7*(4): 36–43.

Goodchild, M. 2007. Citizens as sensors: the world of volunteered geography. *GeoJournal*, 69: 211–221.

Goodchild, M., & Li, L. 2012. Assuring the quality of volunteered geographic information. *Spatial statistics*, (1): 110–120.

Gouveia, C., & Fonseca, A. 2008. New approaches to environmental monitoring: the use of ICT to explore volunteered geographic information. *GeoJournal, 72*(3–4): 185–197.

Graham, M. 2010. Neogeography and the palimpsests of place: web 2.0 and the construction of a virtual earth. *Tijdschrift voor Economische en Sociale Geografie, 101*(4): 422–436.

Graham, M., & Shelton, T. 2013. Geography and the future of big data, big data and the future of geography. *Dialogues in Human Geography, 3*(3): 255–261.

Graham, M., Zook, M., & Boulton, A. 2013. Augmented reality in urban places: contested content and the duplicity of code. *Transactions of the Institute of British Geographers, 38*(3): 464–479.

Graham, M., Hogan, B., Straumann, R., & Medhat, A. 2014. Uneven Geographies of User-Generated Information: Patterns of Increasing Informational Poverty. *Annals of the Association of American Geographers, 104*(4): 746–764.

Graham, S. 1998. The end of geography or the explosion of place? Conceptualizing space, place and information technology. *Progress in Human Geography, 22*(2): 165–185.

Haklay, M. 2010. How Good is volunteered geographical information? a comparative study of OpenStreetMap and ordnance survey datasets. *Environment and Planning B: Planning and Design, 37*(4): 682–703.

Haklay, M. 2013a. Citizen science and volunteered geographic information: Overview and typology of participation. Crowdsourcing Geographic Knowledge. The Netherlands: Springer.

Haklay, M. 2013b. Neogeography and the delusion of democratisation. *Environment and Planning A, 45*(1): 55–69.

Haklay, M., Singleton, A., & Parker, C. 2008. Web mapping 2.0: the Neogeography of the Geoweb. *Geography Compass, 2*(6): 2011–2039.

Haklay, M. 2010. How good is volunteered geographical information? A comparative study of OpenStreetMap and Ordnance Survey datasets. *Environment and Planning B: Planning and Design, 37*(4): 682–703.

Hardy, D., Frew, J., & Goodchild, M. 2012. Volunteered geographic information production as a spatial process. *International Journal of Geographical Information Science* iFirst: 1–22.

Harvey, F. 2013. To volunteer or to contribute locational information? Towards truth in labeling for crowdsourced geographic information. Crowdsourcing Geographic Knowledge. The Netherlands: Springer, pp. 31–42.

Hecht, B., & Gergle, D. 2010. *On the "Localness" of User-Generated Content*, Proceedings of the 2010 ACM conference on Computer supported cooperative work. New York, NY, pp. 229–232.

Hecht, B., & Stephens, M. 2014. *A Tale of Cities: Urban Biases in Volunteered Geographic Information*. In Proc. of ICWSM 2014, University of Michigan. Retrieved at http://www-users.cs.umn.edu/~bhecht/publications/bhecht_icwsm2014_ruralurban.pdf.

Howe, J. 2008. Crowdsourcing: Why the Power of the Crowd Is Driving the Future of Business. New York: McGraw-Hill, pp. 401–408.

Hollenstein, L., & Purves, R. 2014. Exploring place through user-generated content: Using Flickr tags to describe city cores. *Journal of Spatial Information Science*, (1): 21–48.

Jiang, B. (2013). The image of the city out of the underlying scaling of city artifacts or locations. *Annals of the Association of American Geographers, 103*(6): 1552–1566.

Jiang, B., & Jia, T. 2011. Zipf's law for all the natural cities in the United States: a geospatial perspective. *International Journal of Geographical Information Science, 25*(8): 1269–1281.

Kitchin, R. 2013. Big data and human geography Opportunities, challenges and risks. *Dialogues in Human Geography, 3*(3): 262–267.

Kitchin, R. 2014. Big Data, new epistemologies and paradigm shifts. *Big Data & Society, 1*(1). Retrieved at: http://bds.sagepub.com/content/1/1/2053951714528481.long (Accessed march 2016).

Levy, P. 1994. L'Intelligence collective. Pour une anthropologie du cyberespace. Paris: La Découverte.

Lussault, M. 2007. L'homme spatial: la construction sociale de l'espace humain. Paris: Seuil.

Pickles, J. 1995. Ground Truth: The Social Implications of Geographic Information Systems. New York: Guilford Press.

Poorthuis, A., & Zook, M. 2013. Spaces of Volunteered Geographic Information. In: Adams, P., Craine, J., & Dittmer, J. (Eds.) *Ashgate Research Companion on Geographies of Media* (in print).

Poorthuis, A., Zook, M., Shelton, T., Graham, M., & Stephens, M. Using Geotagged Digital Social Data in Geographic Research. In: Clifford, N.,

French, S., Cope, M., & Gillespie, T. (Eds.) *Key Methods in Geography*. London: Sage (forthcoming).

Priedhorsky, R., Masli, M., & Terveen, L. 2010. *Eliciting and focusing geographic volunteer* work. ACM *CSCW'10*, February, Savannah: pp. 6–10.

Purves, R. S., & Derungs, C. 2015. From space to place: place-based explorations of texts. *International Journal of Humanities and Arts Computing, 9*(1): 74–94.

Purves, R. S., & Edwardes, A. J. 2008. Exploiting Volunteered Geographic Information to describe Place. *Proceedings of the GIS Research UK 16th Annual Conference*: pp. 252–255.

Putman, R. 1995. Bowling Alone: America's Declining Social Capital. *Journal of Democracy, 6*(1): 65–78.

Raffestin, C. 1988. Le rôle de la ville d'art dans l'avènement d'une économie de la contemplation. *Cahiers de géographie du Québec, 32*(85): 61–66.

Roche, S., Nabian, N., Kloeckl, K., & Ratti, C. 2012. Are 'Smart Cities' Smart Enough. *Global Geospatial Conference*. Available at: http://www.gsdi.org/gsdiconf/gsdi13/papers/182.pdf (Last accessed 10 October 2014).

Shelton, T., Poorthuis, A., Graham, M., & Zook, M. 2014. Mapping the data shadows of Hurricane Sandy: Uncovering the sociospatial dimensions of 'big data'. *Geoforum, 52*: 167–179.

Straumann, R. K., Çöltekin, A., & Andrienko, G. 2014. Towards (Re) Constructing Narratives from Georeferenced Photographs through Visual Analytics. *The Cartographic Journal, 51*(2): 152–165.

Sui, D. Z. 2004. Tobler's first law of geography: a big idea for a small world? *Annals of the Association of American Geographers, 94*(2): 269–277.

Sui, D., & Delyser, D. 2013. Crossing the qualitative-quantitative chasm I: Hybrid geographies, the spatial turn, and volunteered geographic information (VGI). *Progress in human geography, 36*(1): 111–124.

Sui, D., Goodchild, M., & Elwood, S. 2013. Volunteered geographic information, the exaflood, and the growing digital divide. *Crowdsourcing Geographic Knowledge*. The Netherlands,Springer: pp. 1–12.

Surowiecki, J. 2005. *The Wisdom of Crowds*. New York: Anchor.

Tapscott, D., & Williams A. D. 2008. *Wikinomics: How mass collaboration changes everything*. London. Penguin.

Tobler. W. R. 1970. A computer movie simulating growth in the Detroit region, *Economic Geography, 46*(2): 234–240.

Turner, A. 2006. *Introduction to neogeography*. Sebastopol, CA: O'Reilly.

Whatmore, S. 2002. *Hybrid geographies: Natures cultures spaces*. London. Sage.

Wilson, M. W., & Graham, M. 2013. Neogeography and Volunteered Geographic Information: A Conversation with Michael Goodchild and Andrew Turner. *Environment & Planning A, 45*(1): 10–18.

Zook, M., & Graham, M. 2007. The creative reconstruction of the Internet: Google and the privatization of cyberspace and DigiPlace, *Geoforum, 38*: 1322–1343.

CHAPTER 3

Why is participation inequality important?

Mordechai (Muki) Haklay

Department of Civil, Environmental and Geomatic Engineering,
University College London, m.haklay@ucl.ac.uk

Abstract

Participation inequality – the phenomenon that a very small percentage of participants contribute a very significant proportion of information to the total output – is persistent across Volunteered Geographic Information (VGI) and citizen science projects. It has been identified in both online and offline projects that rely on volunteers' effort over the past 20 years and, therefore, can be expected to appear in new projects. This chapter looks at participation inequality (also known as the 1% rule or the 90-9-1 rule), its origins and some of its characteristics. The chapter also explains how participation inequality emerges in a project at both temporal and spatial scales, and also evaluates its implication on the use of VGI and citizen science data. The chapter suggests a generic rule for analysts of VGI and citizen science datasets, in the form: *'When using and analysing crowdsourced information, consider the implications of participation inequality on the data and take them into account in the analysis.'*

Keywords

Participation inequality, patterns of contribution, citizen science, online and offline communities, 1% rule, 90-9-1 rule

How to cite this book chapter:
Haklay, M. 2016. Why is participation inequality important?. In: Capineri, C, Haklay, M, Huang, H, Antoniou, V, Kettunen, J, Ostermann, F and Purves, R. (eds.) *European Handbook of Crowdsourced Geographic Information*, Pp. 35–44. London: Ubiquity Press. DOI: http://dx.doi.org/10.5334/bax.c. License: CC-BY 4.0.

Introduction

One of the most persistent aspects that can be noted in systems which facilitate user-generated content (among them volunteered geographic information and citizen science data) is the inequality in the level of participation that they exhibit. According to Jakob Nielsen (2006), participation inequality was first recognised by Hill and his team (1992) while studying the development of digital documents and analysing the contributions by different people to the final product. It manifests itself in online forums such as mailing lists, discussion forums, games and ecological observations (e.g. Hill et al. 1992; Mooney & Corcoran 2012; Lund et al. 2011; van Mierlo 2014; Silvertown et al. 2015). In each of these cases, the overwhelming majority of people who use the information or are registered to the service do not contribute any information to it. The proportion of registered people who do not contribute can reach 90% or even more of the total number of users. Of the remaining participants, the vast majority contribute infrequently or fairly little – these account for 9% or more of the users. Finally, the last 1% contribute most of the information. This has led to framing the phenomenon as the 90-9-1 rule (Nielsen 2006). However, participation can be very skewed. As Nielsen demonstrates, in Wikipedia, 0.003% of users contribute two-thirds of the content, with a further 0.2% contributing infrequently, making the relationship 99.8-0.2-0.003% (with the increased use of Wikipedia since 2006, the situation has worsened). There is some evidence to suggest that the proportion can be different – for example, Budhathoki (2010) suggests that in OpenStreetMap the proportions are 70-29.9-0.01%. Recent analysis by Harry Wood (2014) provides an indication of this relationships (Figure 1), with the contribution of the first ranked 1,000 participants dwarfing the effort of all other contributors, and only about 300,000 participants contributing more than 10 points of data - although at the time there were 2 million registered users.

Participation inequality has been observed in VGI and citizen science projects such as OpenStreetMap (Budhathoki 2010; Mooney & Corcoran 2012; Neis & Zipf 2012), Galaxy Zoo (Ponciano & Brasileiro 2014) and bird watching (Cooper & Smith 2010). It is especially noteworthy that participation inequality is not only appearing in online projects, but also can be observed in projects that mainly happen offline, such as participation in environmental volunteering or when analysing the levels of contribution of different volunteers in biological observations across London.

In this chapter, we look at the implications of participation inequality and argue that it is among the most significant aspects of VGI and citizen science. We start by noticing what we already know about participation inequality and its manifestations. This is followed by suggesting possible explanations for how it occurs and evolves over time. The fourth section discusses the potential implications on project development and the use of information that emerges from it. We conclude with open research questions and future directions for investigation that are of specific interest to researchers of VGI.

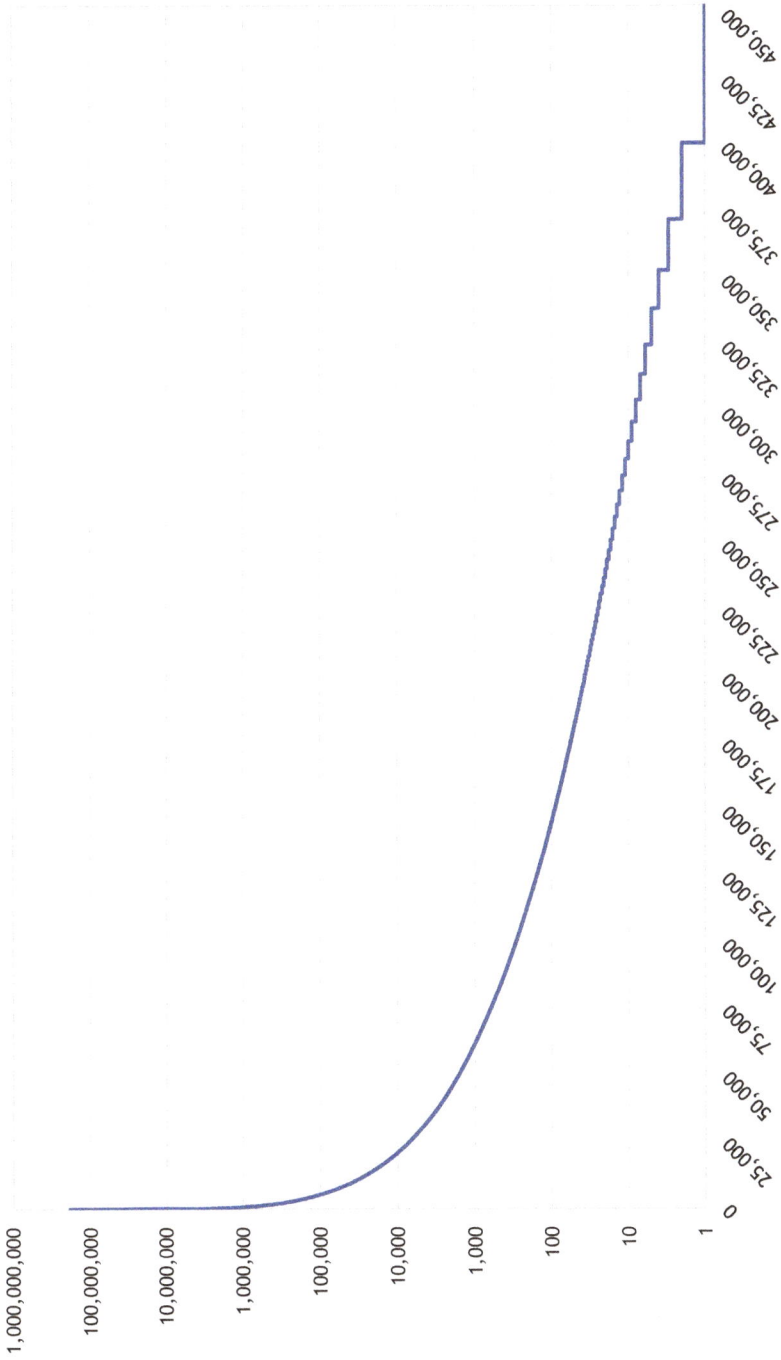

Figure 1: OpenStreetMap contributions (Wood 2014).

Throughout the chapter, OpenStreetMap is being used to demonstrate the nature and implications of participation inequality. While OpenStreetMap have specific characteristics in terms of participants' profiles and social dynamics (Budhathoki 2010; Haklay 2010), it can be used to illustrate the general aspects of the phenomena. Other projects are being used to augment the picture.

Participation inequality – what do we know?

Unlike command and control processes that are common in industrial information creation, VGI and citizen science are produced through a distributed, less coordinated system. Within industrial processes, there is scope for planning of coverage and allocation of resources. For example, when planning the surveying of a city in an industrial process, it is possible to divide the efforts of the surveyors to ensure uniform level of coverage and time allocation to different parts in proportion to the amount of work that is required. Of course, the abilities of the different surveyors will have an impact on the final results but, in general, these can be minimised through quality assurance so the final product is uniform.

Within a system that relies on 'crowdsourcing' – the use of a large group of people with whom there are no direct employment relationships – there is far less ability to dictate to the participants where, when and how they should contribute information. For example, in a system that provides traffic information on the basis of users' satellite navigation devices, there is a co-dependence between the number of users in a given location and the ability to provide information about this place. Moreover, because the devices are used within the context of daily activities, such as the school run or a trip to the local supermarket, there will be more information about places in which many people travel daily (e.g. city centre) and especially during rush hour. While both industrial and crowdsourced systems are socio-technical systems, in the latter the 'socio' requires special attention, particularly to the way it influences the resulting information that emerges from the system.

In the case of participation inequality, since it has been so persistent over the years, it is highly likely to appear in any crowdsourcing project. It has been observed from the pre-Web internet messaging system Usenet (Whittaker et al. 1998) to current large-scale online citizen science (Ponciano & Brasileiro 2014). It is, therefore, part and parcel of VGI and citizen science.

Just as interesting is that the phenomenon repeats itself at various scales (something akin to Power Laws), so analysing the level of participation in OpenStreetMap for the area of London, Europe or across the world will show participation inequality (Haklay 2010; Mooney & Corcoran 2012; Neis & Zipf 2012). Participation inequality also occurs at different temporal scales of weeks, months or years (Neis & Zipf 2012). As can be expected with statistical analysis of this sort, the larger the area or the longer the time frame, the clearer the pattern and the position of various participants.

Another important aspect known about participation inequality is that lowering the barrier for participation does help, but to a limited extent. Even in volunteer computing projects, in which participants download software to their computers that utilises unused processing resources for scientific research, participation inequality persists. IBM World Community Grid serves as an example. This project is an aggregator of volunteer computing projects, and yet few members contributed most of the processing. Of the 350,000 participants, the top contributor has contributed 325 times more than the 250th contributor, and 875 times more than the 1,000th contributor.

The use of a leader board and providing credits to emphasise the position of participants has been shown to encourage competition among contributors, but with a potential to alienate some participants and reduce their motivation (Massung et al. 2013). The assumption that it is always valuable to encourage competition among participants to yield more information should be questioned, and there are alternative, such as the mechanism that encourage collaboration that Silvertwon et al. (2015) offer.

Participation inequality also manifests itself through geographic and temporal patterns. Thus, places that are within the coverage area of highly active participants will have more contributions than areas that do not have many participants. More generally, the geographic distribution of information shows that some places are more popular and receive much more attention than others. Similarly, the temporal pattern of highly active contributors has a disproportionate impact on the temporal patterns of data collection activities as a whole. Thus, the sleeping and working patterns that can be observed within the contributed information will be influenced by the practices of high contributors (Yasseri et al. 2013).

Finally, while high contributors receive a lot of attention, in comparison to the very large group of people who contribute very little both individually and to the overall size of the dataset, we should not forget that they are, statistically, outliers. They are not representative of the overall population, nor should we expect them to be so. There is a need to have the majority of people as consumers of information, as otherwise the producers would lose the *raison d'être* to create and share information.

How participation inequality evolves over time and space

One of the puzzling questions regarding participation inequality is how it evolves. After all, at first look the participants are acting as volunteers and therefore there is no limitation on the number of people who can join a specific activity in citizen science or VGI or how much each of them contributes. Second, arguably, the actions of one participant do not stop another, for example when viewing the same bird or taking a geotagged picture of Big Ben (see Jayaraman 2012). Furthermore, the participants are only loosely coordinated and therefore not necessarily aware of the actions of other participants, and there is no reason for one to compete with another or even be aware of their contribution. However, some of these observations are inaccurate, and a further analysis of the process that created participation inequality can explain the source of the observed patterns.

Firstly, we can start by noticing that, in many VGI and citizen science projects, some resource is finite. For example, in OpenStreetMap or Wikimapia, once a participant has tagged a location and mapped it, this specific place is no longer available to other users to carry out the mapping. This is also true in volunteer thinking projects in which participants help scientists in classifying information online. In such projects, the system allocates the images to participants and, after the image has been viewed by a given number of participants, it is not shown anymore. Therefore, if one participant becomes highly active, they reduce the amount of work that is left to other participants to carry out.

Secondly, the temporal aspects of the project also play their part in generating participation inequality. For example, participants who joined OpenStreetMap early on were facing an empty map, in which it was relatively easy to identify and digitise objects such as motorways. Over time, the ability to digitise objects rapidly diminished as the map became complete. For a volunteer who joins the mapping process today, in many places the effort that is left requires adding more intricate details of building or address information. This is also true in citizen science, for example in the British Trust of Ornithology (BTO) breeding bird survey which started over twenty years ago. A volunteer that joined the project in the early stages will have collected many more records over the years than a person that will join the project today, who will not be able to 'catch up' to such levels of recording.

Thirdly, another side to the temporal aspect is demonstrating the link between participation inequality and other social inequalities. The contributions of participants can be translated into time – for example, one of top contributor to OpenStreetMap in June 2015 (Władysław Komorek) edited over 4.94 million objects in 966 active mapping days over 3 years, contributing on average about 5,100 points in an active day. With an assumption that it is possible to record 2 objects per second, this represents an average investment of about hour in digitising only (without any breaks). This is, of course, a low estimation, since such a participant spends time on mailing lists, meetings and going out mapping. When considering that, across advanced economies, people have about 36.5 hours of leisure a week (OECD 2009), it is clear that, for this participant, OpenStreetMap is the most important leisure activity during that period. However, since leisure time is more available to men, and is reduced in people with major caring responsibilities, it is more likely that men with a well-paid job will be able to become major contributors of VGI and in many citizen science activities. Indeed, many projects have people with such profiles as their top contributors (e.g. Cooper & Smith 2010). However, one should be careful of sweeping generalisations about the profile of top contributors, as they are specific to projects and research area – for example in the EyeWire project, in which participants help brain research by analysing the structure of neurons, 65% of top contributors are women (Kim et al. 2014), while bird watching is dominated by men (Cooper & Smith 2010).

Fourthly, access to financial resources can have an impact on the ability of people to become high contributors. For example, an Australian study of bird-watchers concluded that some of them travel 300 to 1,900 km from home to record an observation (Tulloch & Szabo 2012). Such extensive travel, apart from dedication, also requires financial resources. Other VGI and citizen science activities also involve purchasing specialised equipment and dedicating time to learn how to use it.

Finally, there is a need to consider internal and external motivations of high contributors. The various studies that were mentioned above, and others, demonstrate clearly that the top contributors represent a different demographic group – for example, in EyeWire they are older than the average participant (Kim et al. 2014). Studies show that their internal and external motivations play an important part in maintaining their engagement with a project. For some participants, competition is a significant motivation (Massung et al. 2013) while for others the joint contribution to science is a major one (Nov et al. 2011).

The implications of participation inequality

Based on the analysis above, we can formulate a general rule for crowdsourced geographic information: '*When using and analysing crowdsourced information, consider the implications of participation inequality on the data and take them into account in the analysis.*'

As we have seen, crowdsourced information, either VGI or citizen science, is created through a socio-technical process, which, by necessity, will have impacts on the final outputs. Yet, all too often it is easy to forget the social side – especially when using the information without paying due attention to the metadata of who collected it and when. Even though analysts who use the information are aware that the data source is expected to be heterogeneous because of the nature of the crowdsourced process, it is easy to forget participation inequality and treat each observation as similar to other observations and assume they were all produced in a similar way.

Yet, data is not only heterogeneous in terms of consistency and coverage; it is also highly heterogeneous in terms of contribution, which can have far-reaching implications on quality, coverage and content. As we have explored, the outcome is dependent on the expertise of heavy contributors, their spatial and temporal engagement, and even on their social interactions and conduct.

For example, some of the top contributors of OpenStreetMap naturally concentrate their effort in the city where they live. Knowing where these individuals are active can help in quality assurance processes by comparing novice practices to their actions, potentially changing the number of people that are required to map an area well (Haklay et al. 2010). In some projects, such as iSpot (Silvertown et al. 2015), in which participants help in the identification of a range of species, there are mechanisms to reward high contributors with trust marks and to give their opinions more weight during the identification process.

Another aspect of the impact of high contributors is the social evolution of the project. In some projects, high contributors might exhibit abrasive behaviour towards other participants or protect 'their patch' (the area in which they operate) by aggressively editing any new information to fit their standards. Such conduct is not welcoming to new participants, and can impact on the growth of the project and even its resilience in cases where the high contributor leaves the project.

The specific background and interests of high contributors will, by necessity, impact on the type of data that is recorded. This is especially important in VGI projects where the details of what to record are left to the participants. For example, lack of interest in a class of facilities (e.g. wheelchair accessible toilets) will mean that such information will be lacking from the resulting dataset and might shape the activities of other participants (Stephens 2013).

Interestingly, while some research analysed the biases that are created by high contributors (Haklay 2010; Bégin et al. 2013; Mooney 2013), there is relative lack of attention within the VGI literature to the wider impact that they have on the information and on other participants.

Conclusion

In this chapter, we looked at participation inequality and its implications on VGI and citizen science datasets. We have seen that participation inequality – the phenomenon in which a very small percentage of participants contributes a very significant proportion of information to the total outcome – is persistent. It occurs across spatial and temporal scales and is driven by multiple factors.

Participation inequality impacts on the social and technical outcomes of a project and, because of that, it is critical to remember the impact and implications of participation inequality during the analysis and use of the information. There will be some analysis to which it will have less impact and some where it will have major impact. In either case, it needs to be taken into account. This can be done by including an analysis of participation patterns early on in the analysis of a dataset, and examining the biases that are caused by it.

While we can expect it, we do need to understand more about the process that created it and its impact on the resulting datasets. There is plenty of scope for spatio-temporal analysis to identify the actions of high contributors from their early actions, and evaluate to what degree they impact on other contributors. There is also value in more detailed analysis of how people at different levels of contribution add to the project and whether there are ways to encourage people to move between contribution groups. Finally, the ethical and practical implications of high contributors should be assessed, especially in commercial VGI projects.

Acknowledgements

The author would like to thank Prof Tanya Berger-Wolf for her suggestions on explaining participation inequality and the work of Vyron Antoniou, Valentine Seymour and Gianfranco Gliozzo at UCL on various datasets. Many thanks for Cristina Capineri and Stephen Winter for their comments on an earlier version of this paper. The research was supported by EPSRC (EP/I025278/1, EP/K022377/1) and the EU FP7-ICT (317705).

References

Bégin, D., Devillers, R., & Roche, S. 2013 (May). Assessing volunteered geographic information (VGI) quality based on contributors' mapping behaviours. In *Proceedings of the 8th international symposium on spatial data quality ISSDQ*: pp. 149–154.

Budhathoki, N. R. 2010. *Participants' motivations to contribute geographic information in an online community* (Doctoral dissertation, University of Illinois at Urbana-Champaign).

Cooper, C. B., & Smith, J. A. 2010. Gender patterns in bird-related recreation in the USA and UK. *Ecology and Society, 15*(4): 4. Available at: http://www.ecologyandsociety.org/vol15/iss4/art4/.

Haklay, M. 2010. How good is volunteered geographical information? A comparative study of OpenStreetMap and Ordnance Survey datasets. *Environment and planning. B, Planning & design, 37*(4), 682–703.

Haklay, M., Basiouka, S., Antoniou, V., & Ather, A. (2010). How Many Volunteers Does it Take to Map an Area Well? The Validity of Linus' Law to Volunteered Geographic Information. *Cartographic Journal* 47 (4) 315–322.

Hill, W. C., Hollan, J. D., Wroblewski, D., & McCandless, T. 1992. Edit wear and read wear. In: *Proceedings of the SIGCHI Conference on Human factors in Computing Systems*. ACM: pp. 3–9.

Jayaraman, K. 2012. Tragedy of the Commons in the Production of Digital Artifacts. *International Journal of Innovation, Management and Technology, 3*(5): 625–627

Kim, J. S., Greene, M. J., Zlateski, A., Lee, K., Richardson, M., Turaga, S. C., Purcaro, M., Balkam, M., Robinson, A., Behabadi, B. F., Campos, M., Denk, W., Seung, H. S., & the EyeWirers. 2014. Space-time wiring specificity supports direction selectivity in the retina. *Nature, 509*(7500): 331–336.

Lund, K., Coulton, P., & Wilson, A. 2011 (November). Participation inequality in mobile location games. In *Proceedings of the 8th International Conference on Advances in Computer Entertainment Technology*. ACM: p. 27.

Massung, E., Coyle, D., Cater, K. F., Jay, M., & Preist, C. 2013. Using crowd-sourcing to support pro-environmental community activism. In: *Proceedings of the SIGCHI Conference on Human Factors in Computing Systems*. ACM: pp. 371–380.

van Mierlo, T. 2014. The 1% rule in four digital health social networks: an observational study. *Journal of medical Internet research, 16*(2).

Mooney, P. 2013. *Understanding the activity of contributors to Volunteered Geographic Information projects. How, why, where, and when do they contribute geographic information?* 2nd Meeting of the EU COST Action TD1202, Dresden, Germany.

Mooney, P., & Corcoran, P. 2012. Who are the contributors to OpenStreetMap and what do they do? In: *Proceedings of the GIS Research UK 20th Annual Conference.* Lancaster (GBR), 11–13 April: pp. 355–360.

Nielsen, J. 2006. Participation inequality: Encouraging more users to contribute. *Jakob Nielsen's alertbox, 9*: 2006.

Neis, P., & Zipf, A. 2012. Analyzing the Contributor Activity of a Volunteered Geographic Information Project—The Case of OpenStreetMap. *ISPRS International Journal of Geo-Information, 1*(2): 146–165.

Nov, O., Arazy, O., & Anderson, D. 2011 (February). Dusting for science: motivation and participation of digital citizen science volunteers. In: *Proceedings of the 2011 iConference.* ACM: pp. 68–74.

OECD. 2009. *Society at a glance 2009: OECD social indicators*, OECD.

Ponciano, L., & Brasileiro, F. 2014. Finding volunteers' engagement profiles in human computation for citizen science projects. *Human Computation, 1*: 5–28.

Silvertown, J., Harvey, M., Greenwood, R., Dodd, M., Rosewell, J., Rebelo, T., Ansine, J., & McConway, K. 2015. Crowdsourcing the identification of organisms: A case-study of iSpot. *ZooKeys*, (480): 125.

Stephens, M. 2013. Gender and the GeoWeb: divisions in the production of user-generated cartographic information. *GeoJournal, 78*(6): 981–996.

Tulloch, A. I., & Szabo, J. K. 2012. A behavioural ecology approach to understand volunteer surveying for citizen science datasets. *Emu, 112*(4): 313–325.

Whittaker, S., Terveen, L., Hill, W., & Cherny, L. 1998. The dynamics of mass interaction. In: *CSCW '98 Proceedings of the 1998 ACM Conference on Computer Supported Cooperative Work*: pp. 257–264.

Wood, H. 2014. *The long tail of OpenStreetMap*. Available at: http://harrywood. co.uk/blog/2014/11/17/the-long-tail-of-openstreetmap/#slide18 (accessed July 2015).

Yasseri, T., Quattrone, G., & Mashhadi, A. 2013. Temporal analysis of activity patterns of editors in collaborative mapping project of OpenStreetMap. In: *Proceedings of the 9th International Symposium on Open Collaboration*. ACM: p. 13.

CHAPTER 4

Social Media Geographic Information: Why social is special when it goes spatial?

Michele Campagna

Università di Cagliari, DICAAR, Via Marengo 2, Cagliari 09123, Italy,
campagna@unica.it

Abstract

This contribution introduces the concept of Social Media Geographic Information (SMGI) as a specific type of Volunteered Geographic Information (VGI).
Unlike other kind of VGI, which may originate from geographic measurements
crowdsourcing, SMGI brings in addition a special potential for it may express
community perceptions, interests, needs, and behaviors. Hence, SMGI may
represent an unprecedented resource for expressing pluralism in such domains
as spatial planning, where it may convey the community collective preferences
contributing to enrich knowledge able to inform design and decision making.
In the light of these assumptions, the main issues relevant for SMGI collection and analytics are presented from the perspective of the spatial planning
and governance domain, and a framework for the SMGI analytics in planning,
design, and decision making is proposed.

Keywords

Social Media Geographic Information (SMGI), Spatial Data Infrastructure
(SDI), Spatial Planning and Governance, Geodesign, Planning Intelligence.

How to cite this book chapter:
Campagna, M. 2016. Social Media Geographic Information: Why social is special when
    it goes spatial?. In: Capineri, C, Haklay, M, Huang, H, Antoniou, V, Kettunen, J,
    Ostermann, F and Purves, R. (eds.) *European Handbook of Crowdsourced
    Geographic Information*, Pp. 45–54. London: Ubiquity Press. DOI: http://dx.doi.
    org/10.5334/bax.d. License: CC-BY 4.0.

Introduction

It is often assumed, as a proxy benchmark for the growing amount of information being produced, that in 2009 more data were generated by individuals than in the entire history of mankind through to 2008. In this avalanche of information, social data are increasingly used to build balanced relationships between business and customers. This way, consumers are stimulated by successful web actors to share their data truthfully with the prize of earning the power of being listened to, in addition to other concrete advantages: the online world is beginning to be ruled by users' expectations (Weigend 2009) with major implication for industries and businesses. Likewise, social media are growingly becoming a major arena for politics. We live in a time where the premiere of a candidature to the United States of America presidential elections is expected to be released on Twitter and YouTube channels, and where social movements involving thousands of people are organized on the internet. More and more public authorities are moving online along the stream of a general digital social uptake. Nevertheless, unlike mainstream politics or other sectors of government, in the domain of spatial planning and governance methods and tools to fully exploit the potential of social data, actively (i.e. to create discourse) or passively (i.e. to listen to the community), still are not widely used. This poses questions concerning the actual willingness of citizens to establish such powerful user relationship with public authorities, and what public authorities can do to meet citizens' expectations.

From a technical perspective, nowadays many organizations rely on social media management tools to interact with their customers, which in the case of public authorities include the citizens; however these social engagements tools often remain siloed from other enterprise applications (Oracle 2013). As in many other domains in the private and the public sector, spatial planning and governance is a sub-domain where the integration of social data with authoritative official information may provide opportunities for improving the dialogue with the citizens, by not only listening to their preferences but also by monitoring the social processes they are involved in, towards more pluralist, informed, and community-oriented decision making.

Unstructured social media contents that capture citizens' interests, intentions, perceptions, and needs may enrich traditional institutional and other commercial data sources. Key Performance Indicators can be developed to monitor through real time dashboards the reactions of the community to the public policies and actions helping to respond promptly to citizens' behaviors, moving a step towards a new generation of planning intelligence, thereby contributing to a more sustainable and smart growth.

With this premise the chapter is articulated as follows. In the next section a brief overview of recent advances in digital spatial data sources argue knowledge building in planning is enriched by the availability of social data. Afterwards, a definition of Social Media Geographic Information (SMGI) is given as a special type of Volunteered Geographic Information (VGI). Then, in the core section of the chapter, a novel SMGI analytics is proposed from the perspective of spatial planning and governance. The conclusions briefly summarize and propose issues for further research development.

Social data enrich the planning intelligence

Recent approaches to spatial planning and design propose the concept of Geodesign, emphasizing the role of knowledge about the local territorial context to inform design and decision-making. According to Steinitz (2012) there is no such profession as the Geodesigner, rather a Geodesign process is carried out through collaboration among different experts coming from the design disciplines and from the Geographic (Information) Sciences, as well as stakeholders and other actors from the local communities, or the people of the place. Unlike until a decade ago, such an approach is currently enabled thanks to development both in authoritative and volunteered sources of geographic information.

On the one hand, in an increasing number of countries and regions developments in Spatial Data Infrastructures (SDI) are starting to offer planners dozens, and in some cases even hundreds, of official large-scale spatial data layers, enabling the transition from analogue cartography analysis to digital geoprocessing in the representation and analysis of territorial processes as well as in the environmental impact assessment of design alternatives. This is the common case in Europe, where the Directive 02/2007/EU establishing the INfrastructure for SPatial InfoRmation in Europe (INSPIRE) is promoting public access to official spatial data produced by public authorities at all levels in the Member States. Along the adoption and the implementation of INSPIRE, in a growing number of regions, Advanced Regional SDIs are offering spatial data and services to professionals working on spatial planning and environmental impact assessment and are thus starting to bring innovation into the planning and design practice (Campagna & Craglia 2012).

On the other hand, the wealth of Volunteered Geographic Information (Goodchild 2007) offered by geobrowsers and widespread diffusion of GPS-equipped handheld devices, is starting to represent a novel –but already must-have – sources of information in many fields according to neo-geography or citizens science approaches. OpenStreetMap[4] may be considered one of the most successful example of GI crowdsourcing to create a comprehensive and high quality open spatial dataset as a major alternative to more traditional official or commercial sources. Topography, networks, habitats, biodiversity, diseases spreading, climate change, and hazards are some of the examples of environmental and social processes being mapped by voluntary observers

---

[4] www.openstreetmap.org.

acting as citizens sensors. However, only a fraction of available VGI is purposefully produced and contributed, while an even larger share is made available often as unaware results of the use of social media on web and mobile apps. With regard to the latter share, which is the focus of this paper, the next section argues that social media georeferenced content deserves to be treated individually in research for its peculiar characteristics, and special focus is given to its potential as complementary knowledge base in spatial planning and governance to support design and decision making.

Why social data are special when they are spatial?

Social Media Geographic Information (SMGI) can be defined as any piece or collection of multimedia data or information with explicit (i.e. coordinates) or implicit (i.e. place names or toponyms) geographic reference collected through the social networking web or mobile applications. Social data are acknowledged as a good of major value in the digital economy, and their potential for enhancing more traditional analytics is of the utmost importance. A big part of social data however also features spatial (and temporal) references, thus their integration with more traditional Authoritative Geographic Information (AGI) may enable a further step towards the next generation of geospatial intelligence.

SMGI is a sub-category of VGI and can be active or passive, depending on the type of application with which it is collected: applications purposefully created and/or used to collect SMGI in participatory initiatives (as in Campagna 2014) originate active SMGI, while SMGI harvested by general purpose social media such as Twitter or Instagram are passive, and can be considered more generically as user generated content.

Multimedia content of SMGI may include texts, images, videos, or audios in whatever combination usually aggregated in place marks or posts. Together with spatial references, place marks and posts usually feature a time reference and creator, or user owner.

Application Programming Interfaces (API) can be used by the public to access SMGI. Hence, the data model of each publicly accessible sources depends both on the original data model in the social media owner database and on the API. In general, the publicly available SMGI data model features a subset of the original attributes or multimedia data, implying that the analytical potential is in general greater within the social media companies than for the public (Lazer et al. 2009). Through the APIs, SMGI may be retrieved and accessed by keyword, by space, by time, by user, or by a combination of the former depending on the original social media platform and/or API. Data returned by a query through the API can be converted in a spatio-temporal dataset. Hence, spatial-temporal analyses are supported on SMGI as well as user-behavioral analysis (i.e. the analysis of user's behavior in term of data production and sharing). Spatio-temporal and user-behavioral analyses may be combined with querying and mining techniques on multimedia (e.g. spatio-temporal textual analysis can be defined as the analysis of text in a given area in a given moment or time frame) creating new opportunities for SMGI analytics.

Towards SMGI Analytics for spatial planning and governance

As introduced in the previous sections, it is argued in this paper that SMGI may turn out to be a very valuable source of information to support spatial planning and design. However, a novel analytics is to be formalized for the peculiar data models which make this type of information different from more traditional vector spatial datasets, with which it can be integrated (i.e. AGI; e.g. spatial data layer from institutional SDI) in order to elicit knowledge useful for informing spatial planning, design, or governance.

At the current state of development, common AGI vector datasets available for download in national or regional SDIs as shapefiles or remotely accessible as Web Feature Services, in Italy as in other countries in Europe, feature a geographic and a thematic component or dimension. Hence, they can be represented as follows:

$AGI = <x, y, z; a_i>$ where x, y, z represent the geographic coordinates, and a_i any thematic alphanumeric attributes of the common relational database data types (i.e. text, incl. URL, numbers, or dates).

In contrast, SMGI usually features a richer data model including temporal and multimedia components. Additionally, each piece of information exhibits a user dimension, which may include an identifier as well as other data which convey information on the user's profile. The latter plays a special semantic role for the user who produced and shared the single piece of information becoming a dominant dimension from the analytical perspective. In addition, each piece of SMGI often features a score expressing agreement by or interest for and popularity within the virtual community, due to the functioning of the majority of the social networking apps.

Thus, SMGI can rather be represented as follows:

$SMGI = <x, y, z; t; u; m_i; l>$ where t represents time associated to each element of the set, u the user, m_i the multimedia content (i.e. text, images, video, or audio clip), and l the amount of 'likes and dislikes', the number of 'stars', or any other kind of popularity or agreement score, which indicates consensus on the measure and should be treated accordingly in the analysis. Figure 1 summarizes the main features of the AGI and SMGI different data models.

Therefore, any SMGI analytical framework should include not only traditional spatial analysis but also temporal, multimedia, and user behavioral analyses methods, and these should be tightly integrated in order to fully exploit the knowledge potential embedded in data. From a planning analysis perspective, coupling these methods in an integrated GIS application would be an advantage in as much GIS is (becoming) the common platform for the

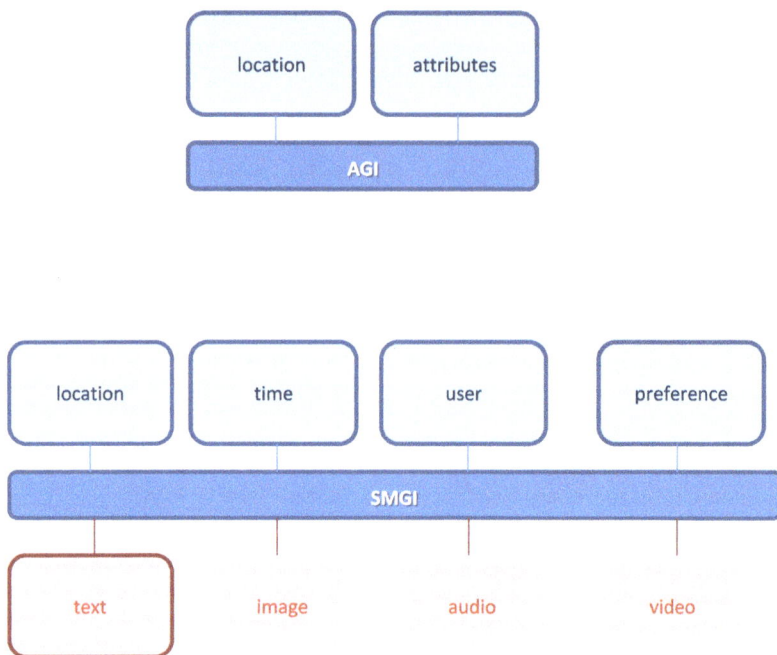

Figure 1: Scheme of the AGI (up) and of the SMGI (down) data models.

planning profession given the role of maps in expressing knowledge and design in this domain. With this consideration in mind, a framework for SMGI analytics have been developed by the author with the objective of exploiting this new resource of information in order to enrich the knowledge about the local context from the social perspective and to better support spatial planning and governance. This framework is under development; nevertheless the current results seem promising and allow proposing a tentative formalization as a guide for further research in this field.

In the remainder of the paper, this tentative framework for SMGI analytics is presented with references to the case studies developed at the University of Cagliari under the supervision of the author. The case studies relate to two major research streams. The first one pertains the development and use of a map-based social networking platform, namely 'Place, I care!' (PIC! 1.0). The latter can be defined as an active SMGI resource where users can create thanks to a user-friendly interface a private or public project. Thanks to an advanced user permissions manager each project can be customized to the contextual use case requirements. This way, accepted users can be allowed (or not) either to post, like/dislike, or comment, enabling different levels of participation in a map based discussion. Simple query functions are available in the map interface and collected data can be easily exported for further analysis with GIS packages. PIC! 1.0 was used in a number of pilot projects (Campagna 2014; Campagna et al. 2015).

The second research stream concerns the collection and use of SMGI produced by major social networking platform such as Twitter, YouTube, Instagram, and Booking.com (i.e. passive SMGI resources). In both cases, SMGI datasets were collected in selected areas at various scales (i.e. from the global, to the regional, to the local) and then integrated with other sources of AGI (e.g. from regional or local SDI) or VGI (e.g. WikiMapia).

When SMGI is integrated in a GIS project with other authoritative and volunteered spatial datasets, the peculiarities of their data models enable the analyst to perform the following analysis on SMGI data:

- **Spatial analysis of user interests**: thanks to the widespread use of social media, the high number of georeferenced posts enables us to investigate the patterns of user interest in space by density (Campagna 2014) and clustering functions (Massa & Campagna 2015). The overlay with topographic AGI such as administrative boundaries, or physical artefacts such as buildings, infrastructure, services or public spaces, may offer useful hints to public authorities to understand not only which places are important to the community and how they are perceived (Campagna 2014), and by whom the community is eventually composed (e.g. local people, commuters, tourist or other);
- **Temporal analysis of user interests**: the temporal reference is often an available attribute in SMGI, enabling to study when given regional destinations, urban districts, public spaces, or other infrastructures and services are used along the year, the months, the week, or the day (an example of this type analysis is given in Massa and Campagna, in this volume);
- **Spatial Statistics of user preferences**: collecting posts by spatial units enables planners to analyze patterns in user interest at different scales. An example is given in Floris and Campagna (2014), where the hot-spot analysis has been used at the regional level to study the distribution by municipality of positive user assessments by user profiles, where the hot-spot analysis has been used at the regional level to study the distribution of positive user assessments by user profiles to discover where young vs. elder, or family vs. solo tourists prefer to go during their trips. In the same study, the hot-spots were then investigated with Spatio-Temporal Textual analysis (see STTx below) and with geographically weighted regression analysis to explore at the local level what physical and locational factors may affect the preferences;
- **Multimedia content analysis on texts, images, video, or audio**: multimedia analysis is well developed in the case of texts analytics. However, it is currently more difficult to automatically extract useful information from images, video, or audio. In the case of text, many software packages can be used to apply simple (i.e. calculating words frequency, or tag clouds) to more advanced (e.g. sentiment analysis) text analysis techniques. These

techniques can be easily applied to subsets of SMGI obtained by spatial, temporal, or user query (see STTx below);

- **User behavioral analysis**: querying SMGI by user enables to study users' behavior in space and time. This information can be used to analyze, for example, whether a public space is visited by local people or by visitors. This information may be useful also for profiling: for the users visiting a certain place or service user Spatio-temporal footprints can be defined to identify people who mainly move locally, regionally, or internationally, and where they come from;

or

- **A combination of two or more of the previous** such as in **Spatial-Temporal Textual Analysis (STTx)**: textual analysis functions can be integrated within GIS applications (Campagna 2014), enabling the application of text analysis techniques to subset of SMGI selected by space and time. Several examples were tested by the authors to analysis the perception of different neighborhoods in the city using Place, I care! data (Campagna 2014), the judgment of tourists on a destination using Booking.com data (Floris & Campagna, 2014), or who is talking about an event of global reach around the world using Twitter and YouTube data (Massa & Campagna 2014). Although not verified yet by a systematic analysis, several case studies on the application of STTx to the same areas with different SMGI sources, thus different type of users, returned similar results, suggesting further research should be devoted to better understand the issue of representativeness.

Conclusions

The tentative framework presented in this paper derives from testing single SMGI sources integrated with AGI in a GIS environment. The research may be considered still in its infancy and the SMGI analytics framework proposed here is likely to evolve substantially in the future, nevertheless the potential seems already be very promising and it offers many issues to be further investigated. Early experiments seem already to demonstrate how it is possible to introduce new dimensions of analysis in order to build useful knowledge for design and decision making in spatial planning and governance. Further research is under development in order to explore the potential of integrating multiple SMGI data sources together. Indeed, each SMGI source has different peculiarities given by specific data models and public accessibility feature. In addition, each SMGI source has different rate of diffusion and usage in different regions making their combined use unique to a given place, making the case for local SMGI–AGI mixes. This issue is further amplified by the fact that also AGI sources may vary differently in different regions and countries. Nonetheless, the possibility of introducing pluralist knowledge on people's perceptions, preferences, or needs is an opportunity of utmost importance to bring innovation towards more sustainable, democratic, and community-oriented design and decision-making in spatial planning and governance.

The case studies referenced have already demonstrated how SMGI data enable us to observe to the community and to establish a dialogue. From the planning perspective, the possibility to tightly couple participatory initiatives to traditional planning knowledge in an integrate environment is a promising frontier to be further investigated. Territorial marketing, urban, and regional as well as sector planning and spatial governance in general have now a new power to inquiry the limit of which seems to be only the ability to look at data from a new perspective and ask smart questions.

Acknowledgments

The results on which part of this contribution is based were developed within the research project 'Efficacia ed efficienza della governance paesaggistica e territoriale in Sardegna: il ruolo della VAS e delle IDT' [Efficacy and efficiency of landscape and environmental management in Sardinia: the role of SEA and of SDI] CUP: J81J11001420007 funded by the Autonomous Region of Sardinia under the Regional Law n° 7/2007 'Promozione della ricerca scientifica e dell'innovazione tecnologica in Sardegna'. The author wishes to thank a number of young researchers for their contribution to this project including Pierangelo Massa, Roberta Floris, Anastacia Girsheva, Konstantin Ivanov, Roberta Falqui e Sara Mura.

References

Campagna, M. 2014. The geographic turn in Social Media: opportunities for spatial planning and Geodesign. In: Murgante, B., et al. (Eds.) *ICCSA 2014, Part II, Lectures Notes in Computer Science* 8580. Springer International Publishing Switzerland: pp. 598610.

Campagna, M., & Craglia, M. 2012. The socioeconomic impact of the spatial data infrastructure of Lombardy *Environment and Planning B: Planning and Design, 39*(6): 1069–1083.

Campagna, M., Floris, R., Massa, P., Girsheva, A., & Ivanov, K. 2015. The Role of Social Media Geographic Information (SMGI) in Spatial Planning. In: Geertman, S., Ferreira, Jr., J., Goodspeed, R., & Stillwell, J. (Eds.) *Planning Support Systems and Smart Cities.* Springer, ISBN 978-3-319-18367-1.

Floris, R., & Campagna, M. 2014. Tourism Planning in Social Media Data: Analysing Tourists Satisfaction in Space and Time *TeMA – Journal of Land Use, Mobility and Environment* Special Issue Eighth International Conference INPUT Smart City Planning for Energy, Transportation and Sustainability of the Urban System, Naples, 46 (June 2014).

Goodchild, M. 2007. Citizens as sensors: the world of volunteered geography *GeoJournal,* 69(4): 211–221.

Lazer, D., Pentland, A., Adamic, L., Aral, S., Barabasi, A.L., Brewer, D., Christakis, N., Contractor, N., Fowler, J., Gutmann, M., Jebara, T., King, G., Macy, M., Roy, D., & Van Alstyne, M. 2009. Computational Social Science. *Science,* 323: 721–723.

Massa, P., & Campagna, M. 2014. Social Media Geographic Information: Current developments and opportunities in urban and regional planning. In: *Proceedings of the 19th International Conference on Urban Planning and Regional Development in the Information Society GeoMultimedia 2014.*

Massa, P., & Campagna, M. 2014. Social Media Geographic Information: recent findings and opportunities for smart spatial planning *TeMA – Journal of Land Use, Mobility and Environment* Special Issue Eighth International Conference INPUT Smart City Planning for Energy, Transportation and Sustainability of the Urban System, Naples, 46 (June 2014).

Massa, P., & Campagna, M. 2015. Integrating Authoritative and Volunteered Geographic Information for spatial planning,. In: Capineri, et al. (Eds.) *European Handbook of crowdsourced information.* London: Ubiquity Press, pp. XX–XX.

ORACLE. 2013. The value of social data *ORACLE White paper.* Available at: https://go.oracle.com/LP=1320 (Last accessed 28 June 2015).

Steinitz, C. 2012. A Framework for Geodesign. Redlands, CA: Esri Press.

Weigend, A. 2009. The social data revolutions. Harvard Business Review. Available at: https://hbr.org/2009/05/the-social-data-revolution (Last accessed 28 June 2015).

PART II

Quality:
Criteria and methodologies

CHAPTER 5

Handling quality in crowdsourced geographic information

Laura Criscuolo*, Paola Carrara*, Gloria Bordogna*,
Monica Pepe*, Francesco Zucca†, Roberto Seppi†,
Alessandro Oggioni* and Anna Rampini*

*CNR-IREA, via Bassini 15, 20133 Milan, Italy, criscuolo.l@irea.cnr.it,
carrara.p@irea.cnr.it, gloria.bordogna@idpa.cnr.it

†Department of Earth and Environmental Sciences, University of Pavia,
Via Ferrata 1, 27100 Pavia, Italy

Abstract

While spatial information quality is an established discipline in traditional scientific geographical information (GI), standards and protocols for representing and assessing the quality of geographic contributions generated by volunteers or by the generic 'web crowd' are still missing. This work offers an analysis of strategies for quality control and describes a simple representation of the components of the quality in crowdsourced GI. In this framework, and based on the research carried out in Criscuolo et al. (2014), we also introduce a methodology for quality assessment, based on the given representation, which goes beyond the limitations of previous methods in the literature defined for a specific purpose, being able to deal with many quality features, GI categories, and types of application. The method is designed as a decision making approach, so flexible as to take into account the purpose of GI analysis, and so transparent

How to cite this book chapter:
Criscuolo, L, Carrara, P, Bordogna, G, Pepe, M, Zucca, F, Seppi, R, Oggioni, A and
Rampini, A. 2016. Handling quality in crowdsourced geographic information. In:
Capineri, C, Haklay, M, Huang, H, Antoniou, V, Kettunen, J, Ostermann, F and
Purves, R. (eds.) *European Handbook of Crowdsourced Geographic Information*,
Pp. 57–74. London: Ubiquity Press. DOI: http://dx.doi.org/10.5334/bax.e. License:
CC-BY 4.0.

as to make explicit the criteria driving to quality evaluation, namely the quality features (e.g. the credibility of the volunteers, or the accuracy of the spatial features, etc.) and their relevance.

Keywords

Crowdsourced Geographic Information, Volunteered Geographic Information (VGI), spatial information quality, quality assessment

Main issues in utilizing crowdsourced GI for scientific purposes

With the development of the geo-web and the increasing popularity of mobile devices and communication technologies, in the last decade many geographic information consumers have extended their role to the most active one of geographic content producers. The geographic information generated so far, is characterized by great heterogeneity – both in semantics, formats, contents, and quality.

In fact, crowdsourced GI is most frequently provided on the web – by both aware contributors within scientific initiatives (VGI) and unaware contributors within social networks – in the form of text commenting events, advice, warnings related with physical locations, geo-tagged photographs, and points of interest (POIs), corresponding for instance to historic, cultural, and naturalistic destinations valuable for touristic or commercial purposes. Contributors frequently provide also a geometric georeferenced representation of the footprints of the POIs in the form of points, polylines, or polygons (e.g. centroid of a building, a polyline for a road or a trail, a polygon for a park boundary). This geometric information can be acquired by GPS, or by sensors, or special equipment.

These contributions arise the interest of the scientific community, historically engaged in the creation and distribution of geographic information, together with concerns on the consequences that such new practices can lead to established scientific disciplines.

In fact, there are many problems related to the creation and use of geographic information coming from non-traditional sources for scientific purposes.

First of all, for a (spatial) dataset to be reusable, it should be coupled with its metadata, which define the domain within which its usage is recommended (temporal and geographic references, spatial resolution, quality and validity, constraints, etc.). Crowdsourced geographic information is often lacking, in whole or in part, meta-information allowing us both to locate it precisely in space and time, and to evaluate the basic parameters for its usage, such as acquisition procedure, measurement accuracy, instrumental precision, time stamps, contact details, etc. (Sui, Elwood & Goodchild, 2013). In some cases, some elements are available to enrich the meta-information of the crowd-sourced contribution, but they are expressed in unusual forms (for instance authors' nicknames, tags and geotags, external links, attached Exif files, etc.).

This issue especially emerges when datasets of user generated content created by their authors for non-scientific purposes (social, promotional, documental, etc.) are retrieved, selected, and exploited in the framework of scientific projects, for public or governmental decision making purposes.

The second critical point arises when processing crowdsourced geographic information. In fact, while gathering large volumes of user generated geographic contributions is relatively easy (typically 15% of social media contents are georeferenced), to spatially overlay and thematically integrate this information could be extremely difficult. This is due to the different – or commonly undefined – instrumental precision, reference systems, spatial and temporal granularity, together with the absence of common attributes and conceptual schemas, which often make the spatial analysis of user-generated georeferenced data a burden.

A third issue is related to the trustworthiness of contributed data. The quality of a crowdsourced contribution indeed is not just a characteristic of the data: it is also related to the author's reliability and experience, i.e. knowledge of the domain and ability in using the tools for data creation. By taking into account these aspects, it is possible to state the trustworthiness of the information.

The concept of trustworthiness suits both the conventional production of expert scientific information, and the crowdsourced contents, even if the latter is more complex, due to several reasons, among which the difficult traceability of authors, their unknown reputation, and the lack of standards and merit systems. In the last decade several studies have been focused on building credibility models (Metzger 2007; Keßler et al. 2013), analyzing quantitatively and qualitatively user generated content fluxes on the web by discussing their intrinsic characteristics, sources, subjects, drives (Eysenbach & Kohler 2002; Coleman et al. 2009; Van Dijck 2009), and currently the issue is still open and debated.

Because of the absence of a systematic procedure for amateurs' data production, it is commonly acknowledged that official data have a greater reliability and usefulness to science, while volunteered and non-specialist data are more affected by inaccuracies and contain less scientific value. Some authors have spent efforts to prove - or contradict - such a hypothesis by comparing datasets of crowdsourced and specialized observations. Dickinson et al. (2010) reports a series of studies in which variations in observer quality are correlated to the author's preparation. Among factors influencing such variations are background and experience (Galloway et al. 2006) together with the type of task (De Solla et al. 2005; Genet & Sargent 2003; Lotz & Allen 2007), the level of training, the company of a specialist in the field (Fitzpatrick et al. 2009), and the age and education of the author (Delaney et al. 2007). On the other hand, several studies have shown that the creative, aggregate use of non-expert contributions can generate new valuable information (De Longueville et al. 2010; Antoniou et al. 2010; Friedland & Choi 2011), and have documented situations in which local knowledge or expertise provide information of greater value than the expert knowledge alone (Fisher 2000). There is evidence of the high potential of crowdsourced geographic information when collected and managed in well-structured contexts, also in the results of the analysis conducted by authors such as Haklay (2010), Girres and Touya (2010), Ciepłuch et al. (2010), who have evaluated the accuracy of OpenStreetMap data against reference sources, and found that sometimes crowdsourced data have a better accuracy than the reference datasets.

Several other sensitive topics can be identified, related to the scientific usage of crowdsourced GI. Since an adequate treatment of these problems would be beyond the scope the paper, we just mention here the complex issue of the reproducibility in the procedures, the one of personal data protection, and those related to the distribution policy (establishment of intellectual property rights, copyrights, and related rights).

In order to make it possible a controlled use of crowdsourced contributions depending on the purpose of the reuse, the authors propose to establish a theoretical framework for a flexible and transparent quality representation and handling (further analysis on this can be found in Criscuolo et al. 2014, Bordogna et al. 2014a and in Bordogna et al. 2014b): flexibility is intended to offer the possibility to customize the criteria of the quality assessment to different purposes and needs; transparency is intended to offer the possibility for a user to know the criteria used for selecting the crowdsourced information.

In the next sections the topic of quality management for generic GI is addressed, firstly by describing a comprehensive model to represent GI quality, then by discussing the approaches for its control, finally by introducing a methodology for its assessment, suitable for both traditional and crowdsourced GI.

Representing quality in crowdsourced GI

In this work the types of multimedia geographic information are grouped in the following categories:

- **images**: photographs, video recordings and graphic objects;
- **annotations**: mostly textual reports;
- **features**: spatial entities, mono- or multi-dimensional, with associated attributes (such as Shapefiles or Geography Markup Language files);
- **measurements**: values derived from human or sensor's observations, mainly as numbers.

The contributions expressed in form of rating, i.e. the public evaluation of user-contributed geographic contents (e.g. thumbs up / down, star ratings…), are deliberately excluded; in fact, these expressions are certainly informative, and are widely used too, but are more similar to quality assessment tools than to stand alone geographic information. For this reason in the present work the contributions in the form of ratings will be discussed as mechanisms for quality control, i.e. a kind of quality indicators, and not as a type of crowdsourced GI.

Each GI item, i.e. each informative contribution, can consist of a single piece or be composed of multiple elements. In case of multiple elements, they may belong to the same category (for instance, they can report measurements of various physical parameters from a single measuring station) or be of different categories (for instance a photograph and its textual description).

Once the category and structure of a GI have been described, it is necessary to represent its quality.

The discussion on quality in GI has a long history, which starts from the last century, deepens with the advent of GIS technology (for a comprehensive review refer to Van Oort 2006), and finds a new flourishing in the last decade, with the advent of geo-web and the proliferation of collaborative mapping applications. In fact, although the quality of GI has been widely discussed and has its reference standard in ISO 19157:2013 (ISO/TC 211/2010 - Geographic information/Geomatics), the quality of crowdsourced GI presents some different features, such as to require new indicators to be adequately described and evaluated (Van Exel et al. 2010).

The quality of crowdsourced GI is actually a composite property: it includes not only some aspects dealing with the characteristics of the data, but also aspects dealing with the characteristics of the data producer and with the application context.

In ISO 19113-15 two main categories of quality are taken into account: Internal and External. The first one relates to intrinsic characteristics of information (spatial accuracy, temporal accuracy, semantic accuracy…), while the second one deals with the fitness for use of the information. These categories are certainly necessary to perform quality assessment on single pieces of information or on whole datasets, but they don't cover a third aspect of user generated information quality, which is important especially for crowdsourced resources: the trustworthiness of information.

To take into account this complexity, we choose to describe the quality of GI through three main categories, inspired by the ISO 19113-15 and by the thematic literature:

- **intrinsic quality**, corresponding to ISO internal quality, which depends on the characteristics of the informative content;
- **extrinsic quality**, which depends on the characteristics of the context, and responds to the needs of assessing the credibility both on the information

and on the author (Flanagin & Metzger 2008; Galloway et al. 2006; Genet & Sargent 2003);
• **pragmatic quality**, first described by English (1999) and similar to the ISO external quality, which measures the capability to meet the needs of a user or of a usage.

The features that contribute to determine the intrinsic, extrinsic, and pragmatic quality of a piece of geographic information can be broken down into elementary properties. These properties are many and varied, and can be updated under different project conditions. We list only a few of them, selected from the most important and most frequent in the specialist literature.

Intrinsic quality can be described, for instance, by the following elementary properties:

• **accuracy**, i.e. its conformity to the actual or expected value;
• **precision**, i.e. the repeatability of the observation or of the measurement;
• **correctness**, i.e. the absence of formal errors;
• **completeness**, i.e. the absence of significant omissions;
• **intelligibility**, i.e. the possibility of the contribution to be understood and examined.

The elementary properties relatable to the extrinsic quality can be:

• **reliability** of the information;
• **credibility** of the author.

Finally, the pragmatic quality can be described by the two following elementary properties:

• **pertinence** of the information;
• **fitness for** a particular **use**.

While the elementary properties contributing to define the intrinsic and extrinsic quality can be defined by evaluating elementary quality indicators associated to specific pieces of information constituting the VGI items, the last two properties, pertinence and fitness for use, may both be defined in terms of the extrinsic and intrinsic quality (Bordogna et al. 1914b).

The representation of quality sketched in Figure 1 is independent from the categories of contributions (images, annotations, measurements, features), from the information content and context, and can therefore be taken as a general framework to evaluate – and possibly compare – the quality in any crowdsourced or generic GI project.

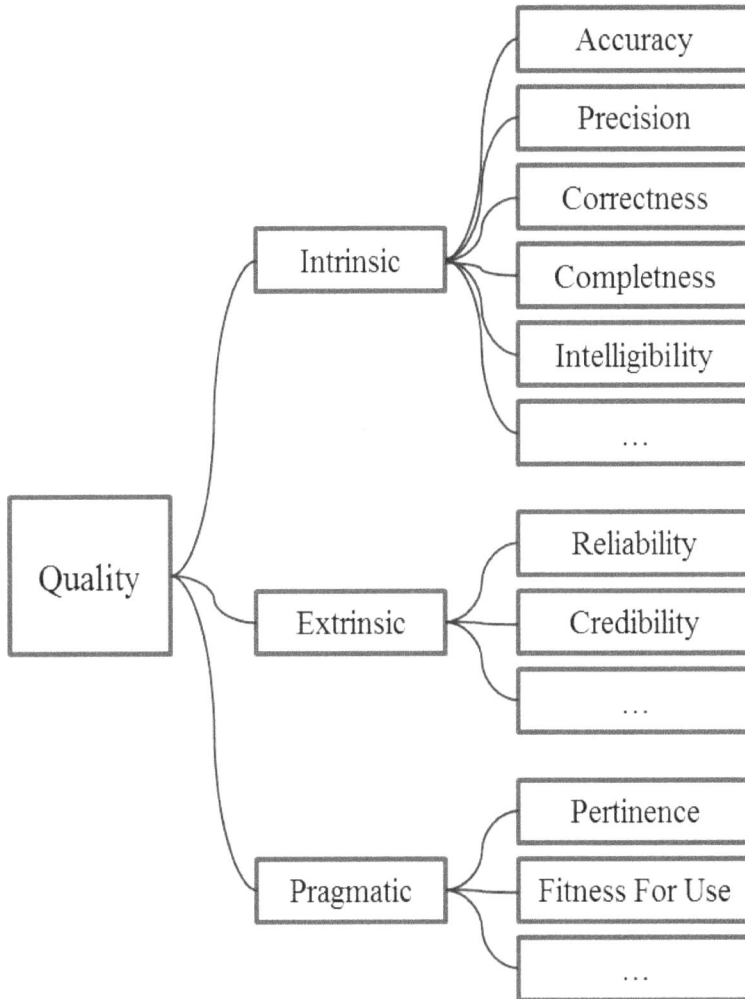

Figure 1: A sketched representation of the proposed categories and the main elementary properties of crowdsourced GI quality.

Approaches to quality control

Once defined the categories of crowdsourced GI and reported its quality properties, we can describe the types of approach to quality control.

In temporal terms the approach to quality control may take place via:

prevention, if it takes place through procedures that precede or are contextual to the submission of information (e.g. learning materials, controlled vocabularies, web forms that guide data producers in compiling their contributions);

correction, if it occurs after the contributions are submitted to the system (e.g. selection of contributions, automatic or manual corrections).

From the point of view of the actors involved, the operations for a quality control can be carried out:

by the **administration team**, if they are performed manually by the project coordinators, a technical staff or a group of experts;

by the **community** of participants, if the group of volunteers itself assesses and validates the information entered;

automatically, if one or more IT components of the system operate the content selection or make some automated edits.

Finally, from the point of view of the remedial action performed, the contributions considered unsuitable may be subject to:

warning, and then be published with an appended message, an alerting symbolism, or a notice;

removal, being excluded from publication and successive processing.

The described approaches to quality control in crowdsourced GI are represented in Figure 2.

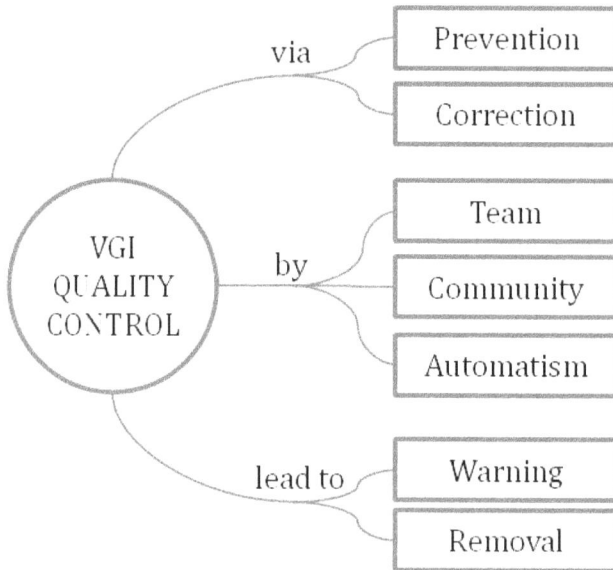

Figure 2: A representation of the approaches to quality control in crowdsourced GI.

For each use and each context of application the best fitting strategy must be carefully designed.

Here follows the discussion of the advantages and disadvantages associated with each option, presenting some possible implementations.

The preventive approach, aimed to facilitate the correct compilation of VGI before its submission, may consist, for example, in simple manuals and handbooks; in assisted completion procedures, with multiple choice fields, wordlists, or auto-completion; in automatic tools for the normalization of contents or for the automatic extraction of metadata; in ontologies and geographic gazetteers (Popescu et al. 2009; Kuhn 2001). Common preventive actions are also the selection and the training of volunteer contributors (Galloway et al. 2006; Crall et al. 2011). All these methods facilitate a uniform and formally correct data entry from contributors (intrinsic quality), but they do not ensure to control the reliability (extrinsic quality) or the fitness for use of the information entered (pragmatic quality).

The corrective methods act instead ex post, by amending or removing the weak VGI contributions. They may include the use of automatic algorithms or geostatistical filters (De Tré et al. 2010; Latonero & Shklovski 2010), but also can apply a human supervision to identify systematic errors and maintain the consistency of the dataset (Dickinson 2010; Huang et al. 2010), or still can monitor in real time the semantic integrity of the collected data (Pundt 2002). The corrective methods are suited to act on the reliability and effectiveness of the information provided (extrinsic and pragmatic quality), as well as on the intrinsic quality characteristics, but since they act for removing, merging, or reshaping inappropriate contributions, they may cause partial or even total loss of information.

A quality assessment performed by a team of experts, or supervisors, offers some guarantees, assuming they use competence, wisdom and fairness in the task. Yet even scientists or professionals cannot always enjoy a full mastery of all the variables, and their judgment can be subjective and approximate. It may happen that local citizens, or specialists in particular activities, or direct observers of phenomena make more detailed and reliable assessments than their scientific supervisors. On some occasions, however, it is hazardous to assign assessment tasks to volunteers. In fact, for lack of expertise, superficiality or bad faith, they could create confusion and even hamper the entire data collection.

The combination of the two methods – the traditional authoritative (or top-down), and the democratic one (or bottom-up), is not only possible, but can also produce significant results. In this context, in fact, the web can be used as a meeting point for a collective assessment: the ongoing access to a web item by a hybrid team of experts, local amateurs, occasional visitors, who are asked for evaluating the content, can give rise to a kind of participative evaluation of quality, with a high potential for selection and judgment (Flanagin & Metzger 2008; Connors et al. 2012).

Finally, the automatic control mechanisms can be extremely useful, especially when crowdsourced contributions form large volumes of data (Spinsanti & Ostermann 2013). At present, however, automatic control systems rarely reach an adequate degree of reliability, comparable to a human validation. They are likely to fail especially as regards to the relevance of the contribution, the pertinence (pragmatic quality), and the intelligibility/correctness of the textual content (intrinsic quality).

The control mechanisms that act for removing flawed contents are of great help to preserve the integrity and the consistency of the data collections. Nonetheless, since they discard information considered inadequate according to pre-set parameters, they lead to the exclusion or to the partial loss of information that, no matter how flawed, might be useful in other contexts. The control mechanisms, which keep the whole submitted information, even if not compliant, but report the flaws, do not lose any entered information and encourage the users to access the data consciously. These warning mechanisms, however, have two adverse consequences: on the one hand the storage process is non-effective, because it allocates some memory to data of doubtful relevance; on the other hand the usage is made more difficult, because the system lets the user decide on the data reliability.

Each one of the described options should be considered and evaluated carefully by the project coordinators; nevertheless, most of the times hybrid methods can help in achieving a proper management of quality, by balancing the pros and cons of the various strategies.

Suggesting a quality estimation method for crowdsourced GI

Whatever the strategies to address the control of quality in crowdsourced GI, subsequently it is useful to define a method to estimate the results. This estimation is important not only to establish the level of quality reached by the single contributions and by the whole dataset, but also to monitor the quality trends over time.

In recent years several efforts have been made to develop procedures for quality assessment. Some of them focus on the credibility issues (Metzger 2007), some others focus on the geographical accuracy (Keβler et al. 2013; Sabone 2009; Goodchild 2008), which is often calculated by comparing different datasets or by validating a data sample with a field survey (Haklay 2010). These proposals, while effective in assessing particular aspects of quality, are useful in their specific context, but do not offer a general or flexible method, nor include the different aspects that characterize the quality of non-traditional GI.

To overcome these limitations, we base on the representation of quality introduced in section 2 which makes it possible to define some elementary quality indices, to be associated with each component of the GI items, and then to aggregate the elementary indices into composite ones, until reaching an overall index of quality for the information item, and, in case, the quality index for a whole dataset. A similar method has been described in Bordogna et al. (2014a), and is proposed here in a simplified form, so as to make it easily applicable and customizable to any provided GI, i.e. traditional, non-expert, volunteered or even unaware.

First, we decompose a generic GI item into its elementary components, which may consist of one or more images, annotations, measurements, geographic features. The overall information of a contribution, which we name GI_{TOT}, is therefore achieved by aggregating the n informative elements GI_i, i=1,... n.

$$GI_{TOT} = \oplus (GI_1, GI_2, GI_3,..., GI_n) \quad \oplus \text{ being a mathematical aggregation operator}$$

An overall quality index Q_{TOT} is then associated to the overall information GI_{TOT}. Q_{TOT} results from the aggregation of the n Q_i indices associated with the n components. In this aggregation step, each index Q_i is associated with a numerical weight K_i, which is properly chosen by the analyst depending on specific design requirements. Also the aggregation is chosen by the analyst, for example it may be a weighted average or a sum.

$$Q_{TOT} = \oplus (K_1 Q_1, K_2 Q_2, K_3 Q_3, ..., K_n Q_n)$$

Each Q_i is in its turn the result of the aggregation of three quality indices – I_i, E_i, P_i – respectively connected to the intrinsic, extrinsic, and pragmatic properties of the GI quality.

I_i, E_i, and P_i, can be in their turn associated with three weights – K_I, K_E, and K_P – that are also set by the analyst, depending on the relevance stated for each property of GI quality.

The overall quality for a GI item results in:

$$Q_{TOT} = \oplus (K_1{}^* \oplus (K_I I_1, K_E E_1, K_P P_1), K_2{}^* \oplus (K_I I_2, K_E E_2, K_P P_2), ..., K_n{}^* \oplus (K_I I_n, K_E E_n, K_P P_n))$$

I_i, E_i, and P_i can be finally decomposed in lower level indices, related to the elementary properties of GI quality: accuracy, precision, correctness, completeness, intelligibility, reliability, credibility, pertinence, and fitness for use.

Even at this level, the comprehensive evaluation of I_i, E_i, and P_i is performed by the aggregation of their lower components:

$I_i = \oplus (accuracy_i, precision_i, correctness_i, completeness_i, intelligibility_i)$
$E_i = \oplus (reliability_i, credibility_i)$
$P_i = \oplus (pertinence_i, fitness for use_i)$

This multi criteria assessment of GI quality depends on both the relevant quality indexes Q_i (those with weight $K_i > 0$), the number of such relevant indexes, and the aggregation operator used to combine them.

The described indices and the progressive levels of aggregation are represented in the Figure 3a.

In order to clarify how the model applies to real cases, we can include as an example a real VGI project and make the quality indices explicit. Let's assume to work as analysts in the famous Wikimapia[1] project, and try to estimate the whole quality of a VGI item, consisting in a polygonal shape with an annexed photo. The quality index associated to the polygonal feature will be named Q_1, and the one associated to the photo will be Q_2.

We choose a sum function to perform the aggregation and set the weights for the two VGI components Q_1 and Q_2, and for the three quality indices I_i, E_i and P_i, depending on our project interests, in the following way:

$K_1 = 1,5$ $K_2 = 1$ assuming more interest in preserving the quality of the polygonal feature than the quality of the photo;

$K_I = 1$ $K_E = 1$ $K_P = 0,5$ assuming more interest in controlling the intrinsic and extrinsic quality than the pragmatic one.

Now we set some numerical values to the elementary quality properties, simulating a likely situation in the Wikimapia project. Let us define the numerical values in the domain $[-1, 0, 1]$:

we set the feature accuracy and the feature precision = 0, assuming, in this example, that is not possible to determine them directly in Wikimapia;[2]

we set the feature correctness, completeness and intelligibility = 1, assuming they are completely fulfilled;

we set reliability = −1 and credibility = 0, assuming that some users from the Wikimapia community commented negatively the entered feature, and assuming the author is a neophyte (corresponding to *user level 0*, or *Unregistered* in Wikimapia);

we set pertinence and fitness for use = −1, assuming the polygonal feature entered is not belonging to the categories requested in the project (for example it could figure out the area in which a temporary event takes place);

we set similarly the values for the elementary quality properties of the photographic component of the VGI item.

---

[1] http://wikimapia.org is a multilingual open-content collaborative map, where volunteers are asked to mark places, add descriptions provided with proof links, give them appropriate categories and upload photos.

[2] In the literature some procedures have been developed and adopted to calculate geometrical accuracy and/or precision of VGI polygonal contributions, usually based on a comparison with the base map images. Nevertheless these procedures can be sometimes challenging or not applicable. This could happen for various reasons: for instance the user generated polygon could refer to a physical element that is not completely visible, or not updated, in the base map images; sometimes it could be difficult to determine which data is more accurate (the polygon from the volunteer contributor or the base image provided by the map application).

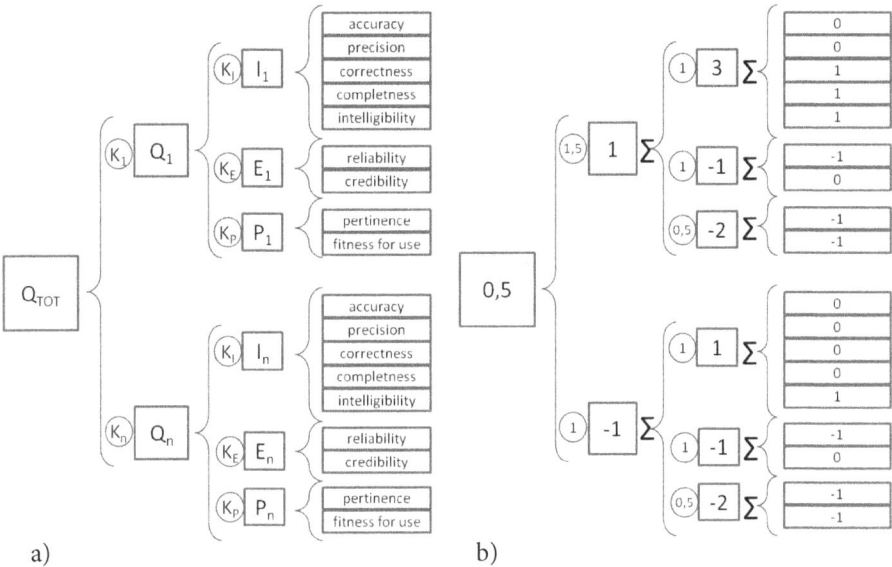

Figure 3: The theoretical model for estimating quality in a generic GI item (a) and its enactment for a plausible VGI case study (b).

The situation described above is represented in Figure 3b, as homologous to the theoretical model in Figure 3a. The numerical score resulting from the simulation (Figure 3b) is a direct consequence of the initial decision – taken by the imaginary analyst – to assign equal weights to the intrinsic and to the extrinsic components of the VGI item, and it is directly connected with the numerical domain stated ($[1, 0, -1]$). These choices lead to a slightly positive total score (0.5), which represents the overall quality index for the volunteered contribution. Some alternative decision – for instance to assign a higher value to the extrinsic quality weight K_E – would lead to different and even negative results. The result, whatever it would be, is meaningful only if compared with analogous ones, belonging to the same dataset, or to different datasets. The procedure indeed doesn't assess itself the quality of the VGI items, but allows normalizing the quality components in order to facilitate their comparison and evaluation. This procedure can be carried out manually, automatically or semi-automatically. The processed items could be finally ranked and possibly filtered, depending on the accomplishment of a minimum quality threshold.

Discussion and conclusion

The issue faced in this work – i.e. to flexibly represent and assess the quality in crowdsourced GI – has led to define a methodology and a set of quality indices suitable to represent and quantify quality.

A synthetic representation of quality control strategies in crowdsourced GI activities has been proposed, as a guide for project design and comparison of existing ones.

Such approaches, in practical cases, are often combined into hybrid strategies. To break them down into their atomic properties helps to describe and normalize such strategies, even in the more complex operational cases. It seems unrealistic to point out one single optimal solution. On the contrary, several effective configurations can coexist, offering suitable solutions for specific use cases. The choice is usually determined by the objective of the crowdsourced GI activity (which can be recreational, social, scientific, professional, experimental, etc.), by the type and the amount of information expected (images, annotations, features, measurements), by the characteristics of the contributors addressed (citizens, unaware web users, trained volunteers, experts, etc.), and by the infrastructure and technologies on which the project leans (geographic databases, web and mobile clients for services, sensors, etc.).

The representation introduced in Figure 2 can be used not only *a posteriori*, i.e. to describe the control strategy performed, but can also be helpful during the design phase, to configure the most effective solution for quality management.

Besides the analysis of strategies for quality control, a simple representation of the components of the quality in crowdsourced GI has been depicted too. It helps in focusing on different aspects of quality and individually evaluating them.

A flexible methodology for quality assessment has been introduced on the basis of the given representation. It can be applied manually or automatically on a wide range of volunteered contributions, and differently weighted according to the needs of the analysts.

The estimation of the quality of crowdsourced GI is a challenge that has been addressed by several authors with different methods. The methods found in literature, however, are usually designed to respond to specific needs, and therefore, as far as valuable and useful in particular cases, appear to suffer from one or more of the following major constraints:

- they aim to quantify the uncertainty of a single quality feature (e.g. the credibility of the volunteers, or the accuracy of the spatial features, etc.);
- they deal with a single category of GI (images, annotations, features, measurements);
- they are suitable only for a specific application (depending on a given technology, or presuming the participation of a certain amount of volunteers, etc.).

The contribution brought by this work is the proposal of a generalized method for estimating the quality that goes beyond these limitations. It is formally defined in terms sufficiently operational to ensure their applicability, but also general enough to ensure its transferability to different application cases.

The proposed method is based on some choices and can be designed as a decision making approach including all the major quality features and allowing the description of each type of GI contribution, under all possible aspects, using different aggregation operators to get to a final decision. Its strength lies in its flexibility: the aggregation operations can be chosen to suit the purpose of the user's analysis, and the decision-making approach allows dealing with any specific case. Even if the analyst does not wish to join to the proposed representation of quality (Figure 1), the method is still applicable, by replacing the suggested indices with alternative properties. Finally, the method can provide a guide to systematize and make explicit the criteria for assessing the quality that are used in an application of crowdsourced or even heterogeneous GI.

Acknowledgments

The present work is based on the research carried out in the framework of a PhD programme of the University of Pavia, Dept. of Earth and Environmental Sciences, and has been supported by the Institute for the Electromagnetic Sensing of the Environment CNR-IREA. The research has been partially founded by the Italian Flagship Project RITMARE.

References

Antoniou, V., Morley, J., & Haklay, M. 2010. Web 2.0 geotagged photos: Assessing the spatial dimension of the phenomenon. *Geomatica, 64*(1): 99–110.

Bordogna, G., Carrara, P., Criscuolo, L., Pepe, M., & Rampini, A. 2014a. A linguistic decision making approach to assess the quality of volunteer geographic information for citizen science. *Information Sciences, 258*: 312–327.

Bordogna, G., Carrara, P., Criscuolo, L., Pepe, M., & Rampini, A. 2014b (November). On predicting and improving the quality of Volunteer Geographic Information projects. *International Journal of Digital Earth*: 1–22. DOI: http://dx.doi.org/10.1080/17538947.2014.976774

Ciepłuch, B., Jacob, R., Mooney, P., and Winstanley, A. 2010. Comparison of the accuracy of OpenStreetMap for Ireland with Google Maps and Bing Maps. In: *Proceedings of the Ninth International Symposium on Spatial Accuracy Assessment in Natural Resources and Environmental Sciences*. Leicester, UK on 20–23rd July 2010: p. 337.

Coleman, D. J., Georgiadou, Y., Labonte, J., Observation, E., & Canada, N. R. 2009. Volunteered Geographic Information: The Nature and Motivation of Produsers. *International Journal of Spatial Data Infrastructures Research, 4*: 332–358.

Connors, J. P., Lei, S., & Kelly, M. 2012. Citizen Science in the Age of Neogeography: Utilizing Volunteered Geographic Information for Environmental

Monitoring. *Annals of the Association of American Geographers, 102*(6): 1267–1289.

Crall, A. W., Newman, G. J., Stohlgren, T. J., Holfelder, K. A., Graham, J., & Waller, D. M. (2011). Assessing citizen science data quality: an invasive species case study. *Conservation Letters, 4*(6): 433–442.

Criscuolo, L. 2014. Monitoring of Italian Glaciers: Official, Volunteered and Incidental Information. Unpublished Thesis (PhD), University of Pavia, Dept. of Earth and Environmental Sciences. Advisors: Zucca F., Seppi R., Pepe M.

Delaney, D. G., Sperling, C. D., Adams, C. S., & Leung, B. 2007. Marine invasive species: validation of citizen science and implications for national monitoring networks. *Biological Invasions, 10*(1): 117–128.

De Longueville, B., Luraschi, G., Smits, P., Peedell, S., & Groeve, T. D. 2010. Citizens as sensors for natural hazards: A VGI integration workflow. *Geomatica, 64*: 41–59.

De Solla, S. R., Shirose, L. J., Fernie, K. J., Barrett, G. C., Brousseau, C. S., & Bishop, C. A. 2005. Effect of sampling effort and species detectability on volunteer based anuran monitoring programs. *Biological Conservation, 121*(4): 585–594.

De Tre, G., Bronselaer, A., Matthe, T., Van de Weghe, N., & De Maeyer, P. 2010. Consistently Handling Geographical User Data Context-Dependent Detection of Co-located POIs. *Communications in Computer and Information Science*, Springer, *81*(1): 85–94.

Dickinson, J. L., Zuckerberg, B., & Bonter, D. N. 2010. Citizen Science as an Ecological Research Tool: Challenges and Benefits. *Annual Review of Ecology, Evolution, and Systematics, 41*(1): 149–172.

English, L. P. 1999. Improving data warehouse and business information quality: methods for reducing costs and increasing profits. J. Wiley & Sons.

Eysenbach, G., & Kohler, C. 2002. How do consumers search for and appraise health information on the world wide web? Qualitative study using focus groups, usability tests, and in-depth interviews. *British Medical Journal, 324*: 573–577.

Fischer, F. 2000. *Citizens, experts, and the environment: The politics of local knowledge*. Duke University Press.

Fitzpatrick, M. C., Preisser, E. L., Ellison, A. M., & Elkinton, J. S. (2009). Observer bias and the detection of low-density populations. *Ecological Applications, 19*(7): 1673–1679.

Flanagin, A. J., & Metzger, M. J. 2008. The credibility of volunteered geographic information. *GeoJournal, 72*(3–4): 137–148.

Friedland, G., & Choi, J. 2011. Semantic computing and privacy: A case study using inferred geo-location. *International Journal of Semantic Computing, 5*(1): 79–93.

Galloway, A. W. E., Tudor, M. T., & Haegen, W. M. V. 2006. The Reliability of Citizen Science: A Case Study of Oregon White Oak Stand Surveys. *Wildlife Society Bulletin, 34*(5): 1425–1429.

Genet, K. S., & Sargent, L. G. 2003. Evaluation of methods and data quality from a volunteer-based amphibian call survey. *Wildlife Society Bulletin*: 703–714.

Girres, J. F., & Touya, G. 2010. Quality Assessment of the French OpenStreet-Map Dataset. *Transactions in GIS, 14*(4): 435–459.

Goodchild, M. F. 2008. Spatial Accuracy 2.0. In: *Proceedings of the eighth international symposium on spatial accuracy assessment in natural resources and environmental sciences*, Shanghai: pp. 1–7.

Haklay, M. 2010. How good is volunteered geographical information? A comparative study of OpenStreetMap and Ordnance Survey datasets. *Environment and planning. B, Planning & design, 37*(4): 682.

Huang, K. L., Kanhere, S. S., & Hu, W. 2010. Are You Contributing Trustworthy Data? The Case for a Reputation System in Participatory Sensing. In: *MSWIM '10 Proceedings of the 13th ACM international conference on Modeling, analysis, and simulation of wireless and mobile systems*. ACM New York, NY, USA c2010, pp. 14–22.

Keßler, C., Theodore, R., & De Groot, A. 2013. Trust as a Proxy Measure for the Quality of Volunteered Geographic Information in the Case of OpenStreet-Map. In: Vandenbroucke, D., Bucher, B., & Crompvoets, J. (Eds.) *Geographic Information Science at the Heart of Europe, Lecture Notes in Geoinformation and Cartography*. Springer International Publishing, Cham: pp. 21–37.

Kuhn, W. 2001. Ontologies in support of activities in geographical space. *International Journal of Geographical Information Science, 15*(7): 613–631.

Latonero, M., and Shklovski, I 2010 "Respectfully Yours in Safety and Service": Emergency Management & Social Media Evangelism. In: *Proceedings of the 7th International ISCRAM Conference*. Seattle, USA: pp. 1–10.

Lotz, A., & Allen, C. R. 2007. Observer bias in anuran call surveys. *The Journal of wildlife management, 71*(2): 675–679.

Metzger, M. J. 2007. Making sense of credibility on the Web: Models for evaluating online information and recommendation for future research. *Journal of the American Society for Information Science and Technology, 13*: 2078–2091.

Popescu, A, Grefenstette, G, and Bouamor, H 2009 Mining a Multilingual Geographical Gazetteer from the Web. In: *2009 IEEE/WIC/ACM International Joint Conference on Web Intelligence and Intelligent Agent Technology, Ieee*: 58–65.

Pundt, H. 2002. Field Data Collection with Mobile GIS: Dependencies Between Semantics and Data Quality. *GeoInformatica, 6*: 363–380.

Sabone, B. 2009. *Assessing Alternative Technologies for Use of Volunteered Geographic Information in Authoritative Databases*. Technical Report, Department of Geodesy and Geomatics Engineering, University of New Brunswick, Canada.

Spinsanti, L., & Ostermann, F. 2013. Automated geographic context analysis for volunteered information. *Applied Geography, 43*: 36–44.

Van Dijck, J. 2009. Users like you? Theorizing agency in user-generated content. *Media, culture, and society, 31*(1): 41–58.

Van Exel, M., Dias, E., & Fruijtier, S. 2010. The impact of crowdsourcing on spatial data quality indicators. In: *Proceedings of the 6th GIScience international conference on geographic information science, 2010*: pp. 213–217.

Van Oort, P. A. J. 2006. Spatial Data Quality: From Description to Application, Thesis (PhD), Wageningen Universiteit.

Sui, D., Elwood, S., & Goodchild, M. 2013. Crowdsourcing geographic knowledge, Volunteered Geographic Information in Theory and Practice, Springer Verlag.

CHAPTER 6

Data quality in crowdsourcing for biodiversity research: issues and examples

Clemens Jacobs

GIScience Research Group, Institute of Geography, Heidelberg University, clemens.jacobs@geog.uni-heidelberg.de

Abstract

The last few years have seen the emergence of a large number of worldwide web portals where volunteers report and collect observations of plants and animals, share these reports with other users, and provide data for scientific research purposes along the way. Activities engaging citizens in the collection of scientific data or in solving scientific problems are collectively called citizen science. Data quality is a vital issue in this field. Currently, reports of species observations from citizen scientists are often validated manually by experts as a means of quality control. Experts evaluate the plausibility of a report based on their own expertise and experience. However, a rapid growth in the quantity of reports to be processed makes this approach increasingly less feasible, creating a need for methods supporting (semi)automatic validation of observation data. This aim is achieved primarily by analysing the spatial and temporal context of the data. Relevant context information can be provided by existing observation data, as well as by spatial data of environmental factors, or other spatio-temporal factors impacting the distribution of species, or the process of observation and contribution itself. It is very important that the

How to cite this book chapter:

Jacobs, C. 2016. Data quality in crowdsourcing for biodiversity research: issues and examples. In: Capineri, C, Haklay, M, Huang, H, Antoniou, V, Kettunen, J, Ostermann, F and Purves, R. (eds.) *European Handbook of Crowdsourced Geographic Information*, Pp. 75–86. London: Ubiquity Press. DOI: http://dx.doi.org/10.5334/bax.f. License: CC-BY 4.0.

specific properties of data emerging from citizen science origins are taken into account. These data are often not produced in a systematic way, resulting in (for instance) spatial and temporal incompleteness. Also, the data structure is not only determined by the natural spatio-temporal patterns of species distribution, but by other factors such as the behaviour of contributors or the design of the citizen science project that produced the data as well.

Keywords

crowdsourcing, citizen science, data quality, data validation, biodiversity

Introduction

Learning more about biodiversity on our planet has become an important challenge as we face climate change and species extinction. Any conservation efforts need to be based on adequate knowledge about distribution, behaviour, and ecology of species. However, the long-term data covering broad geographic regions, which are necessary to gain said knowledge (Dickinson, Zuckerberg & Bonter 2010), cannot be collected using professional data collectors alone. High costs associated with professional data gathering pose another inhibiting challenge. One way of solving this problem is data collection by volunteers. Activities involving citizens in the collection of scientific data or, more general, in scientific research endeavours, are called citizen science. While citizen science itself is not a new phenomenon, we see a growing number of such projects being organized in web portals, revolutionising the way biodiversity data are collected and made available. Recent years have seen a growing number of projects using the possibilities offered by web 2.0 technologies (Dickinson, Zuckerberg & Bonter 2010; Miller-Rushing, Primack & Bonney 2012), where volunteers can upload, manage, and share their own observations of plants and animals, and make them available for scientific research. Opportunities for biodiversity monitoring and ecological research provided by this phenomenon, but also implications for project organisation and management, are extensively discussed in a book by Dickinson and Bonney (2012), and in numerous other publications (e.g., Connors, Lei & Kelly 2012; Chandler et al. 2012; Cosquer, Raymond & Prevot-Julliard 2012; Sullivan et al. 2014). Motivations of initiatives in this field range from furthering public interest in conservation issues and concerns (with data collection as a mere by-product), to systematic generation of such data for specific uses in scientific research, planning or public administration (e.g. monitoring of certain species or groups of species in certain areas or regions). Other projects aim at collection of data about the distribution of species without a predefined, specific goal.

This way of collecting data using the www and the general public is a specific form of crowdsourcing (Howe 2006), i.e. employing the general public to produce web content or to carry out certain labour-intensive tasks (especially tasks that cannot be easily automated using methods of data processing). Other terms are used in biodiversity citizen science depending on data collection procedures employed or goals pursued, such as community-based monitoring or CBM (Conrad & Hilchey 2011). As the data collected always have a geographic reference, they represent a specific type of Volunteered Geographic Information (VGI) (Goodchild 2007; Haklay 2013).

One of the most important concerns with these data is data quality. Assuring data quality is important because a general lack of trust will decrease their use for science or administration (Conrad & Hilchey 2011). A recent study by Theobald et al. (2015) showed that so far only a small portion of biodiversity-related citizen science projects contributed data to peer-reviewed scientific articles. The quality of the output of scientific research depends directly on the quality of the data used (Dickinson, Zuckerberg & Bonter 2010), as does the quality of administrative and planning decisions. On the other hand, citizen science approaches introduce great advantages, considering their ability to provide large amounts of data over broad geographic regions as well as long periods of time, often at relatively low cost (Dickinson, Zuckerberg & Bonter 2010). At the same time, this poses a challenge for data quality assurance: many projects acquire large amounts of observations - often hundreds of observations per day, or even more. Many projects employ manual validation procedures that do not scale well, making (semi)automatic validation methods necessary.

This chapter presents an overview of important issues related to quality of citizen science biodiversity data. Using examples from citizen science projects in the domain of biodiversity, it discusses specific problems and possible avenues to solutions concerning quality assurance for this specific kind of VGI. While there are many commonalities with VGI from other domains, allowing for the adoption of quality assurance approaches and strategies that are also used in other fields of VGI, there are also notable differences or features shared only with few other VGI domains, making adjustments of common approaches and strategies necessary. Most important among these differences are the diversity concerning project design and organisation (from strict monitoring schemes to rather open, opportunistic data collection, resulting in data properties and quality assurance needs varying between projects), and the nature of the information mapped (identification of species requiring some degree of expert knowledge, thereby raising issues of credibility).

Quality of citizen science biodiversity data

When we examine the quality of citizen science data from the biodiversity domain, we need to look at how data quality can be defined, and how it is used and handled in the relevant scientific practice. We approach the term data quality from two different perspectives:

- Data quality in terms of the sum of the data's properties, and
- data quality in terms of the data's fitness for use.

Data quality as the sum of the data's properties

Observations of occurrences of species are geographic data. Therefore, their quality in terms of characteristic properties can be described using the quality features introduced by ISO standard ISO 19113 (ISO 2002). These include the following: completeness, logical consistency, positional accuracy, temporal accuracy, and thematic accuracy. Properties of the data regarding these attributes are determined mostly by the design of the project collecting the data (especially rules and guidelines concerning data collection). Therefore, they are diverse. Considering the aspect of **completeness**, we often find a pronounced spatial and temporal heterogeneity in citizen science biodiversity data. This is especially the case for data collected without using structured monitoring schemes or strict rules - so called casual data collected in an opportunistic way (Chapman 2005). There are many reasons for this heterogeneity, like contributor preference for certain species or groups of species, variable observation effort caused by different (seasonal) weather conditions, or differences in spatial density of observations associated with differences in population density, among many others. Bird et al. (2014) describe approaches to account for the variability in the resulting data caused by such factors. They use several statistical tools to demonstrate effects of certain types of error and bias in citizen science data on modelling results in biology, and describe how to address these issues. Van Strien, van Swaay and Termaat (2013) present a methodology to remedy several types of bias in the data when using them for occupancy models (modelling the distribution of species in space and time).

The **positional accuracy** of citizen science species distribution data depends primarily on the type of location information, e.g. exact point, assignment to an (arbitrary) area or to a map quadrant. The data of many relevant projects are heterogeneous in this respect. The positional accuracy of point data depends (among other factors) on the way the coordinates of an observation are determined, e.g. using a GPS device on site, placing the observation's location on a map or aerial photograph (in a map viewer), deriving the location from a specimen description, etc.

Thematic accuracy refers to the correctness of the classification of objects or of their non-quantitative attributes (Kresse & Fadaie 2004). An important issue regarding thematic accuracy of observational data of animals and plants from citizen science projects is the participants' lack of scientific training and its effect on the reliability or credibility of species identification (Conrad & Hilchey 2011).

The **temporal accuracy** of observational data of animals and plants from citizen science projects is determined by how accurately it can be determined at data collection. The day of observation is mandatory information provided in most cases. Sometimes, the time of day can be specified as well, or is recorded automatically if an observation is reported using a mobile device. Currentness, i.e. the correctness of data in relation to the state of the environment changing over time, is another important aspect of temporal accuracy.

Logical consistency, including aspects like consistency of data structure or compliance with certain rules (Kresse & Fadaie 2004) is usually ensured by adequate design of the reporting tools and data base.

Data quality in terms of fitness for use

Data quality in terms of 'usefulness' can only be assessed for a certain intended use of the data (Devillers et al. 2007). Whether data quality is 'good enough' for a specific use depends on whether the data's properties allow for the question(s) at hand to be answered (Devictor, Whittaker & Beltrame 2010). For example, a precise location in observation data of plants or animals is not important if the data are used for deriving seasonal occurrence for larger regions, but would be important for analysing fine-grained spatial distribution patterns. Bordogna et al. (2014) point to the need for all VGI to assess and improve data quality with respect to the data's intended use and the data user's expectations. They propose a framework to match users' needs and data properties.

Principles of quality assurance for user-generated data

Data quality assurance aims at identifying, correcting and eliminating errors. Chapman (2005) also uses the term 'data cleaning'. On the one hand, this process includes the identification of formal errors, i.e. missing values, typing errors, etc. On the other hand, the suitability of a (formally correct) data set for a particular purpose depends, as we have already seen, on whether the data's characteristics (e.g. position accuracy) are sufficient for this purpose. Such uses can be very diverse and are often not fully foreseen prior to data collection (Dickinson, Zuckerberg & Bonter 2010).

Goodchild and Li (2012) identify three basic approaches to quality assurance for VGI, which are also applicable to citizen science observation data in the field of biodiversity.

The '**crowd-sourcing approach**' builds on the assumption that an error cannot persist if many users work on the same data. Hardisty and Roberts (2013) consider this the best method to identify errors in biodiversity data. Goodchild and Li (2012), however, present a good example where this assumption failed, with a wrong name of a golf course in California persisting for years in

Wikimapia (an online map project collecting information on locations from users). They also conclude that what they call 'obscure' objects (e.g. objects that exist only for short periods of time) may be more susceptible to such errors than others. Observations, especially of more mobile animal species, may well be counted among these.

Another principle, termed the '**social approach**' by Goodchild and Li (2012), uses privileged users as controllers validating the data collected in the project. This approach is widely used in citizen science projects in the biodiversity domain (Wiggins et al. 2011). Data validators are often regional experts for a certain species group (Sullivan et al. 2009), responsible for data validation in a certain area that they know well. The validation process sometimes involves communication between data reviewers and observers, when a reviewer requests more specific information about an unusual report (Bonter & Cooper 2012) that may help to validate it.

In the '**geographic approach**', Goodchild and Li (2012) summarize all methods using rules formalising geographic context. As Elwood, Goodchild and Sui (2012: 580) conclude, '… the richness of geographic context (…) makes it comparatively difficult to falsify VGI, either accidentally or deliberately'. Methods based on this principle allow for automatic verification of data. The necessary geographic context can be gained from observation data already existing in the project in question. This approach requires large amounts of existing data with a relatively high spatial density (Conrad & Hilchey 2011), often not (or not yet) available in citizen science data sets in the biodiversity domain. Consequently there is a need for methods relying on other context sources. Using external context data may provide a solution to this challenge (Elwood, Goodchild & Sui 2012), adding the question of data quality of these context data to the picture. Goodchild and Li (2012) conclude that there is a need for the formalization of relevant geographic context and the rules for describing it.

Using geographic context with distribution data of organisms shows certain methodological similarities with niche or habitat modelling, using known occurrences or absences of a species or of species communities in order to find correlations between these occurrences and a number of environmental factors, with the goal of predicting occurrences (or, at least, finding suitable habitats) in regions without available occurrence data (Engler et al. 2004). Many niche modelling methods need absence data (that is, data about locations where the species in question is definitely not present) to work (Engler, Guisan & Rechsteiner 2004). However, the inability to provide absence data is a notorious weakness of citizen science data in the biodiversity domain, especially if collected as casual data in an opportunistic way (Chapman 2005). This disadvantage can be overcome (or at least mitigated) by using an appropriate project design concerning the protocols and procedures to be followed at observation data collection. A well-established approach is the use of species checklists, allowing to differentiate between species that were observed at a certain place and time and species that were not (for example, the project eBird or the German ornitho.de platform use this method). Certain issues like the detectability of species still need to be taken into account when working with this approach.

Quality assurance for user-generated data from citizen science projects: research and practice, shortcoming and possible solutions

Wiggins et al. (2011) conducted a study analysing the quality assurance mechanisms used in citizen science projects. They found that many projects assure the quality of the data produced by implementing suitable measures before data collection (e.g. project design, training of participants, etc.), while manual validation of observation data by experts is the dominant approach for ex post verification of data. The assessment of correctness (or 'truth') of an observation is based on the plausibility of that observation in the light of the information provided with the observation. The expert's knowledge about the species and the region the observation comes from serve as reference information for the assessment. Also, photographs are often used as evidence.

   Some projects employ automatic assessments of the plausibility of observations. For instance, the project eBird, considered as a 'gold standard' source for bird observations from citizen scientists for use in scientific research, checks the numbers of individuals of species specified by the observer for plausibility, taking into account the location and the season (Sullivan et al. 2009). If the numbers are considered implausible, the observer gets feedback right away. If he or she insists, the observation is passed on to a regional expert for validation. This is also the case for observations that contain species not listed in the species checklist provided to the observer for the location and season (observers can manually add species to the list). eBird now also uses the large amount of data already accumulated in the project to determine parameters for its filter mechanisms, improving filtering results concerning unusual observations (Sullivan et al. 2014). In the German portal 'naturgucker', observers get hints from the system if an observation has certain properties making it implausible. For example, the system checks whether the reported species usually occurs in the region and at the time the observation was made. Another filter checks whether the species has been reported from that region before. Reports of uncommonly rare species will also lead to appropriate feedback to the observer. This project does not flag reports or pass them on to experts for verification, leaving further data quality control entirely up to the crowd. Project Feeder Watch, a North American bird monitoring program, has automatic filters very similar to those of the project eBird, as well using species check lists for regions and seasons, and numbers of individuals observed. Bonter and Cooper (2012) point to the inability of such filters to detect plausible but false reports, and see a need for more research in this area. They expect advances through combining different approaches for plausibility assessment, including assessment of the observers' expertise or experience. Concerning contributors and their properties, Schlieder and Yanenko (2010) explored approaches using social distance between contributors as a confirming factor for the reliability of VGI contributions closely related in space and time. However, such concepts are hardly applicable for citizen science data from the biodiversity domain, as suitable information about contributors to measure their social distance is very rarely available. For an overview of the data quality assurance strategies in projects mentioned in this section, see Table 1.

Many citizen science projects in the field of biodiversity collect observations of plants, animals and fungi in an opportunistic way, producing so called casual data without imposing strict rules or protocols on the contributors. Volunteers contributing to such projects are free to collect and submit observations of a large number of different species at any time and from any place (examples are the Swedish Artportalen project and iNaturalist, an American project with a world-wide scope; see Table 1 for an overview of their respective data quality assurance strategies). This approach has the potential of producing large amounts of data, as the effort required from volunteers is relatively low,

Project	Data quality assurance strategies and options, in terms used by	
	Goodchild and Li (2012)	Wiggins et al. (2011)
eBird (http://ebird.org)	Social approach	Filtering of unusual reports, contacting participants about unusual reports, expert review
Project Feeder Watch (http://feederwatch.org)	Social approach	Filtering of unusual reports, contacting participants about unusual reports, expert review
Ornitho.de (http://www.ornitho.de)	Social approach	Contacting participants about unusual reports, expert review
naturgucker (http://www.naturgucker.de)	Crowd-sourcing approach	Filtering of unusual reports
Artportalen (http://www.artportalen.se)	Social approach	Expert review
iNaturalist (http://www.inaturalist.org)	Crowd-sourcing approach	Filtering of unusual reports

Table 1: Data quality assurance strategies and options employed by the citizen science projects cited in this chapter, in terms used by Goodchild and Li (2012) and Wiggins et al. (2011), respectively. Information about the projects' data quality assurance strategies can be found on their web sites (see table).

encouraging participation and thus furthering high numbers of participants. However, this kind of data has increased needs for ex post quality assurance and suitable data quality parameters, because the usefulness of such projects and their data for science, administration, and planning is often questioned due to a lack of ex ante quality assurance measures (e.g. training of volunteers, implementation of monitoring schemes, etc.).

Most observations consist of at least the species, location, time, and observer, sometimes supplemented with more (project-specific) information. Therefore, methods for quality assurance or plausibility assessment needing only the four basic aspects of an observation have the potential to be useful for many different projects and data sets, but data properties have to be carefully examined in any case. For example, a seemingly exact location in the form of coordinates can have a wide range of accuracy, or even represent different types of locations (i.e. an exact location vs. the centre of a map quadrant).

Conclusion

The scientific studies cited in this chapter, as well as the examples given, provide an overview of the most important aspects of quality of citizen science data from the biodiversity domain and its assurance. They show that manual validation of observations of species by experts based on an assessment of their plausibility in the light of available context information is the dominant approach in citizen science projects in the biodiversity domain. The use of automatic (or semi-automatic) approaches for plausibility assessment is increasing, yet they have important shortcomings as described in section 3. Employing the geographic context for plausibility assessment of crowd-sourced geographic data has high potential for assessing the plausibility of species observations in a (semi)automatic way, despite being rarely used so far. There is a great need for further research on methods to assess the plausibility of citizen science data in the biodiversity domain taking their specific properties into account.

References

Bird, T. J., Bates, A. E., Lefcheck, J. S., Hill, A., Thomson, R. J., Edgar, G. J., Stuart-Smith, R. D., Wotherspoon, S., Krkosek, M., Stuart-Smith, J., Pecl, G. T., Barrett, N., & Frusher, S. 2014. Statistical solutions for error and bias in global citizen science datasets. *Biological Conservation, 173*: 144–154. DOI: http://dx.doi.org/10.1016/j.biocon.2013.07.037

Bonter, D. N., & Cooper, C. B. 2012. Data validation in citizen science: a case study from Project FeederWatch. *Frontiers in Ecology and the Environment, 10*(6): 305–307. DOI: http://dx.doi.org/10.1890/110273

Bordogna, G., Carrara, P., Criscuolo, L., Pepe, M., & Rampini, A. 2014. On predicting and improving the quality of Volunteer Geographic Information projects. *International Journal of Digital Earth*. DOI: http://dx.doi.org/10.1 080/17538947.2014.976774

Chandler, M., Bebber, D. P., Castro, S., Lowman, M. D., Muoria, P., Oguge, N., & Rubenstein, D. I. 2012. International citizen science: making the local global. *Frontiers in Ecology and the Environment, 10*(6): 328–331. DOI: http://dx.doi.org/10.1890/110283

Chapman, A. D. 2005. *Principles and Methods of Data Cleaning – Primary Species and Species-Occurrence Data, version 1.0. Report for the Global Biodiversity Information Facility*. Copenhagen, Denmark: GBIF.

Connors, J. P., Lei, S., & Kelly, M. 2012. Citizen Science in the Age of Neogeography: Utilizing Volunteered Geographic Information for Environmental Monitoring. *Annals of the Association of American Geographers, 102*(6): 1267–1289. DOI: http://dx.doi.org/10.1080/00045608.2011.627058

Conrad, C., & Hilchey, K. 2011. A review of citizen science and community-based environmental monitoring: issues and opportunities. *Environmental Monitoring and Assessment, 176*(1): 273–291. DOI: http://dx.doi.org/10.1007/s10661-010-1582-5

Cosquer, A., Raymond, R., & Prevot-Julliard, A. 2012. Observations of Everyday Biodiversity: a New Perspective for Conservation? *Ecology and Society, 17*(4): 2–16. DOI: http://dx.doi.org/ 10535/8668

Devictor, V., Whittaker, R. J., & Beltrame, C. 2010. Beyond scarcity: citizen science programmes as useful tools for conservation biogeography. *Diversity and Distributions, 16*(3): 354–362. DOI: http://dx.doi.org/10.1111/j.1472-4642.2009.00615.x

Devillers, R., Bédard, Y., Jeansoulin, R., & Moulin, B. 2007. Towards spatial data quality information analysis tools for experts assessing the fitness for use of spatial data. *International journal of Geographic Information Science, 21*(3): 261–282. DOI: http://dx.doi.org/10.1080/13658810600911879

Dickinson, J. L., Zuckerberg, B., & Bonter, D. N. 2010. Citizen Science as an Ecological Research Tool: Challenges & Benefits. *The Annual Review of Ecology, Evolution, & Systematics, 41*: 149–172. DOI: http://dx.doi.org/10.1146/annurev-ecolsys-102209-144636

Dickinson, J. L., & Bonney, R. (ed.). 2012. Citizen Science: Public Participation in Environmental Research. Ithaca and London, Comstock Pub. Associates: 300 pp.

Elwood, S., Goodchild, M. F., & Sui, D. 2012. Researching Volunteered Geographic Information: Spatial Data, Geographic Research, and New Social Practice. *Annals of the Association of American Geographers, 102*(3): 571–590. DOI: http://dx.doi.org/10.1080/00045608.2011.595657

Engler, R., Guisan, A., & Rechsteiner, L. 2004. An improved approach for predicting the distribution of rare and endangered species from occurrence and pseudo-absence data. *Journal of Applied Ecology, 41*(2): 263–274. DOI: http://dx.doi.org/10.1111/j.0021-8901.2004.00881.x

Goodchild, M. F. 2007. Citizens as sensors: The world of volunteered geography. *GeoJournal, 69*(4): 211–221. DOI: http://dx.doi.org/10.1007/s10708-007-9111-y

Goodchild, M.F., & Li, L. (2012). Assuring the quality of volunteered geographic information. *Spatial Statistics, 1*: 110–120. DOI: http://dx.doi.org/10.1016/j.spasta.2012.03.002

Haklay, M. (2013). Citizen Science and Volunteered Geographic Information: Overview and Typology of Participation. In: Sui, D., Elwood, S., & Goodchild, M. F. (Eds.), *Crowdsourcing geographic knowledge: Volunteered Geographic Information (VGI) in Theory and Practice*. New York: Springer, pp. 105–122. DOI: http://dx.doi.org/10.1007/978-94-007-4587-2_7

Hardisty, A., & Roberts, D. 2013. A decadal view of biodiversity informatics: challenges and priorities. *BMC Ecology, 13*(16). DOI: http://dx.doi.org/10.1186/1472-6785-13-16

Howe, J. 2006. The rise of crowdsourcing. *Wired, 14*(6).

International Organisation for Standards. 2002. *ISO 19113 Geographic information – Quality principles*. Geneva, Switzerland: ISO.

Kresse, W., & Fadaie, K. 2004. *ISO Standards for Geographic Information*. Heidelberg: Springer.

Miller-Rushing, A., Primack, R., & Bonney, R. 2012. The history of public participation in ecological research. *Frontiers in Ecology and the Environment, 10*(6): 285–290. DOI: http://dx.doi.org/10.1890/110278

Schlieder, C., & Yanenko, O. 2010. Spatio-temporal proximity and social distance: a confirmation framework for social reporting. In: Proceedings of the 2nd ACM SIGSPATIAL International Workshop on Location Based Social Networks (LBSN 2010), San Jose, CA, USA, November 03–05, 2010: pp. 60–67.

van Strien, A. J., van Swaay, C. A. M., & Termaat, T. 2013. Opportunistic citizen science data of animal species produce reliable estimates of distribution trends if analysed with occupancy models. *Journal of Applied Ecology, 50*(6): 1450–1458. DOI: http://dx.doi.org/10.1111/1365-2664.12158

Sullivan, B. L., Wood, C. L., Iliff, M. J., Bonney, R. E., Fink, D., & Kelling, S. 2009. eBird: a citizen-based bird observation network in the biological sciences. *Biological Conservation, 142*(10): 2282–2292. DOI: http://dx.doi.org/10.1016/j.biocon.2009.05.006

Sullivan, B. L., Aycrigg, J. L., Barry, J. H., Bonney, R. E., Bruns, N., Cooper, C. B., Damoulas, T., Dhondt, A. A., Dietterich, T., Farnsworth, A., Fink, D., Fitzpatrick, J. W., Fredericks, T., Gerbracht, J., Gomes, C., Hochachka, W. M., Iliff, M. J., Lagoze, C., La Sorte, F. A., Merrifield, M., Morris, W., Phillips, T. B., Reynolds, M., Rodewald, A. D., Rosenberg, K. V., Trautmann, N. M., Wiggins, A., Winkler, D. W., Wong, W., Wood, C. L., Yu, J., & Kelling, S. 2014. The eBird enterprise: An integrated approach to development and application of citizen science. *Biological Conservation, 169*: 31–40. DOI: http://dx.doi.org/10.1016/j.biocon.2013.11.003

Theobald, E. J., Ettinger, A. K., Burgess, H. K., DeBey, L. B., Schmidt, N. R., Froehlich, H. E., Wagner, C., HilleRisLambers, J., Tewksbury, J., Harsch M. A., & Parrish, J. K. (2015). Global change and local solutions: Tapping the unrealized potential of citizen science for biodiversity research. *Biological Conservation, 181*: 236–244. DOI: http://dx.doi.org/10.1016/ j.biocon.2013.11.003

Wiggins, A., Newman, G., Stevenson, R. D., & Crowston, K. 2011. Mechanisms for Data Quality and Validation in Citizen Science. In: *Proceedings of the IEEE Seventh International Conference on e-Science.* Stockholm, Sweden, December 05–08, 2011: pp. 14–19.

CHAPTER 7

Semantic Challenges for Volunteered Geographic Information

Andrea Ballatore

Center for Spatial Studies, University of California, Santa Barbara,
aballatore@spatial.ucsb.edu

Abstract

Vast swaths of geographic information are produced by non-professional contributors using online collaborative tools. To extract value from the data, crea-
tors and consumers alike need some degree of consensus about what the entities
of their domain of interest are and how they are related. Traditional information communities, such as government agencies, universities, and corporations,
have devised informal and formal mechanisms to reduce the misinterpretation
of the data they rely on, curating vocabularies, standards, and, more recently,
formal ontologies. Because of the decentralized, fragmented nature of peer production, semantic agreements are more difficult to establish and to document
in volunteered geographic information (VGI), severely limiting the re-usability
and, ultimately, the value of the data. This paper provides an overview of the
semantic issues experienced in VGI, and what potential solutions are emerging from research in geo-semantics and in the Semantic Web. The paradigm of
Linked Data is discussed as a promising route to handle the semantic fragmen-
tation of VGI, reducing the friction between data producers and consumers.

Keywords

volunteered geographic information; data quality; geo-semantics; linked data

How to cite this book chapter:
Ballatore, A. 2016. Semantic Challenges for Volunteered Geographic Information. In:
    Capineri, C, Haklay, M, Huang, H, Antoniou, V, Kettunen, J, Ostermann, F and
    Purves, R. (eds.) *European Handbook of Crowdsourced Geographic Information*,
    Pp. 87–95. London: Ubiquity Press. DOI: http://dx.doi.org/10.5334/bax.g. License:
    CC-BY 4.0.

Introduction

The production of geographic information was, until a decade ago, the exclusive territory of professional surveyors and cartographers, working for governments and private firms. The combination of increasingly powerful, cheap, portable, and interconnected computers has opened up unforeseen possibilities for data collection and sharing beyond professional circles, with already tangible effects (Dodge & Kitchin, 2013). These non-professional mappers and cartographers carry out their efforts on online platforms, producing digital artifacts, such as maps, geo-databases, and gazetteers. This process can be seen as a form of collective communication about some phenomenon of interest (e.g. tourist attractions, noise pollution, or animal behavior). The communication is mediated by machines through the encoding of knowledge from human minds into data and the decoding of data back to knowledge.

To be able to perform this process, the communities that produce volunteered geographic information (VGI) need to devise a shared conceptualization of the portion of the world they want to capture. Questions about what entities exist, what their attributes are, what relationships they have to each other, need answers with some degree of consensus. For a myriad of reasons, such a consensus is often hard to reach. The world and its constituents can be described according to many different, and equally valid, conceptualizations (Smith & Mark, 2001). This problem, rooted in human cognition and communication, is often called 'semantic heterogeneity', and is observable in the ubiquitous vagueness, synonymy, and polysemy in natural languages.

In this chapter, I provide an overview of the semantic challenges that VGI producers and consumers face when describing and interpreting their data. I cover this issue from the perspective of geo-semantics, the discipline at the intersection of geographic information science, computer science, and knowledge engineering. First, I cover relevant work in semantics in the context of geography. Subsequently, I focus on the specific context of VGI, and its peculiar challenges. As a case study, I consider OpenStreetMap and its community. Finally, from a more technological viewpoint, I discuss the emergent Linked Data ecosystem, assessing its promises for more transparent, participatory, and democratic geographic information commons.

Semantics and geographic information

Geography is pervasive in human experience and natural language. On a daily basis, we navigate in and communicate about the geographic world, referring to natural and man-made entities such as roads, cities, mountains, and rivers. Our intuitive understanding of such concepts conceals the complexity that is encountered when trying to encode them in a digital form. The term 'mountain', for example, has a common-sense meaning, but also possesses dozens of local and specialized definitions around the world (Janowicz et al. 2013). When an information system needs to answer the question 'Where is Mount Everest?' there is no single, context-free, cross-cultural way to produce an answer that will satisfy all users. The same consideration can be applied to virtually all natural geographic features, whose boundaries are vague, seasonal, or gradual. Man-made features, while obviously exhibiting more intelligible and crisp organization, are not exempt from heterogeneity, and can be described, categorized, and aggregated through alternative and incompatible conceptualizations.

Geo-semantics, as a subfield of geographic information science, is concerned with providing theoretical and applied means to handle the variations in these concepts, with the purpose of facilitating the creation and processing of information in computationally tractable terms. Standardization of units of measurement, nomenclatures, and other aspects of the geographic domain is indeed an important way to reduce semantic friction, and has been successfully applied to many domains, such as the CORINE nomenclature for land use. However, as Janowicz et al. (2013) argue, geo-semantics is not about imposing standards for what we mean by 'mountain', but should be rather about providing ways to preserve and handle the local definitions across heterogeneous datasets, enabling precise translation mechanisms. These vague geographic concepts are hard to formalize, and their intrinsic cultural grounding makes them poor candidates for long-term universal standardization.

One avenue of research in geo-semantics focuses on ontology engineering in support of conceptual modeling in geographic contexts (Kuhn 2009). Unlike 'big-o Ontology', a branch of Western philosophy interested in the deep structure of reality and being, ontology engineering does not aim at assessing what actually exists in the world outside the human mind, but has the task of constraining the usage of terms in the data towards the meaning intended by their authors. The underlying intuition lies in the usage of formal semantics, such as first order or description logics, to provide machine-readable, less ambiguous descriptions of entities and their relationships, which can be used to support data sharing, integration, and constrained forms of reasoning. Insights from this arena include the formal clarification of *identity, rigidity, role, is-a,* and *part-of* relationships, which wreak havoc when misunderstood in complex information systems (Guarino 2009). This program bears similarities with traditional forms of Artificial Intelligence, with which it shares the formal approach, but differs substantially in that it lacks the ambition to model common-sense knowledge through logic, aiming for more realistic and pragmatic purposes, such as the handling the meanings of 'mountain' in the Himalayas and in Ireland.

VGI and meaning

When even well-funded scientific and corporate organizations struggle to handle semantic heterogeneity, it should come as no surprise that VGI contributors and consumers encounter substantial semantic problems in their work. In online collaborative projects, tensions between alternative conceptualizations of the same portions of reality are common, as a quick exploration of Wikipedia's talk pages would reveal. If they want to create value, VGI contributors interested in cycling are forced to confront, sooner or later, what exactly they mean by 'cycle way', 'cycle lane', and 'road quality', and whether these concepts fit different national and regional contexts different to their own.

Taking a broad stance on the scope of VGI, its semantic structures vary from well-defined and curated geo-referenced datasets, such as GeoNames, to unstructured social media content and blogs. The former are geo-semantically explicit, having the purpose of covering the entire world systematically. By contrast, the latter contain large amounts of geographic information–mainly vague references to place names. Much VGI semantics lies between these two extremes, for example in the case of folksonomies and centralized tagging platforms, such as Tagzania, WikiMapia, and Flickr. Such semantic approaches consist usually of a combination of a top-down, centralized definition of a conceptualization by a small elite, and the emergent semantics of bottom-up, unrestrained tagging.

The most popular VGI project, OpenStreetMap (OSM), deserves separate treatment. This cartographic project is geographically explicit, and produces data substantially more complex than that of competing efforts such as WikiMapia. The main dataset of the project contains an uneven (but impressive) object-based description of the entire planet, including its roads, buildings, parks, forests, lakes, etc. The conceptualization underlying this data is a semi-structured folksonomy, documented on a wiki website,[1] permitting the creation of any new term deemed necessary by users. For example, the term *amenity=university* is used to tag universities. Rather than a fixed ontology, OSM's conceptualization is a transient, evolving product, open to modification and negotiation. The project experiences therefore a tension between the technical need for a stable conceptualization, and the desire of contributors to express their local knowledge without a top-down interpretation of their world being imposed upon them.

Because of its openness, OSM is an ideal resource to study the semantic dimension of crowdsourced cartography. Using a combination of media, including a wiki website, forums, mailing lists, and software tools, contributors negotiate the conceptualization that underpins the data they produce, often disagreeing (Ballatore 2014). By exploring this digital corpus, it is possible to probe the interconnected dimensions of the largely asynchronous negotiation performed by VGI contributors. Most of the observable negotiation in OSM revolves around ontology engineering, i.e. the extraction of an explicit conceptualization from tacit knowledge, but in an informal, online setting (Ballatore & Mooney 2015). The dimensions of this negotiation can be summarized as follows:

---

[1] http://wiki.openstreetmap.org.

Topology and mereology. A topology is needed to represent geographic entities, defining how entities can be connected or contiguous, grounded in a theory of boundaries, interiority/exteriority, and separation. Additionally, a mereology is necessary to encode complex spatial entities, specifying how the parts relate to wholes, for example in the case of large buildings.

Simplification and adaptation. Many domains, such as land cover, road classification, and traffic regulations, have been conceptualized in national and international contexts. However, such conceptualizations are often too complex for the scope of OSM and are filled with technical terminology. In these cases, contributors choose an appropriate subset of the conceptualization, and adapt it to suit their needs. For instance, in France, the European CORINE Land Cover nomenclature has been imported into OSM.

Universalism and localism. A fundamental tension arises between the desire to develop a universal conceptualization that will be applied all over the world, and the need to tap into the heterogeneous and local knowledge of contributors. Initially, contributors attempted an Anglo-centric universal conceptualization, and subsequently, facing an explosion of complexity and spatial variation, fragmented it into regional or national schemas. Notably, the classification of roads has been problematic since the inception of the project, even within the English-speaking world. Similarly, contributors struggle with the complexity of the national road legislation, resorting to translatable national schemas.

Problems of equivalence. The tension between universalism and localism results in problems of equivalence between languages, for example in the conceptualization of restaurants in different countries. As indicated by linguistic translation theory, contributors need to express local concepts that do not have a direct translation in English, such as concepts that depend on local practices, laws, and vocabularies (e.g. courthouse). Specific local entities are often described into more general English terms, losing potentially more precise local knowledge.

Contested definitions. To constrain the intrinsic vagueness of geographic terms, lexical definitions of terms provide an important normative tool to construct a shared conceptualization. As in other domains, lexical definitions can help constrain the intended usage of terms, specifying the necessary and sufficient conditions for their application. Unsurprisingly, conflicts frequently arise about the lexical definitions in OSM. Problems occur when definitions are underspecified, lacking necessary detail, and when they are overspecified, including irrelevant or confusing details. Definitional conflicts result in classification conflicts in the data, when contributors disagree on whether individuals fit a category or not (e.g. is a building a church, a chapel, or a cathedral?).

Conceptual granularity. Information can be expressed at different conceptual granularities, for example describing a geospatial entity as a generic 'tree' or as a 'Pinus roxburghii'. For this reason, contributors often disagree on the level of detail to be included in the conceptualization. In principle, infinite knowledge can be elicited about an entity from multiple perspectives, and the choice of what details should be included is arbitrary, and driven by the desired application. When a category is too generic, its usage is not constrained enough and different conceptualizations emerge. Overly specific categories also cause problems, as they often involve jargon, and are little used. The production of VGI oscillates between different levels of conceptual granularity, in a balancing exercise.

The promises of Linked Data

As I have argued so far, the online production of geographic information faces substantial semantic challenges. VGI communities rely on open tagging and other lightweight semantic approaches to describe their data, which result in frequent inconsistencies, ambiguity, and high terminological heterogeneity, hindering the re-usability and interpretability of the data. The semantic friction encountered in the production and consumption of such data is tackled through different top-down and bottom-up strategies to constrain the usage of terms (e.g. adoption of existing standards, lexical definitions, etc.). For GIScientists, these issues point to exciting research questions. How can we design conceptual models and technologies to support communication about the geographic world, reducing the gap between consumers and producers? How can emergent Web technologies be harnessed to help contributors express their ideas in a clearer, less ambiguous way, without imposing centralized conceptualizations? How can we support the expression and alignment of complex local definitions in intuitive ways?

A promising answer to these questions lies in the Linked Data paradigm (Kuhn et al. 2014). Emerging from 15 years of research in the Semantic Web, Linked Data proposes to express information in an inter-linked data space, built on a triple-based formalism that expresses any data as *subject-predicate-object* statements (e.g. *Dublin is_capital_of Ireland, European_Union is_a Political_entity*). The dominant technologies in this arena are RDF (a simple format to encode triples) and OWL (a logical language to define ontologies, i.e. formal specifications of conceptualizations). The triples are hosted in dedicated triple stores, which are able to index, store, process, and retrieve triples more efficiently than general-purpose database management systems. Unlike traditional datasets, linked entities must have unique Web identifiers (URIs) to enable humans and machines to navigate the data space to retrieve definitions and relations with other entities. To query the triples, SPARQL and its spatial extension GeoSPARQL are currently the most widespread choice, providing a standardized access mechanism, roughly analogous to Web APIs.

As a toy example, let us consider a scenario: a tourist wants to state that they took a picture of the Colosseum in Las Vegas, and simply tags an image file with the string 'Colosseum', which might refer to a Roman building in Rome, to its kitsch replica being photographed, or to an obscure board game. Using the linked data approach, existing entities that match the string in the open knowledge base DBpedia (dbp:Colosseum, dbp:The_Colosseum_at_Caesars_ Palace, where 'dbp:' stands for http://dbpedia.org/resource/), can be suggested to the tourist, who can then select the appropriate entity. The photo can now be described as triples (e.g. photo_001 is_a Photograph; photo_001 represents dbp:The_Colosseum_at_Caesars_Palace), which can be stored and processed automatically, inferring for example that the Colosseum is a theatre designed by the firm Scéno Plus Inc., enabling new avenues for data exploration and reducing the potential misinterpretation of the picture.

This simple idea has proved fruitful in both academic and industrial contexts (Heath & Bizer 2011). Notably, several Linked Open Data (LOD) initiatives have generated an ever-expanding cloud of interconnected datasets containing billions of triples.[2] VGI is a central pillar of this 'online commons' of re-usable open resources, providing the geographic ground for the organization of knowl-edge across domains. Projects like GeoNames, LinkedGeoData, and GeoWord-Net form a constellation of open geo-knowledge bases (Ballatore et al. 2013). Major corporate actors such as Google and Yahoo! have also embraced Linked Data principles, offering increasingly structured search products based on RDF knowledge bases.[3] Media groups including the BBC and the New York Times publish part of their informational assets as Linked Data. Adopting a more lightweight, simpler approach, Microformats promote the semantic annotation of people, places, products, reviews, and organizations in Web pages, support-ing the interpretation of content, without requiring the adoption of more com-plex Semantic Web infrastructure.[4]

Conclusions

In this chapter, I have summarized the challenges faced by VGI from a semantic perspective. First, I discussed the conceptual difficulties intrinsic to the vague-ness of many geographic concepts. Second, OpenStreetMap (OSM) was taken as a case study to highlight the semantic issues that cause friction in the process of VGI production and consumption. VGI contributors coordinate their efforts and express information using a variety of semantic approaches, ranging from open tagging to controlled taxonomies and vocabularies. To produce intelligi-ble data, OSM contributors make choices concerning topology and mereology,

---

[2] http://lod-cloud.net.
[3] https://googleblog.blogspot.co.uk/2012/05/introducing-knowledge-graph-things-not.html
[4] http://microformats.org.

in a tension between universalism and localism. Particularly in transnational contexts, the description of entities encounters problems of equivalence, resulting in contested definitions of geographic concepts, at different granularities. To what degree these aspects extend to other VGI projects is an open research question. As a promising way to support the expression of heterogeneous local knowledge in VGI, I have briefly discussed Linked Data, a technical paradigm that has grown from Semantic Web research. Linked Data aims at constructing a data space in which geographic entities can be defined and described through standardized and precise logical mechanisms.

While I have presented reasons to support the adoption of Linked Data in VGI, the open challenges and current limitations of the approach cannot be ignored. The logic formalisms used in Linked Data, such as OWL, are rather ill-suited for spatio-temporal reasoning and need substantial extensions. The triple model, while conceptually attractive, can be very verbose to describe traditional geographic data such as raster images, and its structural complexity can explode quickly in realistic scenarios. To explore the current limitations of Linked Data, it suffices to take a closer look at the LOD Cloud, whose datasets vary hugely in their interpretability and noise, and whose interlinking is often patchy and uneven.

The approach promotes semantic clarity but cannot enforce it. Without formal constraints, Linked Data can be as obscure and ambiguous as plain text: the halo of clarity fades out as soon as poorly structured datasets are subject to integration and complex processing. Finally, the tools available to produce and process Linked Data often lack usability, and substantial design efforts are needed for deployment in VGI contexts, providing intuitive approaches to encoding local knowledge and alternative truths. None of these issues are insurmountable, and they do not outweigh the enormous potential benefits. Ultimately, the Linked Data paradigm should be considered as a promising technical framework to mitigate semantic problems in data production and consumption, and not as an unlikely fix to ancestral flaws in human communication that are here to stay.

References

Ballatore, A. 2014. Defacing the map: Cartographic vandalism in the digital commons. *The Cartographic Journal, 51*(3): 214–224.

Ballatore, A., & Mooney, P. 2015. Conceptualising the geographic world: The dimensions of negotiation in crowdsourced cartography. *International Journal of Geographical Information Science, 29*(12): 2310–2327.

Ballatore, A., Wilson, D. C., & Bertolotto, M. 2013. A Survey of Volunteered Open Geo-Knowledge Bases in the Semantic Web. In Pasi, G., Bordogna, G., & Jain, L. C. (Eds.), *Quality Issues in the Management of Web Information*. Berlin: Springer, pp. 93–120.

Dodge, M., & Kitchin, R. 2013. Mapping Experience: Crowdsourced Cartography. *Environment and Planning A, 45*(1): 19–36.

Heath, T., & Bizer, C. 2011. Linked data: Evolving the web into a global data space. *Synthesis Lectures on the Semantic Web: Theory and Technology, 1*(1): 1–136.

Guarino, N. 2009. The Ontological Level: Revisiting 30 Years of Knowledge Representation. In: Borgida, A., Chaudhri, V., Giorgini, P., & Yu, E. (Eds.), *Conceptual Modelling: Foundations and Applications. Essays in Honor of John Mylopoulos.* Berlin: Springer, pp. 52–67.

Janowicz, K., Scheider, S., & Adams, B. 2013. A Geo-Semantics Flyby. In: Rudolph, S., Gottlob, G., Horrocks, I., & van Harmelen, F. (Eds.), *Reasoning Web. Semantic Technologies for Intelligent Data Access* . Berlin: Springer, pp. 230–250.

Kuhn, W. 2009. Semantic Engineering. In: Navratil, G. (Ed.), *Research Trends in Geographic Information Science.* Berlin: Springer, pp. 63–76.

Kuhn, W., Kauppinen, T., & Janowicz, K. 2014. Linked Data–A Paradigm Shift for Geographic Information Science. In: *Geographic Information Science.* Berlin: Springer, pp. 173–186.

Smith, B., & Mark, D. M. 2001. Geographical categories: An ontological investigation. *International Journal of Geographical Information Science, 15*(7): 591–612.

CHAPTER 8

Quality analysis of the Parisian OSM toponyms evolution

Vyron Antoniou*, Guillaume Touya[†] and
Ana-Maria Raimond[†]

*Hellenic Military Academy, Greece, v.antoniou@ucl.ac.uk
[†]Laboratoire COGIT, Institut National de l'Information Géographique et
Foréstière, 73 Avenue de Paris, 94165 Saint-Mandé, France

Abstract

The paper presents empirical research on the quality of the toponyms that can be retrieved from OpenStreetMap (OSM) under the purpose of enriching authoritative toponymic databases and gazetteers. An analysis on the volatility of places and points-of-interest (POIs) is presented. We examine how named features behave and change in terms of type, name and location. The challenge is to understand the behavior and consequently the fitness-for-purpose of OSM data when it comes to a possible use and integration with authoritative datasets. We show that, depending on the OSM feature type, the volatility can vary considerably and we elucidate which feature types are consistent, and thus could be used in authoritative gazetteers despite their grassroots nature and if there are spatial patterns behind the location changes of features during their lifespan.

Keywords

Toponyms, geographic names, OpenStreetMap, Gazetteer, VGI, data quality.

How to cite this book chapter:
Antoniou, V, Touya, G and Raimond, A-M. 2016. Quality analysis of the Parisian OSM
  toponyms evolution. In: Capineri, C, Haklay, M, Huang, H, Antoniou, V,
  Kettunen, J, Ostermann, F and Purves, R. (eds.) *European Handbook of
  Crowdsourced Geographic Information*, Pp. 97–112. London: Ubiquity Press.
  DOI: http://dx.doi.org/10.5334/bax.h. License: CC-BY 4.0.

Introduction

Gazetteers are a vital component of any spatial database irrespectively of the level of detail used (i.e. local, national or international). Gazetteers consist of a list of toponyms, a type and their corresponding geography. This geography can be either a point, a bounding box or the footprint of the place. Gazetteers are usually used as an entrance point to a spatial database. People start exploring geographic data by providing a toponym and search for features, relations, maps or events related to that toponym. In other cases, the outcome of a spatial search is accompanied by toponyms that facilitate the understanding of the result. There are also cases where the result of a search is the toponym itself (e.g. reverse-geocoding). Apart from these practical examples, toponyms and gazetteers play a key role in many aspects of everyday life. Examples can be found in explicit geographic applications like routing, mapping and cartography but also in more general cases such as in government, legislation, security and policing etc. (UN, 2006).

However, National Mapping Agencies (NMAs), which are the de facto agencies responsible for creating and updating gazetteers in a national level, are facing difficulties in keeping toponymic databases up to date due to the lack of resources and due to the extensive field work needed for data collection and verification. On the other hand, Volunteered Geographic Information - VGI (Goodchild 2007) can serve as a promising alternative mechanism for collecting toponyms that could enrich and update official gazetteers (Goodchild & Hill 2008). In this context, the aim of this paper is to examine whether OSM can provide consistent toponymic datasets or the grassroots mechanisms alter constantly the spatial features in such a level that hinder their use in gazetteers. More specifically, the research tries to provide empirical evidence on the following questions: i) What is the population and the types of OSM features that have names and can be used as part of a gazetteer? ii) What kind of changes are taking place for these features? iii) What feature types are affected and how much? iv) Are there any underlying spatial patterns for these changes? This study adopts the definition about toponyms that is provided by the United Nations Group of Experts on Geographical Names (UNGEN), and thus the scope of interest includes populated places, civil divisions, natural features, constructed features and unbounded places or areas that have specific local meaning (UN 2006: 9).

OSM urges its contributors to provide names for spatial objects, if applicable, using the *name* key tag. Contributors can add more than one name for spatial features such as international names or old names by using variations of the *name* key such as *int_name* or *old_name*. Moreover, OSM wiki pages provide detailed guidelines on how to correctly assign a name to spatial objects in order to achieve maximum standardization. In our study only the *name* key tag has been examined of the point-based objects of two broad OSM categories: Places[1]

---

[1] http://wiki.openstreetmap.org/wiki/Places

and Points of Interest (POIs)[2]. These categories are in accordance with what United Nations (UN) define as a toponym. The remainder of the paper is structured as follows: Section 2 presents briefly a selection of related work on the subject. Section 3 discusses the methodology used to collect and analyze OSM data. Section 4 presents the results followed by discussion and future work in Section 5.

Related Work

The importance of gazetteers and the challenges posed by the nature of VGI data, and especially of toponyms, has drawn the interest of many researchers and there is extensive literature available. Here, we provide few examples of VGI and authoritative data integration efforts so to highlight that VGI quality and stability is an important factor for this task. Such efforts range from creating a gazetteer by harvesting volunteered big geo-data from Web sources (see for example Gao et al. 2014) to combining both administrative and VGI toponyms. For example, Twaroch et al. (2008) use various web sources to create a surface model of the toponyms' footprints. However, the authors highlight the fact that it is difficult to have crisp boundaries when it comes to VGI data and that there is a need to identify outliers. Similarly, Keßler et al. (2009a) proposed the enrichment of authoritative gazetteers with toponyms extracted from geotags of photos. As the authors support, their approach could benefit from quality indicators of the geotags used. The quality of user-contributed data has been also highlighted as a crucial factor in empirical research with geo-tagged photos (see for example Hollenstein & Purves 2010). Regarding OSM, Hahmann and Burghardt (2010) proposed to link OSM with GeoNames gazetteer using semantic web techniques to produce an enriched, multi-lingual gazetteer and Smart et al. (2010) proposed a methodology for the conflation of toponymic data from multiple sources, including both authoritative and VGI datasets, and taking into account the quality differences of each source. However, as Mooney and Corcoran (2012) explain, developers of location-based services should be cautious when it comes to using OSM data as their research on frequently edited features revealed considerable volatility in the naming process. Moreover, Keßler et al. (2009b) underline the importance that gazetteers should cater both for local and small-scale features, as well as timely and user-centric information. In this context, OSM has the potential to become a valuable source of toponyms. Thus the discussion focuses on the nature, the behavior and the evolution of the toponymic datasets that can be retrieved from OSM and how these factors affect quality elements and their use in gazetteers.

---

[2] http://wiki.openstreetmap.org/wiki/Points_of_interest

OSM Data Extraction

Extending this line of research, this paper focuses on the volatility of OSM features. It goes beyond the naming changes that Mooney and Corcoran (2012) focused and examines also the location changes of OSM features. The area of scope of the research is Paris region (12.012 km²). The study area is large enough to have a great diversity of named features, and is quite complete due to the large number of OSM contributors. In order to collect the necessary data, the Geofabrick[3] shapefile download service was used. The datasets for the area of scope were downloaded at the first week of December 2014. Shapefiles include as an attribute the unique OSM_ID of every OSM feature. These IDs were used in combination with the OSM API to collect and store in a Post-greSQL/Postgis database all the versions of each feature. This method provided a complete timeline of the OSM edits made in the area of scope.

Analysis

Descriptive statistics

A preliminary analysis on the availability of names for the spatial features (grouped by OSM category) was conducted and the results are shown at Table 1.

It can be seen that, depending on the category, there are considerable variations in the presence of names. For example, in the '*Places*' category, OSM contributors have assigned a name at almost all (i.e. except from 3) features. Arguably, this behavior meets the expectations of an OSM user (including the

Category	Total	With names	%
Land use	36,347	2,201	6.1%
Natural	20,138	2,093	10.4%
Places	4,275	4,272	99.9%
Points	192,228	53,052	27.6%
Railways	16,482	4,471	27.1%
Roads	344,870	152,595	44.2%
Waterways	5,190	2,520	48.6%
Total	619,530	221,204	35.7%

Table 1: Total OSM features and OSM features with names for the study area.

---

[3] www.geofabrik.de

author of a gazetteer) as it is generally expected that all point features classified as '*Places*' should have a name. In contrast, this cannot be observed at the
'*Roads*' category. Although, in reality, roads have a name (especially in urban
areas like Paris) or a reference name (e.g. *link to Motorway X*) the percentage
of named features is barely 44.2%. Another category that is of interest for a
gazetteer is the one of 'Points'. This category includes a variety of local features
that OSM contributors deem as Points of Interest (POIs). Here the percentage of named features is just 27.6% but, as it will be explained later, this factor
is not indicative of the completeness of the dataset in terms of names as for
many POIs' subcategories a name tag is not applicable (such as for 'crossing' or
'bench'). Given these results the research focused into two categories that were
deemed as the most interesting when it comes to examining the potential to
create or enrich a gazetteer: OSM Places and OSM POIs.

OSM Places

In terms of changes in type, location and name, a Place point can either remain
stable in its entire life-cycle or undergo a change in one or any combination of
these three factors. In an effort to understand whether the OSM data can serve
as a source of consistent toponyms, the percentage of the features that have
been changed or remained stable has been recorded (Figure 1).

It can be seen that two thirds of the features have never been changed while
the most common change that features undergo is in their geographic location.
In this context, the next issue of interest was to examine the types of places and
their corresponding population versus the location movements that took place
for each '*Place*' type. This classification was used so to examine which types
of features, have been moved by OSM contributors. Again, this can give an
overview of the consistency of OSM Places. The findings are shown in Table 2.

The findings show that, depending on the type of place, there is considerable variation in terms of location change. For example, while only 8% of the
features belonging to the '*locality*' type has been moved, for the features that
belong to the '*town*' type this reaches 80%. Following this observation, the next
step was to examine the magnitude of location change (calculated in meters)
for each type of place. The magnitude of location change is considered as the
distance between two points: i) the centroid calculated taking into account all
positions of the feature during its life and ii) the last position of the features.
The results are shown in Figure 2.

This type of analysis can visualize the volatility in location change of various
place types. It can be observed that entities with large spatial extends (either
crisp of fuzzy) suffer from large changes in their location in contrast with
smaller entities. For example, almost 14% of all '*towns*' have moved over 1,000
m whereas for '*suburbs*' 21% of the features have moved less than 100 m and
65% remained stable (see also Table 2).

Places

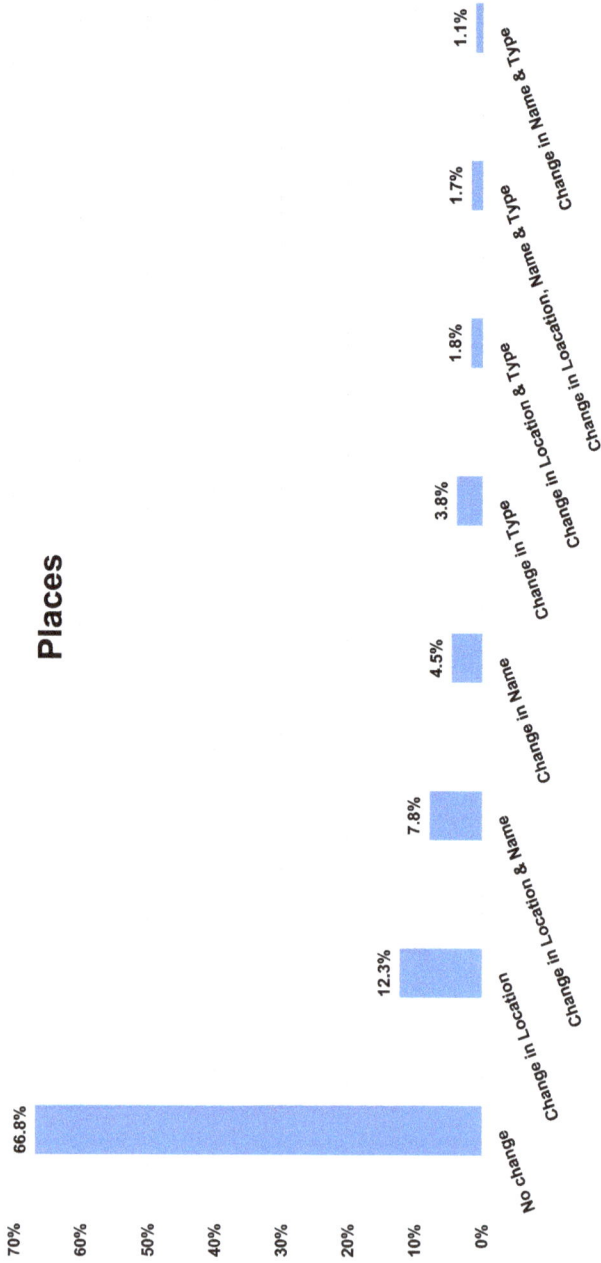

Figure 1: Changes in Places taking into count three factors (location, name and type).

Type	All features	Features moved	%
Allotments	2	2	100%
City	1	1	100%
Hamlet	496	86	17%
Island	9	1	11%
Islet	1	0	0%
Isolated_dwelling	37	3	8%
Locality	2,096	167	8%
Neighbourhood	214	20	9%
State	1	1	100%
Suburb	130	45	35%
Town	248	198	80%
Village	1,039	487	47%
Yes	1	0	0%
Total	**4,275**	**1,011**	**24%**

Table 2: Types of OSM places and number of OSM places that have been geo-graphically moved.

OSM POIs

As noted above, from almost 200K of POIs only 27.6% of them (i.e. 53,052) had a *name* attribute. This is an expected observation as there are types of POIs where the *name* is not an applicable attribute. For example, the POI types of *crossing, bench, traffic_signals* and *survey_points* have in total 75,819 spatial features that account approximately to the 40% of the total population, and less than 0.04% of them (i.e. only 27 features) have names.

Similar to the Places' analysis, the changes of the same elements (i.e. of location, type and name) have been examined also for the POIs. The findings show that about 60% of features have not been changed since their creation while the most common change this time is the change in their name.

In order to examine which POI types are the most volatile in terms of name and location change, a scatter-plot (Figure 4) has been created. The x-axis in Figure 4, shows the percentage of features that had a change in name for the 30 most populous OSM types. Name changes range from minor changes (e.g. alterations in capital letters or blank spaces) up to changes in the entire name. Although it is not clear which OSM feature types should be included in a gazetteer (see also discussion in Section 5), Figure 4 shows that there are types of POIs that have a large rate of name changes and others that remain relatively

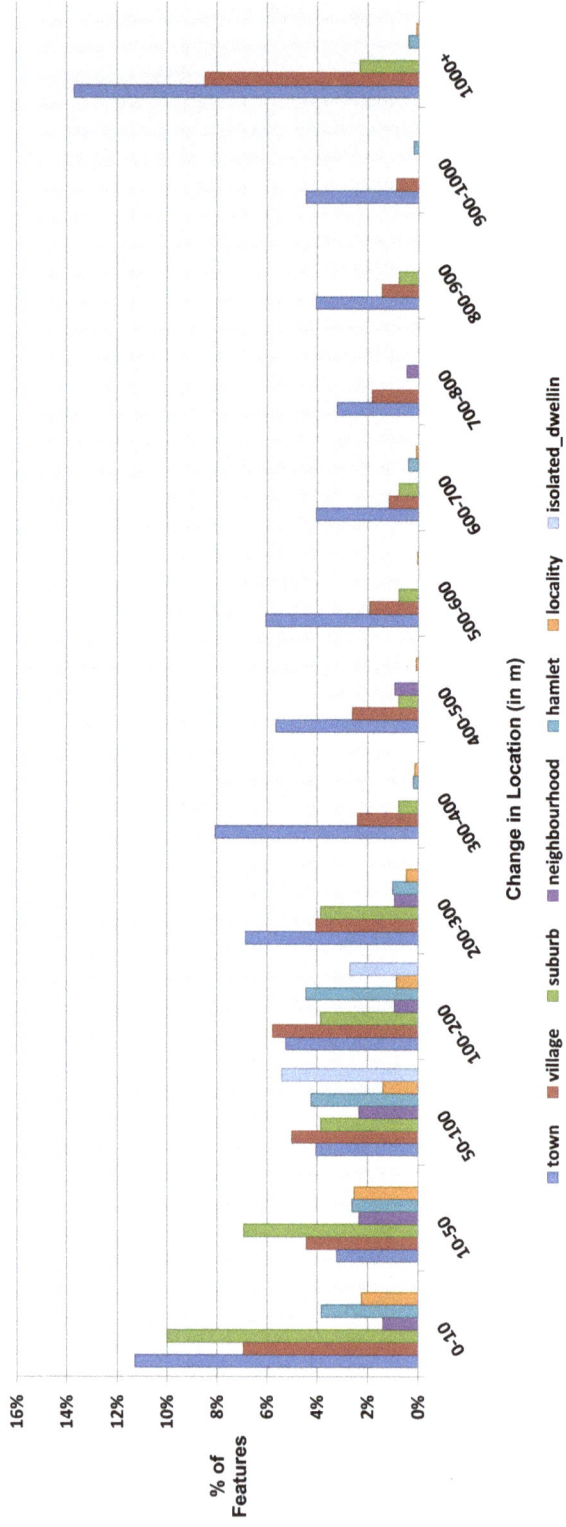

Figure 2: The magnitude of location change (x-axis, in m) per type of OSM 'Place'.

Points of Interest (POIs)

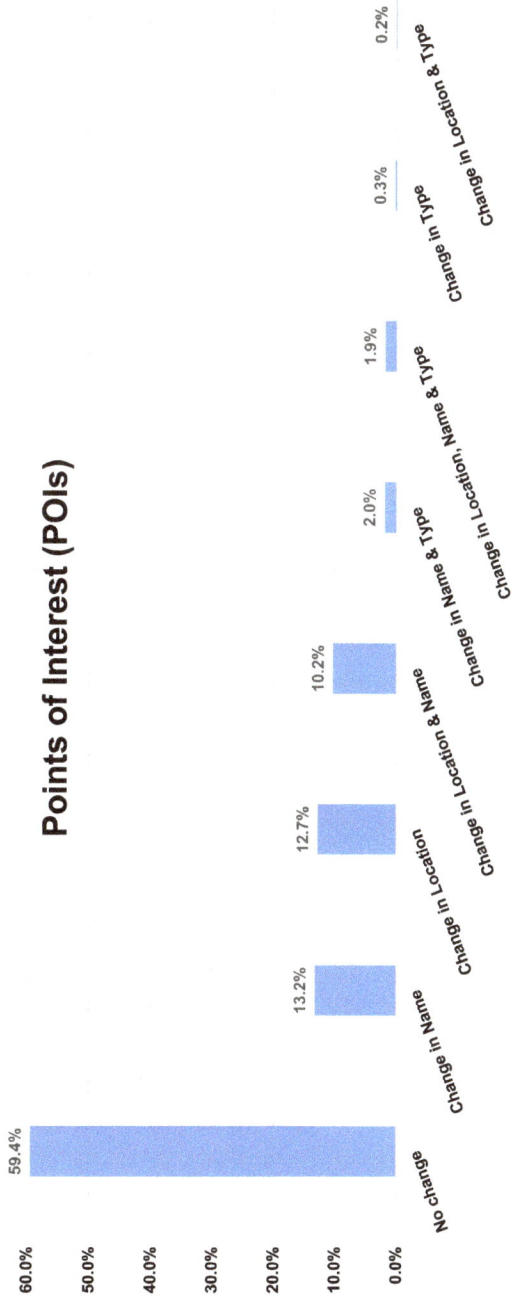

Figure 3: Changes in POIs taking into count three factors (location, name and type).

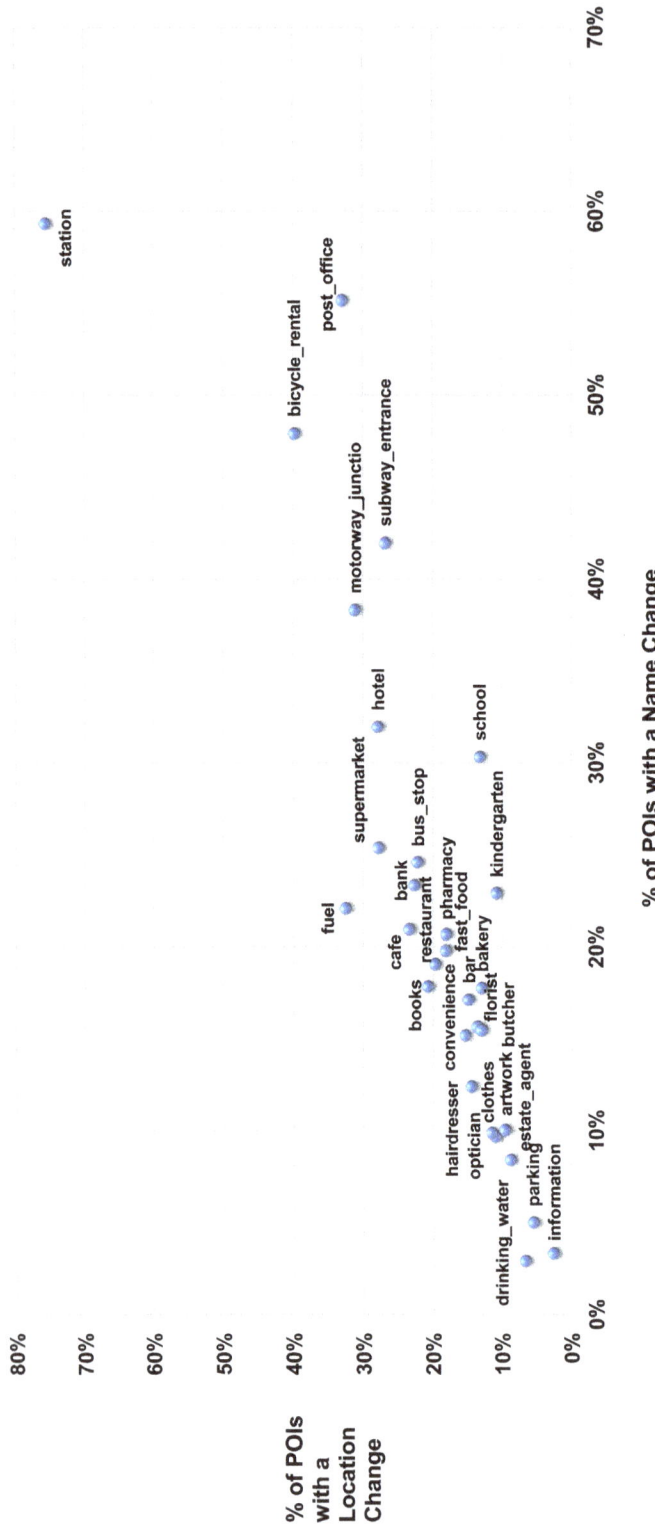

Figure 4: Percentage of POIs that had a name (x-axis) and a location (y-axis) change.

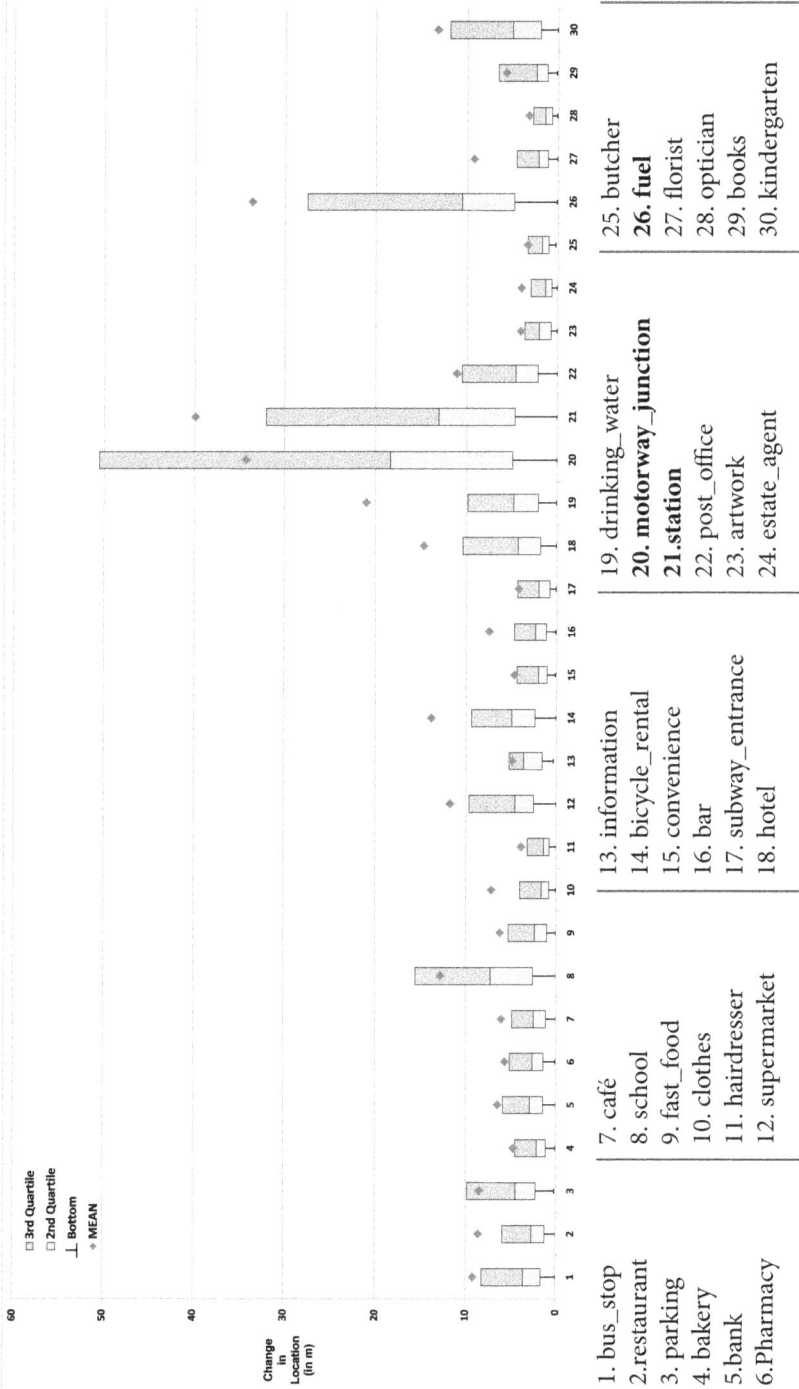

Figure 5: Box-Whisker graph of the location change for each POI type.

1. bus_stop
2. restaurant
3. parking
4. bakery
5. bank
6. Pharmacy

7. café
8. school
9. fast_food
10. clothes
11. hairdresser
12. supermarket

13. information
14. bicycle_rental
15. convenience
16. bar
17. subway_entrance
18. hotel

19. drinking_water
20. **motorway_junction**
21. **station**
22. post_office
23. artwork
24. estate_agent

25. butcher
26. **fuel**
27. florist
28. optician
29. books
30. kindergarten

stable (e.g. lower left corner). The y-axis in Figure 4, shows the % of features that had a location change for each OSM type. Here, again, it can be observed that not all POI types behave the same; certain POI types suffer more than others.

The combined view of changes over these two factors indicates which of these types (if they are to be included in a gazetteer) might be considered as too unstable to populate a gazetteer. However the counter argument can be that the seemingly stable behavior of certain POI types can be explained by poor contributors' attention and thus this stability might indicate obsolete or out-of-date features. In any case, Figure 4 raises awareness of the futures' behavior and gives a better insight on what kind of volatility should be expected per POI type.

After gaining a better understanding on which POI types are volatile in terms of name and location change, the next step was to quantify the latter. In order to better visualize the position change, a Box-Whisker plot has been created in Figure 5 (note that the upper quartile is not marked as outliers in many types make it draw out of scale – for ease of understanding the mean value has been added).

First, this type of analysis can help to understand which spatial features should not be modeled as POIs since the simple geometry of a point appears not to be the best way to model this physical entity. For example, motorway junctions seem not to gather consensus among OSM contributors regarding the position of the POI as the average location change is more than 30 m. On the contrary, there are POI types that despite their location change, the distance between various locations remains well under 10 m (i.e. an arbitrary positional accuracy threshold of hand-held GPS devices). Second, this type of analysis can highlight gross errors and outliers in OSM datasets that might downgrade the overall spatial quality of a dataset. The largest the distance between the mean and the upper level of the second Quartile box (i.e. the 50% of the features), the more outliers and gross positional errors exist in each category. For example, 8% of the features for the *fuel* category have been moved more than 100 m (with a recorded maximum movement of 659 m). Finally, it is made clear that for many POI types a clearer feature extraction guide is needed. For example, when capturing schools or station it needs to be clear for contributors where the point should be positioned: at the entrance of the building, at the centroid of the main building or somewhere else. Let us mention that although there are instructions in the OSM wiki pages how to map each feature, apparently these instructions are not explicit enough and thus inconsistencies occur.

Spatial patterns

Finally, for the entire dataset of POIs a hot-spot analysis was calculated based on the location change for each feature. A visualization based on Z-score is shown in Figure 6. Hot-spot analysis (using the *Getis-Ord Gi* statistic provided in ESRI ArcGIS 10.2.2) can reveal whether a phenomenon is random or not.

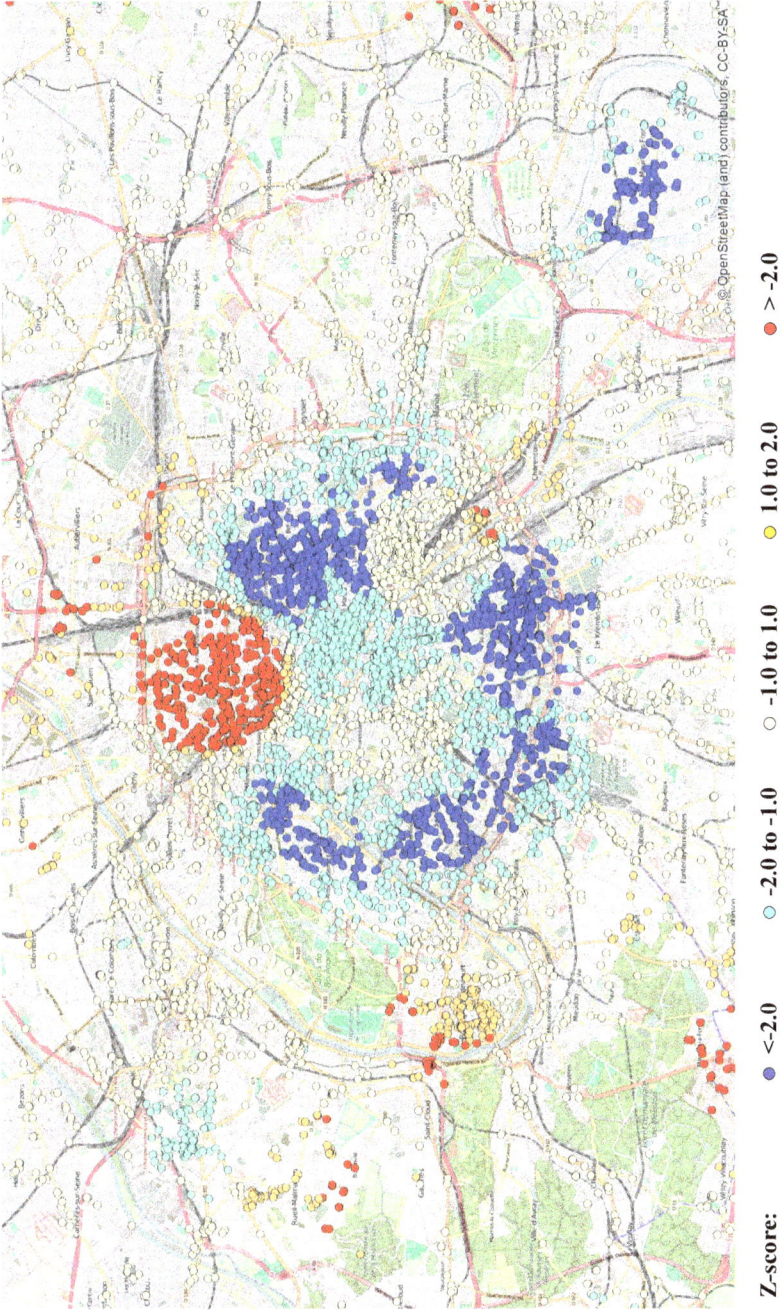

Z-score: ● <-2.0   ○ -2.0 to -1.0   ○ -1.0 to 1.0   ● 1.0 to 2.0   ● > -2.0

Figure 6: Hot Spot Analysis (using the Getis-Ord Gi* statistic) on the location change of POIs.

Here it can be observed that there are concentrated hot colors areas (i.e. areas with not random large movements) and cold colors areas (i.e. areas with not random small movements) for POIs. While a first observation can be made that hot colors areas appear in the popular and touristic area of Montmartre (north of map) and the cold colors areas appear at the periphery of Paris (more residential than touristic), further analysis is needed to fully understand the causes of the phenomenon. For instance, Montmartre is a hill, and the sources, like the ortho-rectified satellite imagery, used for the positioning of the POIs may be less accurate there.

Discussion and future work

VGI datasets are a dynamic source of spatial information. In particular, OSM datasets, which usually function as a proxy in the research on VGI data, have drawn the interest of researchers regarding their use in helping NMAs to complete or update existing geospatial products or even to create new ones (Antoniou 2011). Improved and enriched authoritative products can be toponymic databases and gazetteers. The importance of gazetteers in acquiring accurate results in spatial searches is paramount and thus the update of official gazetteers with local knowledge should be made with caution and meticulous examination of the VGI quality. Unnecessary changes in the names, types or the geographic position (no matter how subtle or small) can introduce problems to authoritative products or location based services. However, once successful, the presence of local and community-level named features and landmarks can considerably enrich and improve gazetteers and geospatial services. A first point of consideration is the decision on which types of user-contributed features should be used in a gazetteer. For example, certain types of POIs are possible to serve as landmarks that can help to provide eloquent and easily understandable routing directions. Although this paper does not delve into the subject of feature type importance, it provides evidence that the selection of OSM types and features should be examined from a quality point of view as well.

What this paper has examined is the behavior and thus the fitness-for-purpose of OSM data as a source of toponymic data. The aim was to examine whether the OSM datasets are a consistent datasets or the grassroots mechanisms alter constantly the datasets in such a level that in practice hinder the use of OSM data. The findings show that VGI and authoritative data conflation is not a straightforward process as they differ considerably in nature. While authoritative toponyms are largely static and hard to change spatial entities, a considerable percentage of VGI toponyms undergo changes. Not all OSM types are fit to support the enhancement of administrative gazetteers as the OSM specification and contribution practices might generate an unwanted volatility in the data. This observation generates a number of questions that could be the aim of future work. First, it is important to understand the nature of these changes. For example, do changes in location serve a better mapping outcome, refer to previous mistakes and thus are a spatial quality improvements or are they simply real-life movements that OSM contributors capture? Relating movement to the geographic extent of the named feature, or to some contributor pattern would be useful to understand how and why the changes occur. Using what Goodchild and Li (2008) call the geographic approach to assess named features movement would also be useful: e.g. check whether a Place feature that refers to a town has been move to the centroid of the town hall. Similarly, it could be examined if there are any time patterns in the changes. For example are these changes concentrated at the early period of the creation of a feature and thus it is an indicator of quality improvement (as discussed in Haklay et at. 2010) or are they happening during the entire life-cycle of each feature and indicate an endogenous volatility of the spatial feature? Nevertheless, contributors might alter OSM features (no matter what the reason) and this change can either be very small and thus authoritative products and services that have integrated OSM data will not be affected or might be large enough to introduce unwanted volatility. Finally, it is of interest to compare, in terms of completeness, the OSM toponyms with authoritative data so to understand at what extend VGI data can help NMAs to improve their gazetteers. Thus, future work will include the comparison between OSM and authoritative toponyms (provided by IGN France, the French NMA).

Acknowledgments

This research took place during a two week Short Term Scientific Mission (STSM), funded by COST TD1202 ENERTGIC Action, in IGN with the help of Dr Bénédicte Bucher. Finally, we are grateful to the reviewers for their helpful comments on the original paper.

References

Antoniou, V. 2011. *User generated spatial content: an analysis of the phenomenon and its challenges for mapping agencies.* Thesis (PhD), University College London.

Gao, S., Li, L., Li, W., Janowicz, K., & Zhang, Y. 2014 (In Press). Constructing gazetteers from volunteered big geo-data based on Hadoop. *Computers, Environment and Urban Systems.* DOI: http://dx.doi.org/10.1016/j.compenvurbsys.2014.02.004

Goodchild, F. M. 2007. Citizens as sensors: the world of volunteered geography. *GeoJournal, 69*(4): 211–221. DOI: http://dx.doi.org/10.1007/s10708-007-9111-y

Goodchild, F. M., & Hill, L. 2008. Introduction to digital gazetteer research. *International Journal of Geographical Information Science, 22*(10): 1039–1044. DOI: http://dx.doi.org/10.1080/13658810701850497

Goodchild, M. F., & Li, L. 2012. Assuring the quality of volunteered geographic information. *Spatial Statistics, 1*: 110–120. DOI: http://dx.doi.org/10.1016/j.spasta.2012.03.002

Hahmann, S., & Burghardt, D. 2010. Connecting Linked GeoData and geonames in the spatial semantic web. In *Proceedings of extended abstracts, 6th International GIScience Conference,* Zurich, Switzerland.

Haklay, M., Basiouka, S., Antoniou, V., & Ather, A. 2010. How many volunteers does it take to map an area well? The validity of Linus' law to volunteered geographic information. *The Cartographic Journal, 47*(4): 315–322. DOI: http://dx.doi.org/10.1179/000870410X12911304958827

Hollenstein, L., & Purves, R. 2015. Exploring place through user-generated content: Using Flickr tags to describe city cores. *Journal of Spatial Information Science,* (1): 21–48.

Keßler, C., Maué, P., Heuer, J. T., & Bartoschek, T. 2009a. Bottom-up gazetteers: Learning from the implicit semantics of geotags. *GeoSpatial semantics, 5892*: 83–102. DOI: http://dx.doi.org/10.1007/978-3-642-10436-7_6

Keßler, C., Janowicz, K., & Bishr, M. 2009b. An agenda for the next generation gazetteer: Geographic information contribution and retrieval. In: *Proceedings of the 17th ACM SIGSPATIAL international conference on advances in Geographic Information Systems,* pp. 91–100.

Mooney, P., & Corcoran, P. 2012. Characteristics of heavily edited objects in OpenStreetMap. *Future Internet, 4*(1): 285–305. DOI: http://dx.doi.org/10.3390/fi4010285

Smart, P. D., Jones, C. B., & Twaroch, A. F. 2010. Multi-source toponym data integration and mediation for a meta-gazetteer service. In *Geographic Information Science, 6292*: 234–248. DOI: http://dx.doi.org/10.1007/978-3-642-15300-6_17

Twaroch, A., F., Jones, B., C., &. Abdelmoty A. 2008. Acquisition of a vernacular gazetteer from web sources. In: *Proceedings of the first international workshop on Location and the web.* ACM: pp. 61–64.

United Nations. 2006. *Manual for the National Standardization of Geographical Names: United Nations Group of Experts on Geographical Names.* Department of Economic and Social Affairs, New York, USA: UN.

CHAPTER 9

Tackling the thematic accuracy of areal features in OpenStreetMap

Ahmed Loai Ali[*,†]

*Bremen Spatial Cognition Center (BSCC), University of Bremen, Germany
†Faculty of Computers and Information, Assiut University, Egypt
loai@informatik.uni-bremen.de, loai.cs@gmail.com

Abstract

With the increasing importance of VGI for GIScience, data quality becomes an issue of high concern. Particularly in collaborative mapping projects, when a group of public participants acts to collect, update and share information about geographic features, aiming to maintain and improve a geo-spatial data-set. OpenStreetMap (OSM) is the most common VGI project that aims to develop free world digital map. Although several studies emphasized the positional accuracy and completeness of the OSM data, particularly in the urban areas, they also highlighted its problematic thematic accuracy. In this chapter, we handle the thematic accuracy quality measure from the facet of classification. This chapter presents an approach for rule-guided classification for VGI projects. The proposed approach exploits the availability of data to learn the distinct characteristics of a set of geographic features. Afterwards, the learned characteristics are used to guide the contributors toward the most appropriate data classes, aiming to improve the data quality. The approach consists of two phases: *Learning* and *Guiding* phases. During the *Learning* phase, data mining algorithms are applied to learn the geographic characteristics of specific features. The learning process results in a set of rules describing these features. The extracted rules are used to develop a classifier. Afterwards, during the *Guiding* phase, the developed classifier is used for several purposes; 1) acts to detect

How to cite this book chapter:
Ali, A L. 2016. Tackling the thematic accuracy of areal features in OpenStreetMap.
In: Capineri, C, Haklay, M, Huang, H, Antoniou, V, Kettunen, J, Ostermann, F and
Purves, R. (eds.) *European Handbook of Crowdsourced Geographic Information*,
Pp. 113–129. London: Ubiquity Press. DOI: http://dx.doi.org/10.5334/bax.i.
License: CC-BY 4.0.

problematic classified entities; and 2) guides and aids the contributors during the classification process. An empirical study followed by an implementation is conducted. The results show the feasibility of the proposed approach and highlight some limitations that could be improved in the future studies. The developed tool generates promising results and improves the classification of OSM dataset as well.

Keywords

Volunteered Geographic Information (VGI), Spatial Data Quality, Thematic accuracy, Spatial data mining.

Introduction

Crowd-sourcing, the advance of web technologies and the availability of location sensing devices empower the public to produce contents associated with implicit or explicit spatial references. This form of User Generated Contents (UGC) has been known as *Volunteered Geographic Information*, in which a group of people voluntary acts to collect, update, and share spatial information (Goodchild 2007). VGI changes the conventional way of mapping activities resulting in collaborative mapping. Those activities were exclusively reserved – for a long time – for mapping agencies and specialized organizations. However, in collaborative mapping, participants are eager to collect information about geographic features producing maps (Gillavry 2004). Among others, Open-StreetMap[1] (OSM), Wikimapia[2] and Google MapMaker[3] are examples of collaborative mapping projects. OSM is the most prominent example of a VGI project; it aims to develop a free digital map of the world editable and available to everyone. During the last decade, several applications and services have been developed based on VGI data including – but not limited to – urban planning, environmental monitoring, crises management, map provision, etc.

Despite the increasing utilization of VGI data, its questionable quality still makes it – in some cases – of limited use (Elwood et al. 2012; Flanagin & Metzger 2008). Among other reasons, contributors' diversities and the fixable contribution mechanisms are resulting in data of heterogeneous quality (Mooney & Corcoran, 2012). Several studies assess VGI data by comparison with authoritative data sources. They conclude the promising completeness and positional accuracy of OSM data, particularly in urban areas (Haklay 2010; Neis et al. 2011). In Hecht and Stephens (2014), the authors highlight the declining of

---

[1] http://www.openstreetmap.org/
[2] http://www.wikimapia.org/
[3] http://http://www.google.com/mapmaker/

data quality with the increasing distance form urban areas. Regarding particular features, Girres and Touya (2010), Haklay and Weber (2008) and Ludwig et al. (2011) emphasize the quality of street networks in France, the UK and Germany respectively. Whereas Arsanjani et al. (2015) and Arsanjani and Vaz (2015) address the promising contributions of land use/land cover features in OSM datasets. The studies highlight the heterogeneous data quality not only regarding the positional accuracy and completeness quality measures, but also regarding the problematic thematic accuracy of data (Haklay 2010; Neis et al. 2011; Vandecasteele & Devillers 2013). Thematic accuracy implies correctness of the assigned classification to a given entity with that entity's characteristics and its geographic context. Hence, this chapter tackles VGI data quality from the *classification* perspective.

In OSM projects, the loose classification mechanisms lead to inappropriate classification of data. Whether a piece of land covered by grass is classified as *park*, *meadow* or *forest*, if a water body is classified as *pond* or *lake*, whether an area is classified from the land use perspective as *residential* or *industrial*, etc. All these classifications mainly depend on contributors' perception (*subjective classification*).

Otherwise, the appropriate classification should reflect the inherent geographic characteristics of an entity (*objective classification*). For example, *park* and *garden* are likely used for entertainment and should contain amusement facilities like a playground, sport area, etc. and a *lake* is likely surrounded by a natural landscape and some facilities like tracks or benches, and is larger in size than a *pond*; whereas *residential* areas mostly cover residential buildings and likely contain some residential services, whereas *industrial* areas usually have industrial properties like a company, factory, etc.

In this chapter, we propose a rule-guided classification approach. The approach aims to improve the data classification; it works to develop a recommendation system able to guide the contributors towards appropriate classification. The approach works to extract the distinct geographic characteristics of a specific feature and encode them in the form of rules. The rules are encoded together into a classifier. Afterwards, the developed classifier is applied to guide the contributors towards appropriate classifications.

As an empirical study, we address the classification of some grass-related features; where a piece of land covered by grass could be classified as *forest*, *garden*, *grass*, *meadow* or *park*. The classification of these features generates a challenge; they are commonly covered by grass, however each class has its distinct characteristics. For example, the *park* and *garden* classes have entertainment characteristics, the *forest* class are usually covered with trees or other woody vegetation and the *meadow* class has agricultural characteristics, etc. The findings are promising and show the feasibility of the approach.

This chapter is organized as follows: the 2nd section gives insights into the classification challenges in an OSM project, while the 3rd section presents the proposed approach of guided classification for VGI. An empirical study is presented in the 4th section. The last section summarizes the findings and points out the future research directions.

Classification Challenges at OpenStreetMap

The OSM project is the most prominent collaborative mapping project: it covers most of the world, has more than 2 million registered users on October 2015[4] and the OSM data is utilized in various services and applications. However, its problematic classifications make its data of limited use (Devillers et al. 2010). In particular, the problematic classification results in inaccurate results and/or incomplete answers. The uncertainty, poor definitions and various individual conceptualizations of geographic features are other reasons behind the problematic classification of data (Fisher 1999; Grira et al. 2010). However, regarding the OSM project, the problematic data classification might come back to the following:

- **Contributors' heterogeneity**: the project harnesses the contributors' diversities to produce rich datasets. However, these diversities influence the resulting data quality (Coleman et al. 2009); contributors have various geographic and cartographic knowledge; this fact results in heterogeneous perceptions of the geographic features and consequently problematic classifications; what is perceived by a contributor as a *park* could be considered by another as a *grass* or *garden* type area.
- **Contribution methodologies**: OSM supports the contributors' heterogeneity by providing different methods of contribution. The most popular contribution methods are either by uploading GPS tracks directly or by editing geographic features over satellite images. The later method is the most common and is known as remote contribution. Figure 1 illustrates, a remote contribution (armchair contribution) process, in which a contributor uses an editor (e.g. JOSM) to contribute information about a specific feature by tracking the feature on satellite images. The contribution method itself generates a challenge during the classification process (Mooney & Corcoran 2012). For example, in Figure 1, the pieces of grass-covered land look similar and their classifications, whether *park, garden, grass* or *meadow*, mainly depend on the contributors perception and need some sense of locality. Moreover, the loose tagging mechanism of OSM also results in problematic classifications. There is no restriction on the number of tags associated with a certain entity; an entity could be associated with no tags or several tags with endless combinations without any integrity checking mechanism. For example, an entity could plausibly be tagged with *leisure=park, natural=grass, landuse=meadow* and *place=garden*.

---

[4] http://osmstats.neis-one.org/.

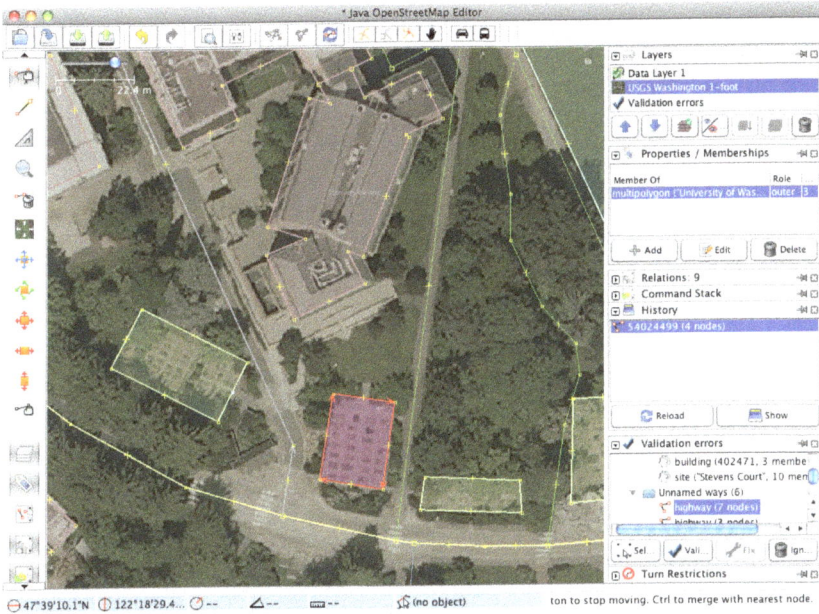

Figure 1: Remote contribution using JSOM editor.

- **Ambiguous recommendations**: the OSM project provides only recommendations for contributors through its Wiki[5] pages. These recommendations resulted from discussions between mappers. However, it is probable that most of the contributors do not spend enough time checking these recommendations. Furthermore, due to ambiguous terminologies (e.g. *wood* or *forest*, *landuse* or *landcover*, etc.), some recommendations might be conceptually misinterpreted, particularly by non-experts. For example, the unclear distinction between *lake* and *pond* classes results in a new class of *lake; pond*.

The previous points summarize the fundamental reasons behind the problematic data classification of OSM. These challenges come up due to the nature of VGI and the OSM project in particular. There exist other reasons due to the nature of geographic data as well. Most geographic features are not well defined; the fact that results in crisp boundaries between classes. In some cases, an identical feature could plausibly belong to multiple classes. However, small details usually exist and distinguish between conceptually overlapping classes. In the case of remote contribution, these details are hardly recognized

---

[5] http://wiki.openstreetmap.org/wiki/Map_Features

by armchair contributors, and consequently, they contribute either imprecise or incomplete data.

Rule-Based Guided Classification Approach

To tackle the classification challenges, we propose a rule-based guided classification approach. Through guiding and recommendations, the approach aims to produce data with appropriate and consistent classifications. The approach consists of two phases: *Leaning* and *Guiding* phases.

Learning Phase

During the *Learning* phase, the approach employs the increasing availability of OSM data in learning the characteristics that distinguish between similar classes. Figure 2 shows a summary of the learning phase. In this phase, the task is to develop a classifier able to distinguish between related classes. Data mining algorithms are used to find the distinct topological characteristics that distinguish between classes. The extracted characteristics have the form of predictive rules. Afterwards, the rules are integrated into a classifier. During the mining process, we depend on qualitative spatial analysis to find the characteristics of a specific class. Topology, direction and distance are the common qualitative spatial relations. In this work, we particularly investigate the topological relations to understand the geographic context of the given classes.

Topological Analysis Based on the first law of geography (Tobler 1970), nearby geographic features are related to each other. For example, the existence of sport's areas and playgrounds inside the *park* and *garden* features, the location of gas stations in a direct access to roads area, etc. Hence, in this work we investigate the topological relations between pairs of entities to find the characteristics that identify each class. Each entity is characterized by its interior and exterior context. At the same time, the *appropriate classification* should reflect the entities characteristics and matches its geographic context.

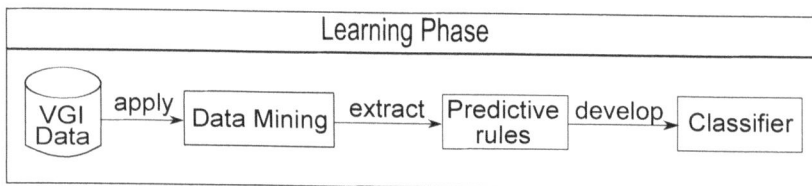

Figure 2: Learning phase of the proposed guided classification approach.

We utilize the 9-Intersection Model (9IM) (Egenhofer 1995), in which the topological relations between pairs of entities are defined as follows: *disjoint, meet, overlap, covers, coveredBy, contains, inside* and *equal*. In this work, *disjoin, meet, overlap, contains* and *coverdBy* relations are considered. While *inside, covers* and *equals* relations are neglected due to two reasons: (a) *inside* and *covers* are the inverse of *contains* and *coveredBy* relations respectively; and (b) the *equal* relation rarely occurs and does not add information to this analysis.

Mining Process The topological analysis aims to find frequent patterns (topological relations) involved between target classes and other geographic features, e.g. *park contains* playground, sport center, etc. Each tag is considered as a new feature. For example, *leisure = playground* and *leisure = sport* are treated differently. We encode them as leisure playground and leisure sport respectively and associate each one with a unique identifier (ID), to facilitate the mining process. The processing is computationally exhaustive and should be done in advance during the preparation for the mining task. Afterwards, the mining process works to extract atomic rules in the following form:

$$Class\ (E,\ C) \leftarrow R(E,\ F) \qquad (1)$$

where E represents an entity, $C \in \{`park`, `meadow`, etc.\}$, R is one of the topological relations where $R \in \{`contains`, `meet`, etc.\}$ and F represents the set of frequent features that mostly involved in a relation R with entities of C.

We apply the Apriori algorithm (Agrawal et al. 1994) to extract the rules. In particular, we use the class association rules mining task, when rules have a predefined class (e.g. *`park`*) as their outcome. Appropriate constraint parameters like *support* and *confidence* should be adjusted to extract and filter the interesting patterns. Afterwards, the extracted rules are integrated into a classifier.

Basically, developing a classifier based on a set of predictive rules consists of the following steps: (1) find all the interesting class association rules from a dataset; (2) filter the extracted rules into a set of predictive association rules; (3) encode the rules into a classifier; and (4) evaluate the classifier on a test dataset.

Guiding Phase

During the *Guiding* phase, the developed classifier could be used in many different scenarios. In this approach, we present three scenarios: the contributing, checking and enriching scenarios. Figure 3 gives brief illustrations of these scenarios as follows:

Contributing Scenario In this scenario, the classifier is embedded into an editing tool. At the contribution time, the classifier checks the validity of a given classification. In case of a problematic classification, the classifier informs the contributor of some recommendations. Then, according to

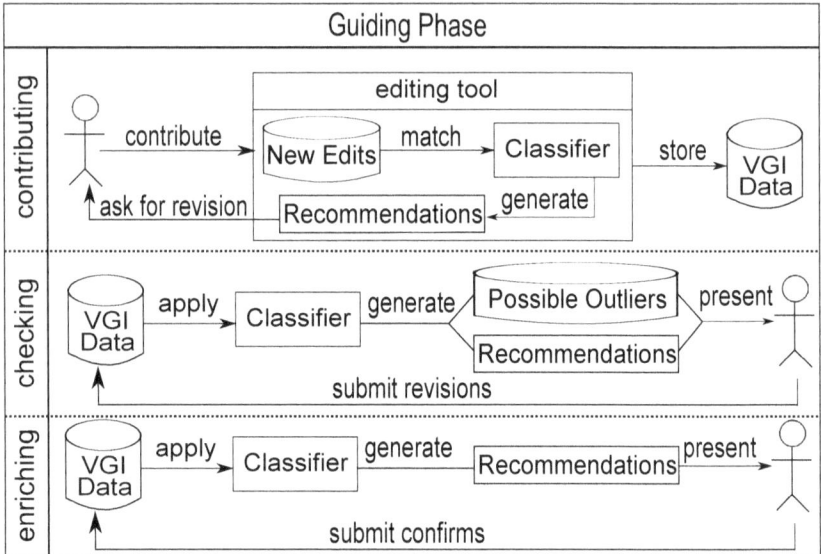

Figure 3: Guiding phase of the proposed guided classification approach.

the given recommendations, the contributor reacts with a correction (if required). The real challenge in this scenario is the computational complexity. During our study, the entities are processed in advance to investigate their geographic context. In contract, this scenario requires on-line processing of contributions at the time of editing.

Checking Scenario This scenario could be directly applied when the developed classifier is applied to an existing dataset generating the potential problematically classified entities. Afterwards, the outliers are presented – associated with recommendations – to crowd-sourcing revision. The contributors act to correct the problematic entities (if required).

Enriching Scenario In which the classifier is applied to a set of unclassified entities. The classifier predicts classifications for these entities and presents them for crowd-sourcing confirmation. The contributors' role here is to confirm the given classification and make corrections (if required). Another enrichment scenario could also be achieved, when the contributor reacts to add more information to satisfy the given recommendations.

In all of the proposed scenarios, the classifier cannot do automatic classification or automatic correction directly on the data source. However, it provides recommendations for directing contributors towards data of appropriate classification. At the same time, developing a global classifier might also be inaccurate. Therefore, the proposed approach is to maintain locality during both the *Learning* and *Guiding* phases. We assumed that a geographic feature should be classified identically, at least on the country level.

Empirical Study: Grass & Green

To evaluate the proposed approach, an empirical study and an implementation are conducted. The study aims to develop a classifier to distinguish between grass-related classes: *forest*, *garden*, *grass*, *meadow* and *park* classes. The classes are the most common grass-related classes within the boundaries of urban cities (the geographic scope of the research). The classification of these features represents a challenge due to the following: (1) in satellite images, they appear similar as a green area; (2) in some cases, a feature could plausibly belong to multiple classes (e.g. *park* and *garden*); and (3) for non-experts, they are all *grass*. Thus, a contributor might be unfamiliar with the characteristics that distinguish be- tween classes. Table 1 shows the OSM Wiki recommendations for these classes. The given recommendations are based on discussion between mapper communities. The given recommendations at Table 1 indicate that there exist unique characteristics that distinguish between classes.

Data Processing

We use an OSM dataset from Germany dated to December 2013. The choice of Germany comes from the following reasons: i) the existence of a large group of active mappers; and 2) several researchers have emphasized the quality of

Class	Recommendations
forest	Some use this tag for land primarily managed for timber production, others use it for woodland that is in some way maintained by humans.
garden	A distinguishable planned space, usually outdoors, set aside for the display, cultivation and enjoyment of plants and other forms of nature. It incorporates both natural and man-made materials. The most common form is known as a residential garden, it is a form of garden and is generally found in proximity to a residence, such as the front or back garden. Residential gardens are usually of human scale, as they are most often intended for private use.
grass	A tag for a smaller areas of mown and managed grass, for example in the middle of a roundabout or verges beside a road. Should not be used where a more specific tag is available.
meadow	Used to tag an area of meadow, which is an area of land primarily vegetated by grass plus other non-woody plants.
park	An area of open space provided for recreational use, usually designed and in a semi-natural state with grassy areas, trees and bushes. Parks are often but not always municipal.

Table 1: OSM recommendations for the target classes.

the data. We extract the entities from the 10 most densely populated cities[6] to ensure active mappers and hence a certain level of quality. The dataset consists of 3,724 *forest*, 3,030 *garden*, 7,336 *grass*, 4,277 *meadow* and 4,445 *park* entities. About 50% of the extracted entities have only one version (edits), which indicates the lower attraction of these entities to mappers. According to Mooney & Corcoran (2012), an increasing number of edits does not usually imply high quality. However, it reflects the heavy collaboration/competition among contributors to improve the data quality. The extracted entities are processed individually by checking the topological relations between each entity and other entities nearby.

Learning Process

During the learning process, the objective is to develop atomic rules per class per topological relation. Due to the uncertainty of spatial context, we take into account that everything is possible. Thus, a 1% *support* threshold is considered sufficient to extract the interesting patterns (frequent topological relations). Each topological relation is processed individually with a given class producing a set of predictive rules of that class. We extracted 8,504 rules: 4,100 describe *forest*, 215 describe *garden*, 745 describe *grass*, 506 describe *meadow* and 2,938 describe *park*. Although a large number of rules have a *confidence* threshold greater than 50%, the rules themselves represent some difficulties in the classification process due to: (1) they have a wide range of *confidence* threshold from 100% to 0.7 %; (2) due to the similarity between some classes, there exist duplicated rules pointing to different classes; and (3) regarding the topological relations, some relations have higher *confidence* thresholds than the others.

Classification Process

During the classification process, each entity is checked against all extracted rules. For example, Figure 4 shows an entity[7] with a *meadow* classification which has *osm_id* = 96279661. The entity matches 46 rules: 26 *park*, 6 *meadow*, 5 *forest*, 5 *garden* and 4 *grass*. Table 2 presents a sample of the matched rules for this entity. The figure illustrates that the entity contains a playground, sport areas and planned footways, which reflect the characteristics of the *park* class.

According to Table 2, considering the maximum *confidence* of the matched rules, the top 20 rules have *confidence* thresholds ranging form 92% to 80% and all of them have the result *Class(E, 'park')*. At the same time, when considering the maximum *confidence* per class, this entity matches with *park*, *meadow*,

---

[6] http://www.citymayors.com/gratis/german_topcities.html
[7] http://www.openstreetmap.org/way/96279661, last accessed April 2015

Figure 4: A entity with osm_id=96279661 classified as 'meadow' (last visit at April 2015).

Rule — Confidence
Class (E, 'park') ← *contains* (E, [1,22,156])) – 92%
Class (E, 'park') ← *contains* (E, [1,15,22, 156])) – 91%
Class (E, 'park') ← *contains* (E, [15,21])) – 89%
Class (E, 'park') ← *contains* (E, [1,15])) – 88%
…
Class (E, 'park') ← *contains* (E, [22])) – 76%
Class (E, 'park') ← *contains* (E, [15])) – 66%
Class (E, 'meadow') ← *containsBy* (E, [128])) – 46%
Class (E, 'park') ← *meet* (E, [15])) – 34%
Where, 1=leisure_playground, 15=highway_footway, 21=sport_soccer, 22=leisure_ pitch, 128=landuse_forest, 156=sport_basketball

Table 2: A sample of matched rules for the entity with osm_id=96279661.

grass, forest and *garden* classes in descending *confidences* of 92%, 46%, 32%, 13% and 12%, respectively. Although the entity is currently classified as *meadow*, its characteristics make it more appropriate to be classified as a *park*. Hence, our recommendation works to guide contributors towards the most appropriate classification.

Evaluation Process

To evaluate the classifier, we do not have a ground-truth dataset for these entities. The available ground-truth datasets cover a higher classification level of land use or land cover. Thus, we depended on manual visual investigation to evaluate the results. Figure 5 presents examples of appropriate and inappropriate classifications, based on the developed classifier and recommendations.

Figure 5(a) gives examples of appropriate classifications. From left to right, the first entity is adjacent to residential houses and other gardens and does not contain much infrastructure. The entity is appropriately classified as *garden*. The second one, located between highways and containing nothing, is most likely to be classified as *grass*. The last entity contains a water body, sports centers, footways and other infrastructure. It is correctly classified as a *park*.

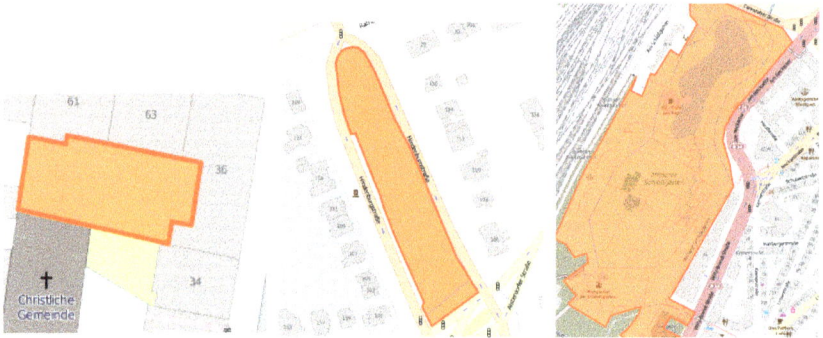

(a) Appropriate Classification of *garden*, *grass* and *park* classes

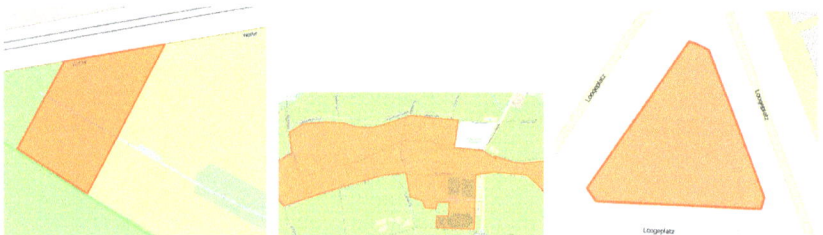

(b) Inappropriate Classification of *garden*, *grass* and *park* classes

Figure 5: Example of appropriate and inappropriate classifications.

In Figure 5(b), the classifier detects these entities as problematically classi-fied entities. From left to right, the first entity is classified as *garden*. The entity meets *meadow* and is located near to a farmland. It does not inherit any plant or decoration characteristics. The classifier recommends *meadow* as an appropri-ate class. Whereas the middle entity is classified as *grass*, despite the fact that it seems too large, contains sports centers, is surrounded by forest areas and is adjacent to a playground. The classifier recommends *park* class for this entity. The entity on the right shows a clear example of inappropriate classification of *park*. The entity is located between roundabouts and does not contain any infrastructure at all. The classifier recommends it to be classified as *grass*.

Grass&Green: a quality assurance web tool

As another way to evaluate the proposed approach, we developed a web tool as a recommendation system called Grass&Green[8] as indicated in Figure 6. The tool presents the generated recommendations for crowd revisions as pro-posed in the checking scenario (see section 3.2). We created social media pages to attract the contributors for revisions: Facebook and Twitter. Moreover, we wrote OSM diaries to announce the tool to the OSM community. In this tool, the user logs in via his/her OSM account and contributes directly to the project. The tool presents entity by entity, combined with the recommended classes and

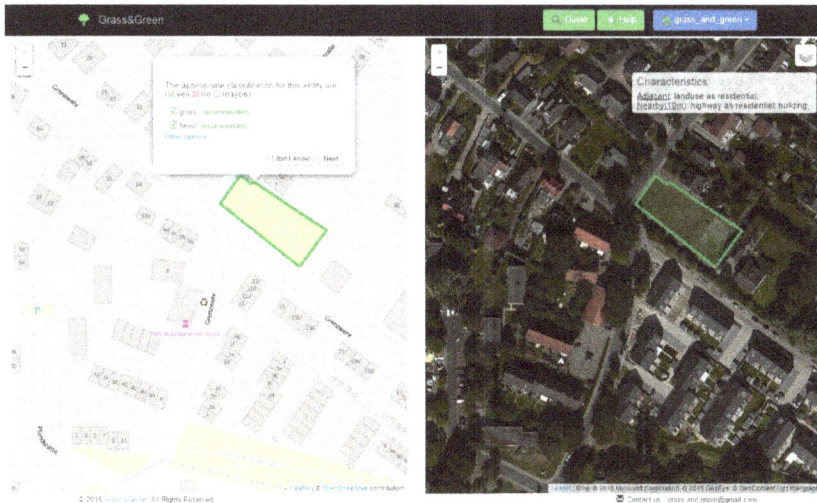

Figure 6: Grass&Green: the main contribution interface (last visit at September 2015).

[8] http://opensciencemap.org/quality

other potential classes. Due to the ambiguity of grass-related features, we provide the users with the two most recommended classes. The user has the ability to press an "I don't know" (inactive participation) in unclear cases. Moreover, the user could also change the recommendations or choose between (yes, no or maybe) in case of high ambiguity. The tool provides users with detailed textual and visual descriptions about the target classes and other grass-related classes.

Eleven days after launching the tool, we obtained promising results. We had around 80 users from various countries. They checked 560 entities: 485 active and 75 inactive participations. They agree with the generated recommendations as follows: 30.10% full agree, 60.84% partial agree and 9.05% disagree. The findings indicate the feasibility of the approach and the tool acts perfectly to improve the classification of OSM data.

Discussion and Conclusions

Conceptualization of geographic features has long been a topic of debate (Frank 1997). However, with the increasing role of public participants in collecting geospatial data, it becomes a crucial issue. GIS applications exploit VGI as an auxiliary data source. That means the data developed by public participants is used to provide services for others. A fact that raises more attention to the resulting data quality.

In particular, how do the participants perceive the space? How do they group and categorize the geographic features? How do they find the commonalities and differences between conceptually overlapping classes? All these questions might be addressed by utilizing the developed geospatial ontologies. Frank (1997) discussed the vital role of ontology in GIS applications, to achieve a better understanding of the space and to build more efficient information systems. For example, the OWL2 ontology that has been developed for structuring city information modeling with respect to land use mapping (Montenegro et al. 2012). OSMonto is another ontology, which has been developed to enrich the semantics of OSM tags, without correcting or modifying any conceptual mistakes in the taxonomy of OSM tags (Codescu et al. 2011). However, the link between ontologies' producers and consumers, in the GIS domain, still needs more research.

In VGI, the data is classified following the bottom-up approach; where the participants contribute data based on their local knowledge. They translate their observations into classes and categories. While in professional methods, the data is classified based on a top-down approach; where a pre-defined model is developed based on strict measures defining the classes. The difference of VGI approach leads to questionable data classification. Therefore, guiding amateur participants is needed for enhanced data classification. For example, designing intelligent data capturing interfaces is one possibility, among others, to support the contribution of enhanced data classification. This chapter calls for the development of intuitive interfaces for VGI projects; negotiation, exemplifications and comparisons are human-centered approaches that could be used to support the VGI participants during the contribution process.

In this chapter, we addressed the VGI quality from a classification perspective. We investigated the classification correctness of an entity with respect to its inherent geographic characteristics. We proposed an approach for guided classification. The approach tackles the classification challenges in the OSM project by guiding the contributors during the classification process. The approach has two phases: the *Learning* and *Guiding* phases. During the *Learning* phase, the approach utilizes the OSM dataset to learn the distinct topological characteristics that distinguish between similar classes. Data mining algorithms have been used to develop a classifier. Afterwards, the developed classifier is used in different scenarios (contributing, checking and enriching) during the *Guiding* phase. The approach aims not only to improve the classification of data, but it could be used to enrich the data source as well.

We conducted visual investigations and an implementation to evaluate the proposed approach. We developed a classifier to distinguish among a set of grass-related classes. The selected classes have some similarity, but each one has its unique characteristics. The findings emphasized the feasibility of the proposed approach. The developed tool shows the positive response of the crowds towards the data quality. The presented results are preliminary indicators of an enhanced data classification. We will keep investigating the tool results and check the enhanced data classification in more details. In future work, the research would investigate how to generalize the developed classifier. In addition, the intuitive user interface would be studied to develop human-centered guided classification that corresponds to the nature of VGI.

Acknowledgments

We gratefully acknowledge support provided by the German Academic Exchange Service (DAAD), as well as the hosted research group in Bremen Spatial Cognition Center (BSCC) at University of Bremen. Furthermore, acknowledge for ICT COST Action IC1203 for Short Term Scientific Mission (STSM) support. We also thank all anonymous users of the developed tool (Grass&Green).

References

Agrawal, R., Srikant, R., et al. 1994. Fast algorithms for mining association rules. In: *Proc. 20th int. conf. very large data bases*, VLDB, Volume 1215: pp. 487–499.

Arsanjani, J. J., & Vaz, E. 2015. An assessment of a collaborative mapping approach for exploring land use patterns for several european metropo-lises. *International Journal of Applied Earth Observation and Geoinformation, 35*(Part B): 329–337.

Arsanjani, J. J., Mooney, P., Zipf, A., & Schauss, A. 2015. Quality assessment of the contributed land use information from openstreetmap versus authoritative datasets. In OpenStreetMap in GIScience,. Springer: pp. 37–58.

Codescu, M., Horsinka, G., Kutz, O., Mossakowski, T., & Rau, R. 2011. OSMonto-an ontology of OpenStreetMap tags. State of the map Europe (SOTM-EU) 2011.

Coleman, D. J., Georgiadou, Y., Labonte, J., et al. 2009. Volunteered Geographic Information: the nature and motivation of produsers. *International Journal of Spatial Data Infrastructures Research, 4*(1): 332–358.

Devillers, R., Stein, A., B´edard, Y., Chrisman, N., Fisher, P., & Shi, W. 2010. Thirty years of research on spatial data quality: achievements, failures, and opportunities. *Transactions in GIS, 14*(4): 387–400.

Egenhofer, M. J. 1995. On the equivalence of topological relations. *International Journal of Geographical Information Systems, 9*: 133–152.

Elwood, S., Goodchild, M. F., & Sui, D. Z. 2012. Researching Volunteered Geographic Information: Spatial data, geographic research, and new social practice. *Annals of the Association of American Geographers, 102*(3): 571–590.

Fisher, P. F. 1999. Models of uncertainty in spatial data. *Geographical information systems, 1*: 191–205.

Flanagin, A. J., & Metzger, M. J. 2008. The credibility of Volunteered Geographic Information. *GeoJournal, 72*(3–4): 137–148.

Frank, A. U. 1997. Spatial ontology: A geographical information point of view. In: *Spatial and temporal reasoning*. Springer: pp. 135–153.

Gillavry, E. M. 2004. Collaborative Mapping: By the People, for the People. *Society of Cartographers Bulletin, 37*(2): 43–45.

Girres, J.-F., & Touya, G. 2010. Quality assessment of the french Open- Street-Map dataset. *Transactions in GIS, 14*(4): 435–459.

Goodchild, M. F. 2007. Citizens as sensors: the world of volunteered geography. *GeoJournal, 69*(4): 211–221.

Grira, J., B´edard, Y., & Roche, S. 2010. Spatial data uncertainty in the VGI world: Going from consumer to producer. *Geomatica, 64*(1): 61–72.

Haklay, M. 2010. How good is Volunteered Geographic Information? A comparative study of OpenStreetMap and Ordnance Survey datasets. *Environment and planning. B, Planning & design, 37*(4): 682.

Haklay, M., & Weber, P. 2008. OpenStreetMap: User-generated street maps. *Pervasive Computing, IEEE, 7*(4): 12–18.

Hecht, B., & Stephens, M. 2014. A tale of cities: Urban biases in Volunteered Geographic Information.

Ludwig, I., Voss, A., & Krause-Traudes, M. 2011. A comparison of the street networks of Navteq and OSM in Germany. In: *Advancing Geoinformation Science for a Changing World*. Springer: pp. 65–84.

Montenegro, N., Gomes, J. C., Urbano, P., & Duarte, J. P. 2012. A land use planning ontology: Lbcs. *Future Internet, 4*(1): 65–82.

Mooney, P., & Corcoran, P. 2012. The annotation process in OpenStreetMap. *Transactions in GIS, 16*(4): 561–579.

Neis, P., Zielstra, D., & Zipf, A. 2011. The street network evolution of crowd-sourced maps: OpenStreetMap in Germany 2007–2011. *Future Internet, 4*(1): 1–21.

Tobler, W.R. 1970. A computer movie simulating urban growth in the detroit region. *Economic geography, 46*: 234–240.

Vandecasteele, A., & Devillers, R. 2013. Improving volunteered geographic data quality using semantic similarity measurements. ISPRS-International Archives of the Photogrammetry, *Remote Sensing and Spatial Information Sciences, 1*(1): 143–148.

CHAPTER 10

Enhancing the management of quality of VGI: contributions from context and task modelling

Benedicte Bucher*, Gilles Falquet†, Claudine Metral†
and Rob Lemmens‡

*Université Paris Est, IGN, France, Benedicte.Bucher@ign.fr
†University of Geneva, Switzerland, Gilles.Falquet@unige.ch,
Claudine.Metral@unige.ch
‡University of Twente, The Netherlands, R.l.g.Lemmens@utwente.nl

Abstract

This chapter presents contributions to managing the quality of Volunteered Geographical Information (VGI) and of crowd sourced geographical information (CSGI) brought by the representation of specific knowledge items: task and context. Task and context modelling have been studied in different communities. We propose an approach for integrating their results with the perspective of improving the quality management of VGI and CSGI.

Keywords

Task modelling, context, quality, specification, user-generated

Introduction

The ENERGIC COST ACTION targets the usage of Volunteered Geographical Information (VGI) and of crowd sourced geographical information (CSGI) in

How to cite this book chapter:
Bucher, B, Falquet, G, Metral, C and Lemmens, R. 2016. Enhancing the management of quality of VGI: contributions from context and task modelling. In: Capineri, C, Haklay, M, Huang, H, Antoniou, V, Kettunen, J, Ostermann, F and Purves, R. (eds.) *European Handbook of Crowdsourced Geographic Information*, Pp. 131–142. London: Ubiquity Press. DOI: http://dx.doi.org/10.5334/bax.j. License: CC-BY 4.0.

scientific applications. One challenge addressed in this action is data quality management.

Data quality management has been addressed for several years with respect to classical geographical data, by operational bodies like the National Mapping Agencies and by scientists mainly from the geographical information science community. The issues they tackle are also relevant in the context of volunteered and user generated data, for example managing the lack of a universal data model about geographical space and the unavoidable heterogeneities between geographical data. Section 1 lists findings about geographical data quality management from the literature and from the experience of the French National Mapping Agency, IGN. These findings relate to the management of data specifications, the definition and documentation of quality criteria, the assessment of which inherent characteristics of the data will impact the output of a given application result and the communication of quality to the user.

Section 2 specifically studies the potential of context and tasks modelling to implement these findings in the context or VGI and CSGI. Context can account for much heterogeneity in VGI and CSGI. Tasks are useful pieces of knowledge to plan the usage of relevant resources to achieve an objective. Context and tasks modelling are studied by communities tackling information management and exchange between implemented components and humans, like distributed architectures, interoperability, ubiquitous mapping, location based services and human–machine dialogue interfaces.

An approach to integrate the context and tasks models to address part of the research questions expressed in the beginning of this chapter is discussed at the end of this chapter.

Geographical data quality management

External quality

Quality is defined in ISO 9000 as the degree to which a set of inherent characteristics fulfils some requirements. This definition of quality is relative to an application. For example, important inherent characteristics of 3D data for visualization applications refers to accurate and realistic textures as well as consistency of visible shape elements and very low level of detail for elements non-visible in the current scene. Important characteristics of the same data for firemen access application requirements refers to the exhaustiveness and geometric accuracy of specific features like windows, electricity cables and tramway cables. This quality is referred to as 'fitness for use' or 'external quality' (Devillers & Jeansoulin 2006).

Most applications considered in ENERGIC action share some common data requirements:

• the ability to discover and reuse the data,
• the ability to combine the data with other data,

- the ability to ground a result on data, i.e. to control the gap between data and interpretation.

The two first criteria are recurrent and are promoted by initiatives like the INSPIRE directive or by the W3C vision for Linked Open Data where data should be produced once and made available (Bizer, Heath & Berners Lee 2009). The third criterion relates to having a documentation of uncertainty related to the data and how it propagates to the result. Besides, when it comes to volunteered content, it is also useful to consider requirements expressed by the Web community. W3C propose a ranking scheme for open data that list important quality criteria from their perspective: publication on the web (one star), in a machine readable format (two stars), in a non-proprietary format (three stars), compliant to RDF standards – using dereferenceable URIs to name things (four stars) and publish links to other URIs (five stars).

With respect to the above requirements, external quality of data will very much depend on metadata and documentation.

Besides, user requirements will eventually be met by an application involving software and data. Hence, geographical data quality assessment is closely related to geographical software quality assessment.

Internal quality

A specific intermediate quality concept is needed to document inherent characteristics of geographical data that will be useful for every user to evaluate their ability to fulfil their application requirements. This is the 'internal quality' (Devillers & Jeansoulin 2006). The data producer should distribute its data together with the description of this internal quality and the users can at the end use this description (and the data) to assess external quality of the data for their application. Indeed, many geographical data (base maps for instance) have seldom been acquired for one specific application but rather to be reused in several applications, and possibly by users who sometimes are far from the production of the data and who did not express their quality requirements, hence did not express which inherent characteristics are important for them.

Internal quality has been traditionally documented by national mapping agencies, and in current ISO/OGC metadata standards (ISO TC211 2014) based on three elements:

- the targeted description, called the data specifications,
- quality criteria describing some distance between the produced data and an imaginary flawless data sets compliant with these specifications,
- the lineage metadata.

In other words, data producers have considered that the characteristics that will help future users assess the 'external' quality of their data are globally the assessment of how far the geographical data are from the reality they aim at describing. These three items are described more precisely hereafter.

The specifications are the scope of the observation process that will lead to the data: when, where, what objects, what level of detail. It is crucial to explicitly describe specifications because there is no natural such abstraction even though one may intuitively think so. Our world is heterogeneous in multiple ways whereas an abstraction will provide only one specific set of categories and classification schemes. The most satisfying solution so far is to provide abstractions that are relevant for a certain spatial and temporal scope and also for a given point of view on reality. Not only are specifications an important item for describing quality, but also they improve the homogeneity of the data during data production. It has always been a necessity and a challenge to share specifications among operators involved in the acquisition process for national mapping agencies producing topographical data over a national territory (Sheeren, Mustière & Zucker 2004). Indeed, geographical database specifications should refer to a common ontology of reality which does not exist (Abadie 2009).

Quality criteria are measures of distance between produced data and what is called 'terrain nominal', i.e. data that would have been produced strictly considering the specifications (and in real time). When a quality criteria is attached to a product (and not to a specific data set), it means a commitment of the producer to respect a certain thresholds during data production. Quality criteria describing 'uncertainties' and 'errors' possibly introduced during the actual production have been standardized after four fundamental dimensions: positional accuracy, attribute accuracy, logical consistency, completeness (Goodchild & Li 2012). Usually a product description includes some commitments of the producer about these criteria threshold. For example such metadata for a road data product can be: the product should describe every road longer than 50 m, thanks to a series of points acquired at the axis of the road, with 10 m precision, with time accuracy of 6 months and exhaustiveness of 98% on the national territory. Whereas these examples refer to explicit attributes and entities, it is also important in geographical data to consider some implicit spatial properties and relationships. An important paradigm of geographical data is that many relations and properties are not explicit in the data but can be computed based on the coordinates. Several authors study the evaluation of some spatial properties and relationships, usually referred to as spatial consistency rules (Servigne et al. 2000). A recurrent quality criteria referring to consistency is the topology.

Last, the lineage metadata refers to sources data and processes that led to the data. It is somehow comparable to the 'source code' of a software that will be useful to debug i.e. if something unexpected happens in the application to investigate if it can be explained by the geographical data production process.

Quality challenges intensified by VGI and CSGI context

Whereas by essence, VGI and CSGI production process can be seen as an opportunity to improve some dimensions of internal quality like the update frequency, this production process also makes it more difficult to handle certain dimensions of internal quality.

The main flaw in our opinion is the weak definition and documentation of expected specifications of the produced data. Besides, Linus's law 'given enough eyes, all bugs are shallow', referring to the ability of the crowd to converge on the truth, does not always work for VGI. Goodchild and Li (2012) and Haklay et al. (2010) have shown that users do not always agree on a value. This is also a motivation for explicitly stating quality specifications. Brando and Bucher (2010) focus on the definition of such specifications for user generated geographical content prior to the production of the data. They proposed a method to instantiate specifications based on OSM tags, Wikipedia infoboxes and the NMA product specifications (Brando, Bucher & Abadie 2011). Yet, their work does not address the issue of acceptability of these specifications by contributors and of evolution of such specifications.

One aspect of the proposal concerns the establishment of explicit consistency rules between the user generated content and reference data provided by the French national mapping agency, a public funded professional organizations who commit to reach specific level of quality criteria for some image data and topographic themes. Acquiring external rules that can be used to evaluate consistency of geographical content now is more generally an important domain of research (Goodchild & Li 2012) and has led to the creation of the organization OSMGB in UK which aims at listing such rules and setting up a formal quality insurance model to improve the trust of local administration in collaborative geodata.

Another flaw is the lack of explicit commitment to follow these specifications and reach quality criteria thresholds, e.g. of any update frequency, and the lack of assessment of quality criteria to document the gap between acquired data and the specifications. So far, documentation of quality criteria is done in punctual studies, like for research about the quality of VGI data comparing OSM data with data whose quality already is documented, like Haklay (2010) in the UK and Girres and Touya (2010) in France. Goodchild and Li (2012) and Haklay et al. (2010) also showed evidence that there are not always enough people interested in a particular area or feature.

To conclude this first section of the chapter, managing quality of VGI and CSGI can benefit from knowledge gained about the management of quality of geographical data. The quality of a data set is documented either according to a dedicated application or in a more generic way as a distance between a flawless ideal representation of a geographical space conforming to a given abstract model and a data set produced by remote sensing, in situ sensing and symbolic knowledge production. It is highly recommended to define explicitly a targeted abstraction of a geographical space and it is easier to try and provide some that is 'locally' relevant. A relevant abstract model should not only be composed of classification schemes but also of consistency rules. Documenting the distance between a targeted abstraction and a dataset cannot be done exactly but is approximated by: quality criteria (exhaustiveness of a feature or of an attribute, and so on), lineage information (also known as provenance metadata).

When it comes to VGI and CSGI specific stakes are:

- the actual description and maintenance of data specifications,
- the shareability of specifications among the contributors, among users and the possibility to compare and align the model it with other data models,
- producers' commitment to quality criteria.

Contributions brought from context and tasks modelling

This section lists some contributions to address the objectives of quality management listed just before.

Context modelling

Firstly, since there is no such thing as a universal widely shared abstract model of reality, we advocate it is better to keep the data as close as possible to their production process (typically to keep sensor data) with context information that explain the data (see Chapter *Enquiring VGI*) then trying to merge every contribution into a pivot model. In this perspective, context modelling is an important metadata to account for much heterogeneity in VGI and CSGI.

Some *context* elements are already studied in the literature about VGI to infer quality and trust metadata like the contributor profile, his status within the VGI system (normal/advanced user in Wikimapia, normal/sysop in Wikipedia, ordinary/Data-working-group in OSM), his motivation and level of quality requirements with respect to data (Coleman et al. 2009), the places they live in (Goodchild 2009) (Bishr & Kun 2007), their relationships with other contributors (Bishr & Kuhn 2007).

Other relevant elements are studied in the domain of location based services and ubiquitous mapping where *context* is an important element to understand how someone may mentally interact with an abstract representation – usually accessible through a visual representation- of his surrounding, which are the time of the day, the season, the user age, nationality, gender (Jakobsson 2002) and culture (Edsall 2007).

Another very important *context* element is the contributor intention. In collaborative content edition, it is described through the effects of the contribution on the content at the moment when the contribution was defined (Sun et al. 1998), for example: refining a shape, fixing an alignment between two features, adding a missing building. Describing the intended effect on the representation requires some abstract model of reality that must be as close as possible to the model the contributor had in mind. This refers to the possibility for the contributor to annotate contributions with a shareable abstract model. To enhance the interoperability of abstract models, it is now encouraged to publish them as 'vocabularies' on the web of data, i.e. as RDF schemas available online thanks to dereferenceable URIs. RDF vocabularies to distribute and share geodata are studied in the geographical information domain and in the semantic web community (Goodwin, Dolbear & Hart 2008; Vilches-Blázquez et al. 2010; Atemezing et al. 2014).

Task modelling

Tasks models organize knowledge about the usage of relevant resources to achieve an objective. In the context of VGI and CSGI quality management, this is useful with respect to modelling three kinds of tasks:

- the usage of space by a citizen when he is producing data, for instance going to work – and producing a GPS track,
- the collaboration or cooperation between citizens to produce data, for instance the organization of edition during a mapping party,
- the user task that requires geographical data, for instance evaluating the impact of a new road on the local biodiversity.

The first kind of task is an element of context that is useful to elicit the abstract model people have in mind when they produce data –the last *context* element mentioned in section above-. As demonstrated by (Gibson 1979), people see the landscape through his functional relevance to their goals. In other words, if a contributor rides a bike he will see the street from a different perspective than if a contributor is in a wheeling chair.

The second kind of task has been studied by (Das et al. 2014) who experimented with a task assignment model to organize the production of one content among several contributors to optimize exhaustiveness, cost and precision. The production is modelled as a task decomposed into subtasks that can be assigned to people. The system requires user profiles to make the assignment based on user expertness and availability, and define the reward they need. There exists relevant work in the literature to guide strategies for collaborative geographical data production. Wilkinson and Huberman (2007) study the nature of the collaboration that will impact the quality of the produced content. Maué and Schade (2008) propose a solution where contributors ask themselves for reviewers when they lack confidence in their own contributions.

In the domain of model collaborative edition, some authors have proposed a model were user contributions are directly expressed as operations and not as a new content, in order to be as close as possible to contributor intention. In Brando, Bucher and Abadie (2011), user edition can be expressed as the enforcement of relationships (i.e. implicit information) instead of geometries because the authors thought users may be more expert in assessing relationships between objects than geometries.

Rehrl et al. (2013) proposed a task/operation based model to analyse user contribution to a collaborative geographical content.

Last, task-based application design can be useful to express external quality criteria. The application is modelled as a task which has pre and post conditions, input and output data (Sun & al. 2012). A task also has a method to decompose high level tasks into elementary tasks, noting that these can be either machine tasks (computation) or user tasks (e.g. finding a geographic feature on a map). As an example let us consider the task 'find a restaurant'. This task is associated to subtasks such as (1) 'consult the list of all restaurants in a given area', which requires the completeness in the area, with an accuracy of 10 m, (2) 'find route to address', which requires a traffic network representation that is topologically correct and complete. The evaluation of fitness for use can benefit from the development of typologies and ontologies of tasks performed on spatial data. Several researchers have already worked in this direction. For instance, von Hunolstein and Zipf (2003) define a task typology in map-based mobile guides: high-level tasks have been associated to subtasks and a mapping between goals and tasks has also been defined. For example the task 'Navigation' is associated to subtasks such as 'routing from point A to B' and to goals/purposes 'navigation, exploring, planning, education. Park, Yoon and Kwon (2012) present a task ontology for intelligent tourist information service, based on travelers' needs and activities. Lemmens (2006) proposed an ontology to support the chaining of operations in geographical information architectures. Bucher and Jolivet (2008) demonstrated the difficulty to document pre and post-conditions of an elementary task (Bucher & Jolivet 2008). Beyond defining a vocabulary to express pre and post conditions, a major bottleneck is the acquisition of their value because it requires setting up benchmarks simulating all possible specific cases of geometrical configurations.

Discussion and conclusion

Quality management traditionally requires the documentation of specifications, the control of quality criteria value, and the description of lineage metadata.

An important challenge raised by VGI CSGI quality management is ambiguities, inconsistencies and heterogeneities due to different abstractions of the geographical space involved in production. These are not limited to features classifications; they should also include important relationships between elements used in consistency management, affordances of features in contributor activities and rules to encode the perceived reality in data. Another challenge is to manage the quality of data products, hence to somehow commit to some thresholds for the quality criteria.

In section 2 we advocated that it is very relevant to tackle these issues from the perspective of knowledge engineering. The derivation of usable information from raw, heterogeneous and distributed acquisitions would greatly benefit from enhanced model of the context in which a contribution is produced. The modelling of information derivation from raw acquisition can be seen as a flexible process where the integration is done when it is needed and where the sources are preserved as much as possible in order not to lose any meaningful information. The notion of context comprehends many elements which have already been studied in various domains like VGI quality assessment, ubiquitous mapping and ecology. Task models can also contribute to this knowledge engineering project in several ways: to clarify how users perceive the space they will describe, to get external quality criteria, and to improve the coordination of citizen and their interactions towards the production of a common content.

There is still work to be done to integrate the different findings in context modelling and in tasks modelling. An interesting perspective is to improve the description of user intention when they contribute. Rehrl et al. (2013) paves the way for a relevant approach of the problem. Their low-level tasks categorization, such as create/update a geographic feature or a relation, could be extended to conceptualize higher level intentions, such as for instance to reflect a change of navigation restriction that occurred in the reality, to propose a more detailed description of the cross-road geometry and topology, to update an attribute value to reflect a change in the specifications, to fix an inconsistent misalignment of buildings in the data. Other typical VGI tasks need modelling such as selecting, evaluating, integrating existing data, assigning sensor task to contributors, evaluating user capacities with respect to quality criteria. The examination of data quality issues and the literature shows, in our mind, an opportunity to define an ontology of 'human sensing' tasks that would describes capacities to produce pieces of data by a given human agent or several human agents together, with explicit objectives assigned and in a given observation context.

References

Abadie, N. 2009. Schema Matching Based on Attribute Values and Background Ontology. In: *Proceedings of the 12th AGILE International Conference on Geographic Information Science (AGILE'09)*. Hanovre (Germany), June. Available at: http://www.agile-online.org/Conference_Paper/CDs/agile_2009/AGILE_CD/pdfs/138.pdf.

Atemezing, G., Abadie, N., Troncy, R., & Bucher, B. 2014. Publishing Reference Geodata on the Web: Opportunities and Challenges for IGN France.

In: *Proceedings of Terra Cognita, the 6th International Workshop on the Foundations, Technologies and Applications of the Geospatial Web in Conjunction with the 13th International Semantic Web Conference*, October, Trentino (Italy). Available at: http://event.cwi.nl/terracognita2014/terra2014_1.pdf.

Bishr, M., & Kuhn, W. 2007. Geospatial Information Bottom-Up: A Matter of Trust and Semantics. In: Fabrikant, S., & Wachowicz, M. (Eds.) *The European Information Society – Leading the Way with Geo-information, Proceedings of the 10th AGILE International Conference in Geographic Information Science.* Aalborg (Denmark), Springer Verlag, LNGC, pp. 365–387.

Bizer, C., Heath, T., & Berners Lee, T. (2009). Linked data – the story so far. *International Journal on Semantic Web and Information Systems,* 5: 1–22.

Brando, C., & Bucher, B. 2010. Quality in User Generated Spatial Content: A matter of specifications. In: Painho, Yasmina Santos, & Hardy (Eds.) *Geospatial Thinking, Proceedings of the 13th International Conference on Geographic Information Science (AGILE'10),* Guimarães (Portugal). Available at: http://www.agile-online.org/Conference_Paper/CDs/agile_2010/ShortPapers_PDF/105_DOC.pdf.

Brando, C., Bucher, B., & Abadie, N. 2011. Specifications for User Generated Spatial Content. In: Geertman, S., Reinhardt, W., & Toppen, F. (Eds.) *Advancing Geoinformation Science for a Changing World, Springer-Verlag Lecture Notes in Geoinformation and Cartography.* Utrecht, Netherlands, CD.

Bucher, B., & Jolivet, L. 2008. Acquiring service oriented descriptions of {GI} processing software from experts. In: *Proceedings of 11th AGILE International Conference on Geographic Information Science.* Girona (Spain). Available at: http://plone.itc.nl/agile_old/Conference/2008-Girona/PDF/94_DOC.pdf.

Das, M., Thirumuruganathan, S., Amer-Yahia, S., Das, G., & Yu, C. 2014. An expressive framework and efficient algorithms for the analysis of collaborative tagging. *VLDB Journal, 23*(2): 201–226.

Devillers, R., & Jeansoulin, R. (2006). Spatial Data Quality: Concepts. In: Devillers & Jeansoulin (Eds.) *Fundamentals of Spatial Data Quality.* ISTE, London (UK), p. 312.

Edsall, R. 2007. Globalization and cartographic design: implications of the growing diversity of map users. In: *Proceedings of the 23rd International Cartographic Conference.* Moscow. Available at: http://www.academia.edu/2324974/Globalization_and_cartographic_design_Implications_of_the_growing_diversity_of_map_users.

Gibson, J. J. 1979. The Ecological Approach to Visual Perception. Boston: Lawrence Erlbaum Associates.

Goodchild, M., & Li, L. 2012. Assuring the quality of volunteered geographic information. *Spatial Statistics, 1*: pp. 110–120.

Goodwin, J., Dolbear, C., & Hart, G. 2008. Geographical Linked Data: the Administrative Geography of Great Britain on the Semantic Web. *Transactions in GIS, 12*(1): 19–30.

Haklay, M. 2010. How good is OpenStreetMap information? A comparative study of OpenStreetMap and Ordnance Survey datasets for London and the rest of England. *Environment and Planning*, *37*(4): 682–703.

Haklay, M., Basiouka, S., Antoniou, V., & Ather, A. 2010. How many volunteers does it take to map an area well? *The validity of Linus's Law to volunteered geographic information*, *4*(4): 315–322.

ISO TC211 2014 19115-1:2014. Geographic information — Metadata — Part 1: Fundamentals.

Jakobsson, A. 2002. User requirements for mobile topographic maps, GiMoDig deliverable IST-2000-30090 D2.1.1. Available at: http://lib.tkk.fi/Diss/2006/isbn9512282062/article5.pdf.

Lemmens, R. 2006. Semantic interoperability in distributed geo-service, PhD thesis, ITC, Enschede, Netherlands. Available at: http://repository.tudelft.nl/view/ir/uuid:31b0eae6-c411-4bbd-a631-153498889671/.

Maué, P., & Schade, S. 2008. Quality of Geographic Information Patchworks. In: *Proceedings of 11th AGILE International Conference on Geographic Information Science*. Girona (Spain). Available at: http://ww.w.agile-online.org/Conference_Paper/CDs/agile_2008/PDF/111_DOC.pdf.

Park, H., Yoon, A., & Kwon, H-C. 2012. Task model and task ontology for intelligent tourist information service. *International Journal of U-and E-Service, Science and Technology*, *5*(2): 43–58.

Rehrl, K., Gröechenig, S., Hochmair, H., Leitinger, S., Steinmann, R., & Wagner, A. 2013. A Conceptual Model for Analyzing Contribution Patterns in the Context of VGI. In: Krisp (Ed.) *Progress in Location-Based Services, Lecture Notes in Geoinformation and Cartography*. Springer-Verlag, Berlin. Available at: http://flrec.ifas.ufl.edu/hochmair/pubs/Rehrl_LBS2012_analyzingContributionPatterns.pdf.

Servigne, S., Ubeda, T., Puricelli, A., & Laurini, R. 2000. A Methodology for Spatial Consistency Improvement of GeographicDatabases. *Geoinformatica*, The Netherlands, *4*(1): 7–34.

Sheeren, D., Mustière, S., & Zucker J-D. 2004. Consistency Assessment Between Multiple Representations of Geographical Databases: a Specification-Based Approach, In: Proceedings of the 11th International Symposium on Spatial Data Handling (SDH'04), Leicester (UK)

Sun, C., Jia, X., Zhang, Y., Yang, Y., & Chen, D. 1998. Achieving convergence, causality preservation, and intention preservation in real-time cooperative editing systems. *ACM Transaction CHI*, *5*(1): 63–108.

Sun, Z., Yue, P., Lu, X, Zhai, X., & Hu, L. 2012. A Task Ontology Driven Approach for Live Geoprocessing in a Service-Oriented Environment: A Task Ontology Driven Approach for Live Geoprocessing. *Transactions in GIS*, *16*(6): 867–884.

Vilches-Blázquez, L., Villazón-Terrazas, B., Saquicela, V., de Leon, A., Corcho, O., & Gómez-Pérez, A. 2010. GeoLinked Data and INSPIRE through an Application Case. In: *Proceedings of the 18th ACM SIGSPATIAL*

International Conference on Advances in Geographic Information Systems. ACM SIGSPATIAL GIS 2010, San Jose, California, USA.

Von Hunolstein, S., & Zipf, A. 2003. Towards task oriented map-based mobile guides. In: *Proceedings of the International Workshop "HCI in Mobile Guides".* 5th International Symposium on Human Computer Interaction with Mobile Devices and Services, Udine (Italy), pp. 8–11

Wilkinson, D., & Huberman, B. 2007. Assessing the value of cooperation in Wikipedia. *First Monday, 12*(4). Available at: http://firstmonday.org/article/view/1763/1643

PART III

Data analytics

CHAPTER 11

A methodological toolbox for exploring collections of textually annotated georeferenced photographs

Ross S. Purves* and William A. Mackaness†

*Department of Geography, University of Zurich, Switzerland,
ross.purves@geo.uzh.ch
†School of Geosciences, University of Edinburgh, UK

Abstract

This chapter provides a brief overview of some methodologies used to extract meaning from the analysis of geotagged images. Broadly they draw from research in natural language processing and statistical and exploratory techniques. The confidence we attach to outputs from such analysis depends upon the questions we ask, our ability to take account of both the behaviour and motivation of the users contributing to user generated content, and the close relationship between how the data are spatially aggregated and the meanings associated with descriptions of images.

Keywords

geo-tagged, images, user behaviour, spatial aggregation, spatial analysis, natural language processing

How to cite this book chapter:
Purves, R S and Mackaness, W A. 2016. A methodological toolbox for exploring collections of textually annotated georeferenced photographs. In: Capineri, C, Haklay, M, Huang, H, Antoniou, V, Kettunen, J, Ostermann, F and Purves, R. (eds.) *European Handbook of Crowdsourced Geographic Information*, Pp. 145–156. London: Ubiquity Press. DOI: http://dx.doi.org/10.5334/bax.k. License: CC-BY 4.0.

Introduction

We continue to witness phenomenal growth in the production of user generated content (UGC). Some of that content comes in the form of photographs. Many are either annotated or tagged in a manner that may reveal aspects of users' conceptual understanding of place. In this article we concern ourselves with methods to extract meaning from large collections of textually annotated georeferenced photographs. Such collections have been the subject of considerable attention over the last decade, for a number of reasons. Firstly, and perhaps most importantly, the data are accessible. For instance, both Flickr[1] and Panoramio[2] provide application programming interfaces which make it possible for researchers to scrape images and associated metadata, while Geograph[3] content is provided under a Creative Commons Attribution-ShareAlike licence. Secondly, unlike other social media, the link between position, annotation and content is often relatively direct and closely linked to people's sense of place. People take pictures of things and events that happen somewhere, at some time, and describe them accordingly.

A wide range of applications have been developed that variously utilise this data in order to extract information and meaning:

- Automatic generation of gazetteer data (Kessler et al. 2009)
- Extraction and delineation of vernacular place names (Hollenstein & Purves 2010)
- Tag recommendations for images based on location (Rattenbury & Naaman 2009)
- Adding information to existing spatial databases (Antoniou et al. 2010)
- Extraction of place semantics at a range of scales (Feick & Robertson 2014; Purves et al. 2011; Rattenbury & Naaman 2009)
- Summarising and aggregating properties of the semantics of space (Ahern et al. 2007; Dykes & Wood 2009; Purves et al. 2011)
- Exploring movement of groups of individuals in space (Girardin et al. 2008)
- Identification and prediction of locations in text (O'Hare & Murdock 2012)
- Extraction of events using space-time clustering (Andrienko et al. 2010)

All of these approaches require methods which go beyond analysing spatial patterns associated only with the locations of photographs. This is the province of an established toolbox of geostatistical techniques for point pattern analysis able to describe spatial distributions and multi-scale patterns (O'Sullivan & Unwin 2003: chap. 4). Additionally we may wish to infer place semantics from other metadata associated with images (e.g. user, annotation, and time as well

---

[1] www.flickr.com
[2] www.panoramio.com
[3] http://www.geograph.org.uk/

as location). In this article we are only concerned with annotations written by the user: the person who uploaded the photograph, and typically, but not always, took it. This user information allows association of a set of photographs with an, pseudo-anonymous, individual and annotations can take the form of a title, a narrative (often descriptive text), and a set of tags. Tags are lists of key words selected freely by a user (Rattenbury & Naaman 2009) and, like all annotation associated with photographs in user generated content, may have a number of different motivating factors, including organisation of content for personal reasons, providing informative descriptions and making photographs findable by others. Locations may reflect the scene photographed, but with the advent of smart phones capable of automatically annotating images with GPS coordinates, more commonly reflect the photographer's position. Finally, temporal information often reflects both time of upload to the database and the time at which a photograph was recorded by a camera as having been taken.

This set of properties allows us to formulate a set of basic questions which can be asked of a collection of annotated, georeferenced photographs:

1) What language is used to describe photographs?
2) How can structured knowledge be extracted from annotations?
3) What influence do users have on information extracted from annotated georeferenced photographs?
4) How can we capture the relationship between language and location?
5) How do descriptions extracted from annotations vary according to scale and region definitions?

In the following, we introduce a methodological toolbox, drawn from a representative set of literature working on georeferenced annotated images, which allows us to explore these questions. As argued above, our focus goes beyond purely spatial analysis, and in particular focuses on textual annotations. In fact, many of the methods applied come from the domains of statistical natural language processing and information retrieval and focus on extracting information from a corpus (Manning & Schütze 1999). Common to all corpora are the basic notions of documents (in our case represented by annotations related to an individual photograph). Information about authorship (in our case in the form of unique users) is somewhat less common, and explicit links to spatial locations are what make our collections of georeferenced photographs particularly interesting. Thus, in the following, we will firstly introduce some **global analysis methods** – and ignore potential stratifications of the data by user or location (Questions 1 & 2). We will then discuss the link between **user behaviour and language** (Question 3) before finally looking at the explicit link between **language, location and scale** (Questions 4 & 5).

In this paper we use as exemplary data two examples of UGC: Geograph and Flickr. Our analyses are based on previous work reported in Purves et al. (2011). We focus on two forms of text input associated with georeferenced images in the British Isles: firstly short descriptive texts from Geograph, and secondly, tags associated with images in Flickr.

Global analysis methods

The first question that we can ask of any corpus concerns its composition. These are simple questions of frequency – what words occur and how often, and how are frequencies distributed in a corpus. A second, often neglected question is to ask, are the answers to the former in any way surprising?

For narrative text, function words (prepositions, conjunctions, pronouns) will typically be most frequent in any corpus and only by filtering out such terms (often called stop words) or exploring specific parts of speech (for example the use of nouns and proper nouns) can peculiarities of a collection with respect to general language be explored (Manning & Schütze 1999; Purves et al. 2011) (Table 1). Word frequencies in a corpus typically broadly follow Zipf's law – frequency is inversely proportional to rank. This implies in turn that a small number of different words account for a large proportion of the total word count in any corpus, and many words occur rarely in a given corpus.

It is important to note that tag lists are typically shorn of much the accoutrement of narrative text, and consist of relatively informative, freestanding terms (O'Hare & Murdock, 2012; Purves et al. 2011; Rattenbury & Naaman 2009). Thus, frequency counts of tags may already be informative with respect to semantic content, with for example around 80% of the Flickr tags analysed by Purves et al. (2011) taking the form of generic nouns (e.g. **church**[4], **hill**, **wedding**) or proper nouns (e.g. **tom, monday, nikon, edinburgh**) (Table 1). Hollenstein and Purves (2010) reported an average of 25% of tags as referring to locations and Rattenbury and Naaman (2009) identified some 12-16% of tags as being 'place tags'. Place tags still typically show Zipfian distributions.

In the above we effectively ignore the semantics or meaning of individual terms or tags. Thus, **forest** and **woods** are treated as entirely independent terms, as are **New York** and **Big Apple**, despite their obvious overlapping meanings. The first step in dealing with this problem is tokenisation – that is parsing some given input text to a set of meaningful units. This, at first glance, trivial problem is anything but. Approaches to tokenisation can have significant impacts on results (for example, is **New York** one token or two?) (Manning & Schütze 1999: chap. 4). The second step typically involves the use of more advanced methods such as lemmatisation and tagging of parts of speech, which fall firmly into the domain of natural language processing. Once again, the popularity of tags can be attributed to their simple structure, but it is important to note that

---

[4] We refer to tags in the text thus: **tag**

	Geograph (Top 10)			Geograph (Top 10 nouns)			Flickr (Top 10)	
Rank	Count	Word	Rank	Count	Word	Rank	Count	Tag
1	426936	the	13	45768	road	1	187605	london
2	275878	of	21	24085	view	2	97696	england
3	189089	to	24	21119	farm	3	96622	uk
4	184705	a	32	17242	lane	4	40528	2007
5	179553	in	36	16232	hill	5	34032	scotland
6	171429	and	37	16157	church	6	29654	unitedkingdom
7	153707	on	38	15815	bridge	7	24525	2006
8	152091	is	43	14737	river	8	21535	edinburgh
9	141579	from	45	14150	square	9	20215	ireland
10	132451	this	48	13690	house	10	17596	dublin

Table 1: Most frequent terms from narratives of 912874 Geograph photographs and tags of 759638 Flickr photographs for data collected in a bounding box corresponding to the British Isles in April 2008 (more details in Purves et al. (2011)).

this does not remove problems of, for example, ambiguity (e.g. does the tag **bath** refer to a town in England or a place to wash oneself?).

One approach taken to explore in more detail how words or tags are semantically related to one another is the use of co-occurrence to identify meaningful collocations – an 'expression consisting of two or more words that correspond to some conventional way of saying things'" (Manning & Schütze 1999: 151). The key task here is to disentangle expressions which co-occur by chance from those whose co-occurrence is statistically and semantically meaningful.

A surprisingly effective and efficient approach to this is adding some form of structure to words or tags found in a collection through annotation. Such annotation tasks often take the form of the formulation of a set of rules, applied independently by a group of annotators, in which final decisions about class membership is based on some majority decision (e.g. Purves et al. 2011; Rattenbury & Naaman 2009). Thus, for example, Purves et al. (2011) generated a simple taxonomy classifying words or tags as elements (things that are visible in an image), qualities (properties which might modify an element or suggest feelings or moods) and activities. Using this taxonomy it was then possible to explore co-occurrence, and identify both meaningful collocations or co-occurrences (e.g. **steep hill** or **city park**). Annotation tasks such as those described here can be seen as substituting specialised task-defined term dictionaries for more commonly available, but less specific, semantic resources such as WordNet (Miller 1995).

User behaviour and language

Other chapters in this book concern themselves with issues of participation inequality – the basic notion that a small number of users contribute much of the content to most examples of user generated content. The importance of this observation in analysis of georeferenced annotated photographs is straightforward – are we analysing the way in which many people have described a particular type of photograph (and their locations) – or the behaviour of only a few? Thus, for example, tags describing **trucks** and **lorries** were the 21st and 22nd most frequent in a collection of 450,272 photographs contributed by a total of 12,682 users, but only used by 15 and 7 users respectively. By contrast, the most frequent tag, **edinburgh**, was used by a total of 7,427 users, and the 20 most frequent tags were all used by more than 300 users. However, simply being used rarely does not *per se* indicate that a tag is not meaningful. In this particular case **trucks** and **lorries** are presumably the subject of interest of a small group, but this does not mean that the locations where they were photographed are unrepresentative. Considering the influence of individual users on tag semantics is therefore an important, and ongoing research challenge, in the analysis of annotated georeferenced photographs.

Purves et al. (2011) explored tagging behaviour by binning all photographs contributed to a collection, sorted by user prolificness. Histrograms of individual tag usage then showed the proportion of tags contributed by more or less prolific users, along with z-scores provided a summative value indicating whether a tag was used in similar ways by all contributors to a collection. This approach has the advantage of allowing exploration of individual tags, rather than contributions, and their influence through user behaviour. Furthermore, it provides a way of dealing with bias caused by, for example, bulk uploads, at the level of individual tags, rather than users.

Language, location and scale

In a book on Volunteered Geographic Information it is of course the location of information which is of primary interest. Georeferenced images were adopted very rapidly by researchers in this area because not only were locations explicitly recorded, but the assumption that the content was linked to a location is more immediate and seems more realistic in describing images taken *somewhere*. However, issues of granularity quickly become apparent, with for example the most frequent three tags in a collection of 1,520,212 images captured within the bounding box of Scotland being **scotland**, **edinburgh,** and **glasgow** respectively (Figure 1). Clearly **scotland** is not wrong, but neither is it informative. This problem is identical to that illustrated by the top ten words from Geograph in Table 1 – **the** is indeed a very frequent word, but it isn't terribly interesting!

One approach to identifying more interesting terms is to home in on those which more effectively characterise a document by comparing frequency of a chosen term in a given document to frequency across a corpus as a whole. This approach is known as term frequency- inverse document frequency (tf-idf) and is a baseline ranking method in information retrieval. It can be applied in a geographical context by counting the number of images with a particular tag within a prescribed region (or cell) and comparing this with frequency over a larger geographic region (Ahern et al. 2007; Rattenbury & Naaman 2009). The basic effect of geographical applications of tf-idf is to privilege locally common, but globally rare tags over globally common tags. Recognising the nature of user generated content and the issues relating to user behaviour described above, many researchers have added a term to capture user frequency in this characterisation, typically ranking tags used by many higher within in a region (Ahern et al. 2007; Feick & Robertson 2014; O'Hare & Murdock 2012; Rattenbury & Naaman 2009).

Obviously the size and form of the regions within which frequencies are calculated will have an influence on the results. The former property, size effectively captures notions of scale, while the latter, form, is closely related to the classical Modifiable Areal Unit Problem (MAUP). To capture notions of scale it is important to characterise tag semantics at multiple scales (c.f. Ahern et al. 2007; Feick & Robertson 2014; Rattenbury & Naaman 2009). Dealing with MAUP has led to a number of approaches. Rattenbury and Naaman (2009) and Ahern et al. (2007) generated regions bottom up, by clustering on photograph positions themselves using K-means. Feick and Robertson (2014) imposed a multi-scale hexagonal tessellation, which they is argued is better able to capture the complex geometries of real world regions. They explored similarity between tag characterisation of connected hexagons to identify larger semantic regions.

Figure 1 illustrates some of these notions for a dataset consisting of 1,520,212 photographs, containing a total of 53,842 unique tags and captured by 31,292 unique users. The ten most common tags were: **scotland, edinburgh, glasgow, uk, united kingdom, geotagged, england, music, uploaded:by=flickr_mobile,** and **highlands**. Seven of these are toponyms, but contain little or no useful information (the images were all from within Scotland's bounding box, and Edinburgh and Glasgow are simply the two most populous cities). Two (**geotagged** and **uploaded:by=flickr_mobile**) refer to properties of the data which are self-evident in the first case and refer to an application used to deliver data in the second. Finally, **music** reflects Flickr's popularity as a platform for describing leisure activities (Antoniou et al. 2010). Figure 1 ranks tags using three methods discussed above for a square grid. Firstly, tags are ranked using only frequency and, as was the case in Table 1, simply reflect characteristics of the collection as a whole (note the predominance of **scotland**). Secondly, tf-idf, filtered for multiple users gives back a much more local picture, and is dominated by more local toponyms, with the exception of larger cities, where activities and their locations (e.g. **fringe festival**, **murrayfield** (rugby), and **bongo**

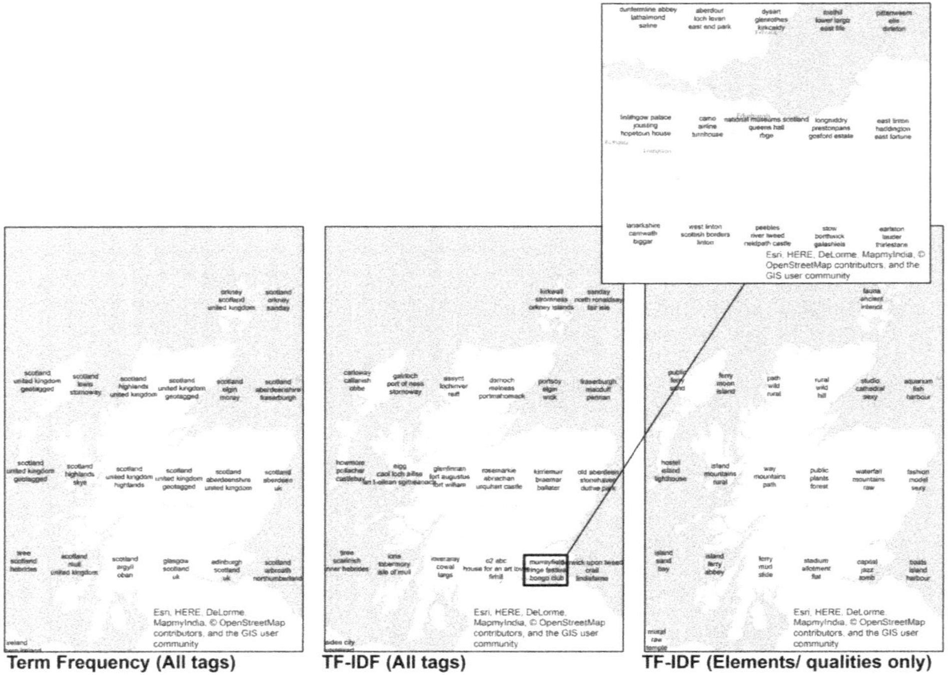

Figure 1: Top ranked tags for Flickr images in Scotland's bounding box according to term frequency, TF-IDF and TF-IDF filtered using list of elements and qualities according to [ref]. Only terms used by a minimum of 3 users are shown. Regions are defined as 1 degree x 1 degree.
Inset map shows top three terms calculated by TF-IDF for region around Edinburgh (Region defined as 0.2 degree x 0.2 degree).

club (Nightclub, gig, and events venue) for Edinburgh) become visible. Zooming in to a more detailed grid using tf-idf reveals finer granularity toponyms. To start to explore not only the names of the locations in grid cells, but what sorts of places these might be, tags are filtered according to a structured list from Purves et al. (2011). The resulting tf-idf values show locations associated with, for example, outdoor activities (**rural, wild, hill**) or more urban locations and activities (**stadium, allotment, flat**).

The techniques described so far focus on tags independent of one another. But, as discussed above co-occurrence can reveal more semantically rich information (e.g. **castle ruin** or **tall building**) and by using (most profitably) interactive visualisations such co-occurrence can be geographically located (Dykes

Figure 2: Top ten Geograph terms describing elements and their co-occurrence with one another presented as a spatial treemap Dykes & Wood (2009). The size of a rectangle indicates the overall count of co-occurrences for a particular term, while the nested rectangles indicate the relative predominance of individual collocates, and the colours link these to location – thus, for example, the most common terms used with **farm** are **hill**, **house**, **lane** and **road**. Figure adapted from data published in Purves, Edwardes & Wood (2011).

and Wood 2009), for example by using spatial treemaps (Dykes & Wood 2009). Spatial treemaps are hierarchical structures which can show 1) the overall occurrence of an individual term, 2) the most commonly co-occurring terms associated with each term, and, when linked to a key using colour 3) the locations of the co-occurrences. Figure 2 shows co-occurrences of the top ten most frequent elements, together with a colour legend linking the distribution for co-occurring terms across the British Isles in the Geograph dataset. Such visualisations allow us to start to explore the link between particular sorts of locations and their properties, for example the relative importance of **river** and **road** with respect to **bridge** compared to the importance of **land, house, road,** and **hill** with respect to **farm**.

Recommendations

The motivations for analysing geotagged imagery are as varied as its contributors. Thus the challenge lies not in the analysis per se, but in the initial processing of the data and in the interpretation of the results. Consensus need not be a prerequisite in extracting semantics; just because a prolific user contributes images of a highly thematic form does not make that contribution biased. However, some basic understanding of what properties in a collection might be surprising and a related awareness for the spectrum of existing approaches are both indispensable. In this short chapter we have scratched the surface of available methods – however we hope this material and the related references will prove a useful starting point for researchers new to the area.

Of course, the astute reader is still waiting for a silver bullet – but the reality is that all techniques should be seen as exploratory, and that great care is required in the interpretation of these qualitative outputs. Nonetheless, we recommend the following basic considerations, which we link here to the questions set out in the introduction:

- Global views on datasets allow an initial quick view of datasets (Q1)
- Consideration of the meaning of tags, and an understanding of potential ambiguities can be aided by simple methods such as co-occurrence (Q2)
- User behaviours can lead to significant biases, for example through bulk uploads and users with particular thematic interests (Q3)
- Purely frequency-based methods are unlikely to reveal interesting spatial patterns – however, simple methods such as tf-idf can rapidly increase the amount of information available in collection (Q4)
- When analysing geographic data basic notions such as scale and MAUP cannot be forgotten (Q5)
- Novel visualisation techniques can provide useful insights and lead to the generation of new hypotheses (Q5)

Acknowledgements

Much of our contribution in this paper derives from work undertaken with a wide variety of colleagues over the years. We would like to in particular acknowledge the contributions of Alistair Edwardes, Livia Hollenstein and Jo Wood to the research reported in this paper, though all errors are of course our own. We would also like to acknowledge contributors to Geograph British Isles, see http://www.geograph.org.uk/credits/2008-04-16, whose work is made available under the following Creative Commons Attribution-ShareAlike 2.5 Licence (http://creativecommons.org/licenses/by-sa/2.5/). RSP gratefully acknowledges the support of the Swiss National Science Foundation project PlaceGen (200021_149823).

References

Ahern, S., Naaman, M., Nair, R., & Yang, J. H.-I. 2007. World explorer: visualizing aggregate data from unstructured text in geo-referenced collections. In: *JCDL'07: Proceedings of the 2007 Conference on Digital Libraries*. ACM, New York, NY, US, pp. 1–10.

Andrienko, G., Andrienko, N., Mladenov, M., Mock, M., & Pölitz, C. 2010. Discovering bits of place histories from people's activity traces. In: *Visual Analytics Science and Technology (VAST), 2010 IEEE Symposium on*. pp. 59–66.

Antoniou, V., Morley, J., & Haklay, M. 2010. Web 2.0 geotagged photos: Assessing the spatial dimension of the phenomenon. *Geomatica, 64*: 99–110.

Dykes, J., & Wood, J. 2009. The geographic beauty of a photographic archive. In: Segaran, T., & Hammerbacher, J. (Eds.) *Beautiful Data*. O'Reilly, Sebastapol, CA, pp. 85–102.

Feick, R., & Robertson, C. 2014. A multi-scale approach to exploring urban places in geotagged photographs. *Comput. Environ. Urban Syst.*

Girardin, F., Calabrese, F., Fiore, F.D., Ratti, C., & Blat, J. 2008. Digital Footprinting: Uncovering Tourists with User-Generated Content. *Pervasive Comput. IEEE, 7*: 36–43.

Hollenstein, L., & Purves, R. 2010. Exploring place through user-generated content: using Flickr to describe city cores. *J. Spat. Inf. Sci, 1*: 21–48.

Kessler, C., Maué, P., Heuer, J.T., & Bartoschek, T. 2009. Bottom-Up Gazetteers: Learning from the Implicit Semantics of Geotags. In: *GeoS'09: Proceedings of the 3rd International Conference on GeoSpatial Semantics*. Springer-Verlag, Berlin, Heidelberg, pp. 83–102.

Manning, C. D., & Schütze, H. 1999. Foundations of statistical natural language processing. MIT press.

Miller, G. A. 1995. WordNet: a lexical database for English. *Commun. ACM, 38*: 39–41.

O'Hare, N., & Murdock, V. 2012. Modeling Locations with Social Media. *J. Inf. Retr., 16*: 13–62.

O'Sullivan, D., & Unwin, D. J. 2003. Geographic information analysis. Wiley, Hoboken, New Jersey.

Purves, R. S., Edwardes, A. J., & Wood, J. 2011. Describing place through user generated content. *First Monday, 16*.

Rattenbury, T., & Naaman, M. 2009. Methods for extracting place semantics from Flickr tags. T*WEB, 3*: 1–30.

CHAPTER 12

Gaining Knowledge from Georeferenced Social Media Data with Visual Analytics

Gennady Andrienko* and Natalia Andrienko

Fraunhofer Institute IAIS, Sankt Augustin, Germany, and City
University London, UK
*gennady.andrienko@iais.fraunhofer.de

Abstract

Analysis of the collections of geographically referenced posts published in social media, such as Twitter, Flickr, and YouTube, can bring new knowledge about places, geographical objects, and events interesting to people, and about people's mobility behaviours. Gaining knowledge from large data collections requires combining computational analysis with human interpretation, judgement, and reasoning, which, in turn, require appropriate visual representations of the data and analysis results. Visual analytics integrates computational analysis techniques with interactive visual interfaces to support collaborative human–computer analytical activities. We give a brief overview of visual analytics approaches to extracting various kinds of information and knowledge from georeferenced social media data.

Keywords

Visual Analytics, Social media analysis, trajectories, movement data, temporal data, spatio-temporal clusters

---

How to cite this book chapter:
Andrienko, G and Andrienko, N. 2016. Gaining Knowledge from Georeferenced
Social Media Data with Visual Analytics. In: Capineri, C, Haklay, M, Huang, H,
Antoniou, V, Kettunen, J, Ostermann, F and Purves, R. (eds.) *European Handbook of
Crowdsourced Geographic Information*, Pp. 157–167. London: Ubiquity Press. DOI:
http://dx.doi.org/10.5334/bax.l. License: CC-BY 4.0.

Introduction

Microblogging services, such as Twitter, and services for sharing photo and video, such as Flickr and YouTube, allow the users to supply their posts with geographic coordinates. The high popularity of these services in conjunction with the widespread proliferation of devices capable of providing location information has led to great and constantly increasing volumes of location- and time-referenced data produced by myriads of users. By analysing these data, it is possible to extract interesting new information about various places and events as well as about people's interests, mobility behaviours, and life styles.

Analysis of social media data is currently a popular topic in visual analytics, a research discipline that aims to support synergistic human–computer analytical workflows by combining computational analysis techniques with interactive visual interfaces supporting human interpretation, judgement, and reasoning (Keim et al. 2010). We give a brief overview of the published literature that describes visual analytics approaches to extracting different kinds of information from georeferenced social media data. Most of the works do not focus on extracting a single type of information but deal with several types.

Analysis of georeferenced photo data

The photos published at Flickr, Panoramio, and other photo sharing services are supplied with metadata, which include the dates and times of the shots and may also include titles and/or text tags indicating the contents of the photos. For many photos, the metadata include the coordinates of the locations where the photos had been taken. Collections of metadata records including geographic coordinates were analysed in multiple ways according to the possible analysis foci (space and place or people) and respective tasks (Andrienko et al. 2009). The photo data were considered from two distinct perspectives: as spatial events (independent points in space and time) and as trajectories of people (i.e. of the photo authors).

Analysing photo taking events

In analysing the data as spatial events, spatial density-based clustering was used for identifying popular places attracting much attention of the photo authors. Visualisation of the times when the photos had been taken in these places revealed different seasonal patterns of the place visits. To study the spatial distribution of the photos over a territory and compare the temporal patterns of visiting different parts of it, the territory is divided into compartments, e.g. by a regular (Andrienko et al. 2009) or irregular (Jankowski et al. 2010; Andrienko et al. 2012) grid, and the photo taking events are aggregated by these compartments and time intervals. The resulting time series of the event counts are visualised on a map (Andrienko et al 2009) or on a time graph (Jankowski et al. 2010; Andrienko et al. 2012), which is linked to a map display through interactive techniques, including synchronous highlighting, selection, and filtering of corresponding visual objects. By analysing the time series using either mostly interactive (Jankowski et al. 2010) or computationally supported (Andrienko et al. 2012) techniques, the researchers detected places with interesting temporal patterns of visits, such as periodic peaks at particular times of the year, very high irregularly occurring peaks, and significant increase of place popularity starting from a particular time. To understand the reasons for these patterns, the researchers extracted frequently occurring words and word combinations from the titles of the photos that had been taken in the places and times of the peaks or sudden increases of attendance. In most cases, the extracted words referred to various public events (festivals, open-air shows and concerts, etc.), but also to interesting natural phenomena, such as cherry tree blossoming or abundant snowfalls. A different approach to identifying public events and other happenings attracting people's attention is by using spatio-temporal clustering of the photo taking events (section 6.2.3 of Andrienko et al. 2013a) which finds occurrences of multiple photos taken closely in space and time, i.e. spatio-temporal clusters. For the clusters, frequently occurring words and word combinations are extracted and investigated using a text cloud display linked to a map (Figure 1).

Sections 7.2.1-7.2.5 of the book Andrienko et al. (2013a) present an example of an in-depth analysis of time series of the presence of distinct photographers by regions of Switzerland. The analysis includes, among other techniques, visually supported clustering of the time series and interactive generation of models for predicting the number of photographers that can be expected to visit the regions in the future at different times of a year. The time series can also be viewed from a different perspective: as a sequence of spatial distributions of the photographers' presence in different time intervals. To study the temporal patterns of the occurrence of similar and dissimilar spatial distribution patterns, the distributions are clustered by similarity, summarized by the resulting clusters, and compared using multiple map displays and special interactive operations supporting comparisons (section 8.1.1 of Anrienko et al. 2013a). The temporal distribution of the clusters is visually represented on temporal displays. The provided example demonstrates how the analysis reveals an interaction between temporal periodicity and temporal trends in the sequence of the spatial distributions of the presence of Flickr photographers over the territory of Switzerland.

Analysing trajectories of photo authors

Trajectories of people can be constructed from georeferenced photo data by arranging the records of each individual photographer in a chronological

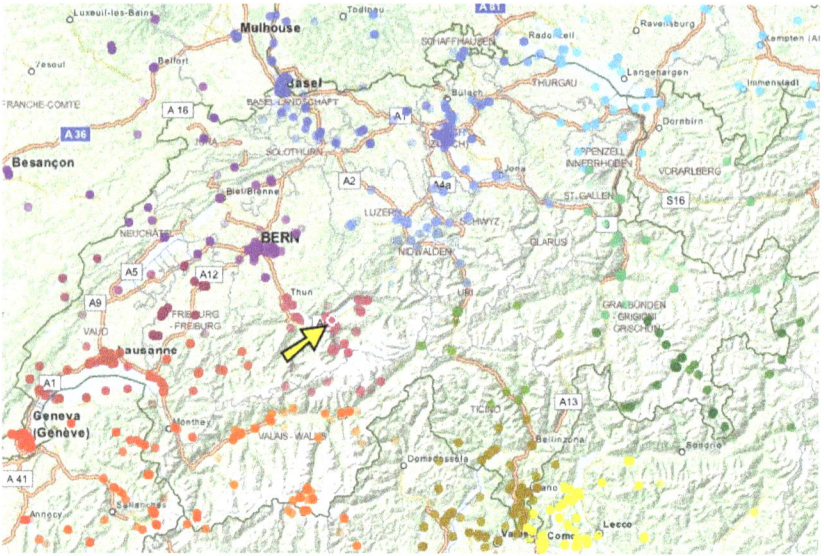

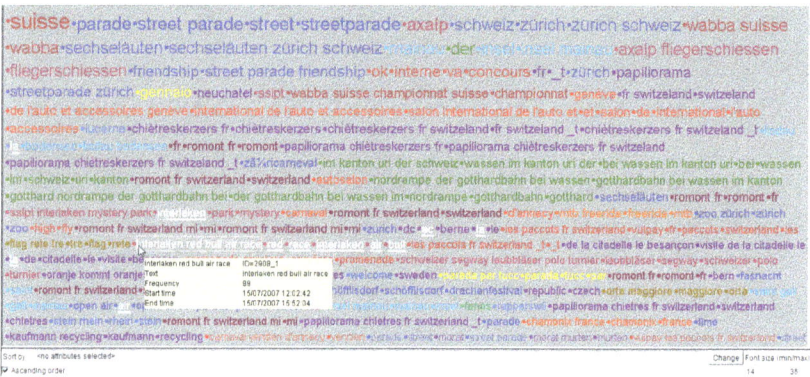

Figure 1: Top: the frequent occurrences of words and combinations in the photo titles within spatio-temporal clusters of Flickr photos are represented on a map by point symbols coloured according to the spatial positions of the clusters. Bottom: the words and combinations are represented in a text cloud display, the font sizes being proportional to the frequencies and the colours corresponding to the spatial locations, as in the map. One of the word combinations ('Interlaken red bull air race') is selected in the text cloud view by mouse-pointing; the corresponding point is highlighted on the map (marked with an arrow). Source: Andrienko et al. 2013a.

sequence (the same idea applies to any kind of georeferenced data that include identifiers of individuals, in particular, to data from YouTube, Twitter, and other social media). Trajectories of individuals can be aggregated into flows between compartments of a territory division and visualised on flow maps to enable studying of mass movement patterns (Andrienko et al. 2009). The aggregation of trajectories into flows can be done by time intervals for studying seasonal differences between the mass movement patterns (Jankowski et al. 2010). A set of trajectories can also be analysed for discovering frequent sequences of place visits (section 7.3.4 of Andrienko et al. 2013a). The extracted frequent sequences can be explored using a text cloud display combined with an interactive map and a space-time cube. By analysing people's trajectories, one can also detect meetings of two or more individuals, including repeated meetings of the same pairs or groups of individuals, and joint trips of two or more photographers (Andrienko et al. 2009); however, performing such analyses may be unethical, as they may compromise the personal privacy of the individuals.

This overview gives an idea about the diversity of the possible approaches to analysing georeferenced photo data and the kinds of information and knowledge that can be extracted from such data. The same range of approaches is also applicable to georeferenced microblogging data, such as data from Twitter. The types of information that can be extracted from the two different sources of data are the same but the interpretation may be different. Thus, people mostly take photos when they encounter interesting places, objects, or events; besides, not all taken photos but only the best or the most interesting ones may be published. It should also be taken into account that photos are rarely taken in low light conditions, and that there are situations and places in which taking photos is prohibited. Therefore, the photo data cannot be considered representative of people's presence and movements over a territory and of people's everyday activities. Figure 1 shows that photo data may reflect people's leisure activities and touristic travels. However, it would be wrong to assume that this is always the case. The possible relation of the published photos to the author's leisure time, travels, or professional activities can be judged from the temporal frequency and regularity of the photos and from their spatial distribution.

Analysis of georeferenced microblog data

Posting microblog messages from mobile devices may occur more frequently and spontaneously and in a wider range of places and situations than taking and publishing photos. Besides, there is no time gap between producing and publishing a message, while photo authors may not publish their photos immediately after taking but may do this after some (often quite long) time. Therefore, unlike photos, microblog data are suitable for real time analysis, which may discover information about currently happening events, in particular, abnormal and disastrous events, such as earthquakes or storms (Chae et al. 2012; Andrienko et al. 2014). This requires processing of the message texts. One of the approaches is pre-filtering of the messages for selecting only those that contain analysis-relevant keywords, such as terms denoting extreme weather conditions (Andrienko et al. 2014). Another approach is extracting significant terms, i.e. such words that do not occur frequently in microblog messages in general or in the times (seasons) or places where they have occurred (Chae et al. 2012; Bosch et al. 2013). Each occurrence of a significant term is treated as a separate spatial event. A spatio-temporal concentration (cluster) of events with the same term may indicate that something is happening in this place and time, and the term gives an idea of what may be happening. The significant terms from such spatio-temporal clusters are shown on a map display using the text cloud technique, with the font size being proportional to the number of the term occurrences. The map is constantly updated in real time as new messages appear. By means of an interactive tool called Content Lens, the user can select a particular area and explore in more detail the term occurrences in this area. To increase the relevance of the information that is shown to the user, various user-constructed filters can be applied to the data (Bosch et al. 2013). In the other approach (Anrienko et al. 2014), the message texts are only used for the selection of potentially relevant messages and not used in the further analysis. The work focuses on real time detection of spatio-temporal clusters of relevant events, taking into account only the event locations and times but not the texts, and on tracing the cluster evolution (growing, shrinking, moving, merging, and splitting) over time (Figure 2). The individual events making the clusters and their message texts can be accessed on demand.

An example of an offline investigation of microblog posts related to a disastrous event (an epidemic) is presented in section 6.3.2 of Andrienko et al. (2013a). Although it uses data generated synthetically (based on real data), it shows the principal possibility of using microblog data for identifying the origin and possible cause of an epidemic, the ways of disease propagation, the spatial spread, and the evolution over time.

However, detecting and investigating disastrous or abnormal happenings is not the only possible use case for microblog data. Georeferenced microblog posts, at least those from active bloggers, may to some extent be considered as representative of the people's daily lives and used for studying people's behaviours. Thus, an analysis of a collection of Tweets posted by residents of the Seattle area (USA) revealed interesting patterns of collective and individual behaviours (Andrienko et al. 2013b). For this analysis, the Tweets were classified according to their topics, such as family, work, education, food, sports, etc. based on the occurrences of topic-specific keywords (for example, the topic 'family' is associated with the terms denoting family members: mother, mom, father, daddy, and so on). The researchers explored how much the Tweet topics are related to the locations from which the Tweets were posted and to the times when this happened. For this purpose, they aggregated the Tweets by the topics, areas in space, and time intervals and visually explored the results using maps and time histograms. It was found that there are areas where particular topics

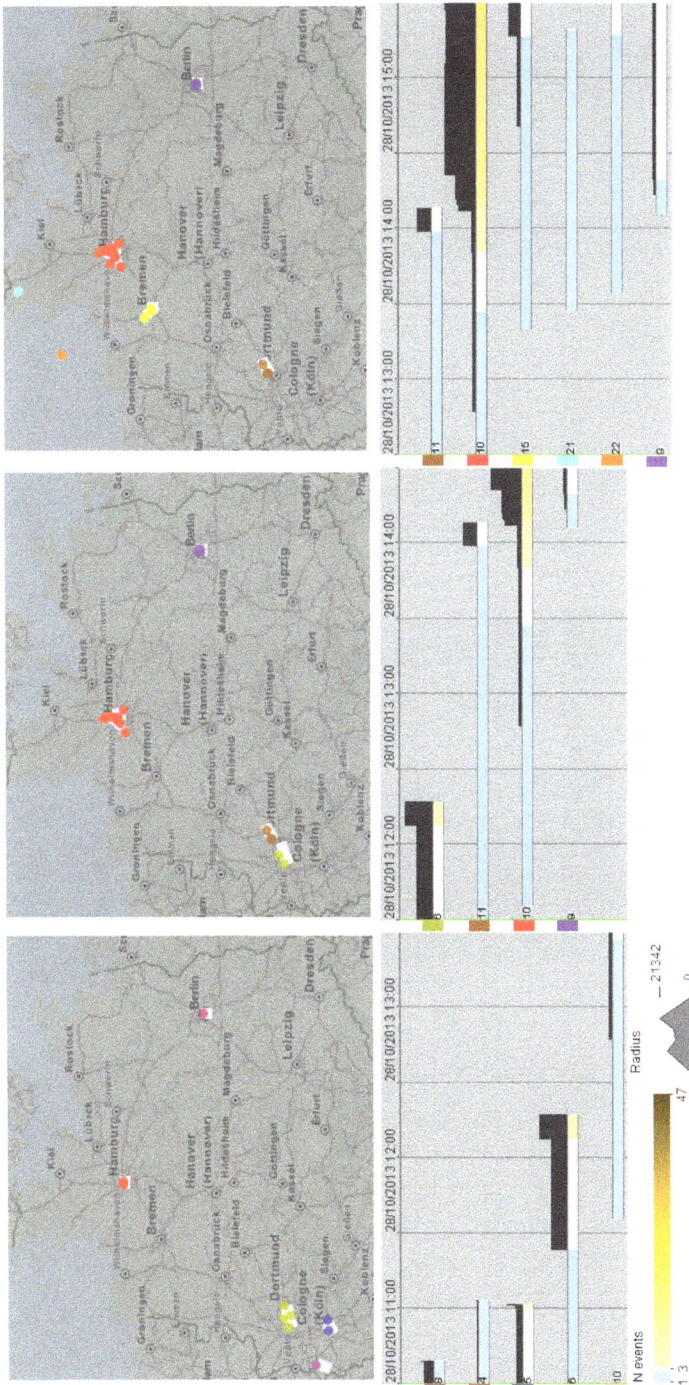

Figure 2: Emergence and evolution of spatio-temporal clusters of georeferenced Tweets related to a hurricane on October 28, 2013. Source: Andrienko et al. (2014).

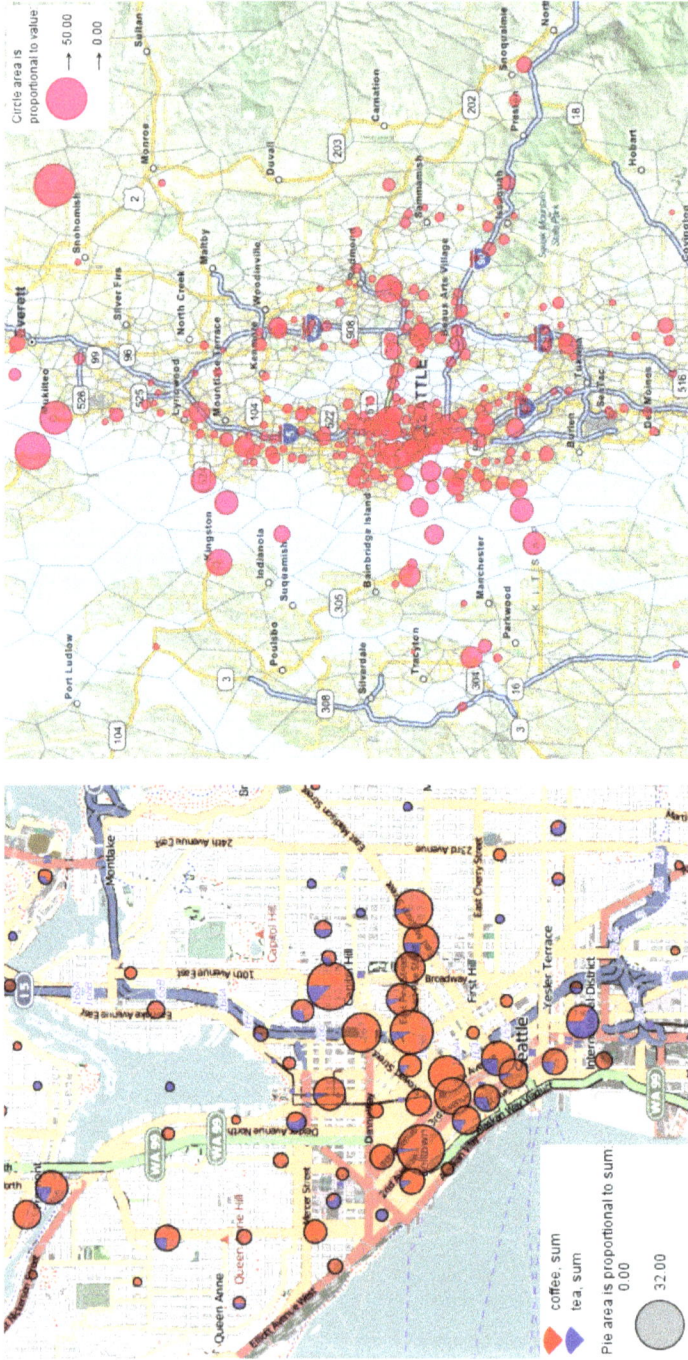

Figure 3: Left: the spatial distribution of the Tweet topics 'coffee' and 'tea' in the central area of Seattle. Right: the spatial distribution of the topic 'transportation'. Source: Andrienko et al. (2013b).

prevail, which may be related to the kinds of objects or facilities located in the areas (e.g. a university or a stadium) or to the characteristics of the population (e.g. an international district; see Figure 3, left). The researchers also looked at the spatial distributions of the different topics and found that some of them are correlated with the distribution of certain kinds of objects or facilities. Thus, the topic 'transportation' occurs along the main transportation corridors (Figure 3, right). Regarding the temporal distributions of the Tweet topics, the researchers found several very interesting patterns of *when* certain topics occupy the peoples' minds. Thus, 'food' occurs more frequently during lunch and dinner times, 'coffee' during/after breakfast and over the forenoon, 'transportation' during working day rush hours, and 'sports' and 'alcohol' in the evenings and over the weekend.

Although the study shows that the contents of some microblog posts are related to the places the authors visit and/or the activities they perform, these data in general contain a large proportion of noise, which includes texts with unidentifiable topics and texts with topics that are not relevant to the places of message posting (thus, a person may Tweet about work while being at home or about food while travelling in public transport). In fact, the proportion of noise outweighs the proportion of potentially relevant data. Therefore, it makes sense to analyse the topic distribution in space and time at the level of a large population of microbloggers, to have a sufficiently large amount of potentially relevant data and to be able to use valid statistical summaries. At the level of individuals, the message texts can hardly be indicative of the individuals' activities or purposes for visiting different places.

In analysing mobility behaviours of individuals, it is reasonable to look not at the message texts but at the temporal patterns of visiting different places (Andrienko et al. 2015). Significant (repeatedly visited) personal places are extracted from the collection of posts of each individual by spatial clustering of the post locations. Place semantics (i.e. the meanings, purposes for visiting, or activities performed in the places) can be determined based on the times over the weekly cycle when the individuals were present in the places. Thus, a place where a person is present in the evenings and nights of all days can be identified as the person's home place. However, separate consideration of the data of each individual is unfeasible and harmful for the personal privacy. The paper of Andrienko et al. (2015) proposes a privacy-respecting approach, in which data of a large number of Twitter users are analysed all together using a combination of computational techniques and visualisations presenting the data and analysis result in aggregated form. After extracting personal places and identifying their meanings in this manner, the original georeferenced data are transformed to trajectories in an abstract semantic space. The semantically abstracted data can be further analysed without the risk of re-identifying people based on the specific places they attend. The paper presents an example of analysing mobility behaviours of Twitter users in the area of San Diego (USA).

Conclusion

To summarise, georeferenced data from social media can be analysed as spatial events (i.e. independent points in space and time) and as trajectories of people. To analyse such data, visual analytics proposes a number of approaches combining computational techniques (clustering, aggregation, statistical summarisation, pattern detection, etc.) with interactive visualisations. With these approaches, it is possible to extract interesting information and gain new knowledge about places, events, and people's interests, behaviours, and habits. Metadata of the photos published through photo sharing services can reveal people's interests to tourist attractions, public events and other happenings, or natural phenomena and patterns of touristic behaviour. Georeferenced microblog posts can be analysed in real time for early detection of abnormal or disastrous events. It may also be useful to analyse the evolution of such events by looking at the spatio-temporal distribution of the event-related posts. Besides the information concerning unusual happenings, microblog data may be a source of knowledge about everyday mobility and activities of people. As both the popularity of the social media and the interest to analysing social media data are growing, we can expect the appearance of new analysis methods and new use cases for information that can be extracted by these methods.

Acknowledgements

This research was supported by European Commission within projects SoBig-Data RI (grant agreement 654024) and VaVeL "Variety, Veracity, VaLue: Handling the Multiplicity of Urban Sensors", grant agreement 688380.

References

Andrienko, G., Andrienko, N., Bak, P., Kisilevich, S., & Keim, D. 2009. Analysis of community-contributed space- and time-referenced data by example of Panoramio photos. In: *Proc. VMV – Vision, Modelling, and Visualization Workshop*, Braunschweig, Germany, November 2009.

Andrienko, G., Andrienko, N., Mladenov, M, Mock, M., & Poelitz, C. 2012. Identifying Place Histories from Activity Traces with an Eye to Parameter Impact. *IEEE Transactions on Visualization and Computer Graphics (TVCG)*, *18*(5): 675–688.

Andrienko, G., Andrienko, N., Bak, P., Keim, D., & Wrobel, S. 2013a. *Visual Analytics of Movement*. Springer.

Andrienko, G., Andrienko, N., Bosch, H., Ertl, T., Fuchs, G., Jankowski, P., & Thom, D. 2013b. Discovering Thematic Patterns in Geo-Referenced

Tweets through Space-Time Visual Analytics. *Computing in Science and Engineering*, 15(3): 72–82.

Andrienko, N., Andrienko, G., Fuchs, G., & Stange, H. 2014. Detecting and Tracking Dynamic Clusters of Spatial Events. In: *Proc. IEEE Visual Analytics Science and Technology (VAST)*, Proceedings, pp. 219–220. DOI: http://dx.doi.org/10.1109/VAST.2014.7042499

Andrienko, N., Andrienko, G., Fuchs, G., & Jankowski, P. (2015). Scalable and Privacy-respectful Interactive Discovery of Place Semantics from Human Mobility Traces, *Information Visualization, 15*(2):117–153. DOI: http://dx.doi.org/10.1177/1473871615581216.

Bosch, H., Thom, D., Heimerl, F., Püttmann, E., Koch, S., Krüger, R., Wörner, M., & Ertl, T. 2013. ScatterBlogs2: Real-Time Monitoring of Microblog Messages Through User-Guided Filtering. *IEEE Transactions on Visualization and Computer Graphics (TVCG), 19*(12): 2022–2031.

Chae, J., Thom, D., Bosch, H., Jang, Y., Maciejewski, R., Ebert, D., & Ertl, T. 2012. Spatiotemporal Social Media Analytics for Abnormal Event Detection using Seasonal-Trend Decomposition. In: *Proc. IEEE Visual Analytics Science and Technology (VAST)*, 143–152. DOI: http://dx.doi.org/10.1109/VAST.2012.6400557

Jankowski, P., Andrienko, N., Andrienko, G., & Kisilevich, S. 2010. Discovering Landmark Preferences and Movement Patterns from Photo Postings. *Transaction in GIS, 4*(6): 833–852.

Keim, D. A., Kohlhammer, J., Ellis, G., & Mansman, F. (eds.) 2010. *Mastering the Information Age – Solving Problems with Visual Analytics*, Eurographics.

CHAPTER 13

Head/tail Breaks for Visualization of City Structure and Dynamics

Bin Jiang

Faculty of Engineering and Sustainable Development, Division of GIScience,
University of Gävle, SE-801 76 Gävle, Sweden, bin.jiang@hig.se

Abstract

The things surrounding us vary dramatically, which implies that there are far
more small things than large ones, e.g., far more small cities than large ones
in the world. This dramatic variation is often referred to as fractal or scaling.
To better reveal the fractal or scaling structure, a new classification scheme,
namely head/tail breaks, has been developed to recursively derive different
classes or hierarchical levels. The head/tail breaks works as such: divide things
into a few large ones in the head (those above the average) and many small
ones (those below the average) in the tail, and recursively continue the division process for the large ones (or the head) until the notion of far more small
things than large ones has been violated. This paper attempts to argue that
head/tail breaks can be a powerful visualization tool for illustrating structure
and dynamics of natural cities. Natural cities refer to naturally or objectively
defined human settlements based on a meaningful cutoff averaged from a massive amount of units extracted from geographic information. To illustrate the
effectiveness of head/tail breaks in visualization, I have developed some case
studies applied to natural cities derived from the points of interest, and social
media location data. I further elaborate on head/tail breaks related to fractals,
beauty, and big data.

How to cite this book chapter:
Jiang, B. 2016. Head/tail Breaks for Visualization of City Structure and Dynamics, In:
    Capineri, C, Haklay, M, Huang, H, Antoniou, V, Kettunen, J, Ostermann, F and
    Purves, R. (eds.) *European Handbook of Crowdsourced Geographic Information*,
    Pp. 169–183. London: Ubiquity Press. DOI: http://dx.doi.org/10.5334/bax.m.
    License: CC-BY 4.0.

Keywords

Big data, social media, natural cities, fractals, head/tail breaks, Ht-index

Introduction

The things surrounding us vary dramatically, which implies that, instead of more or less similar things, there are actually far more small things than large ones, e.g., far more small cities than large ones in the world. This dramatic variation is often referred to fractal or scaling, and is well captured by geographic information of various kinds. Reflected in the points of interest (POI), there are far more POI in cities than in countryside; in terms of social media, there are far more users in cities than in the countryside. The new kind of geographic information constitutes what we now call big data (Mayer-Schonberger and Cukier 2013) in contrast to conventional small data. Unlike small data (e.g., census or statistical data), which are often estimated and aggregated, geographic information in the big data era is accurately and precisely measured at an individual level. This kind of geographic information due to its diversity and heterogeneity is likely to show the scaling pattern of far more small things than large ones. To better reveal the scaling structure, a new classification scheme, namely head/tail breaks (Jiang 2013a), has been developed to recursively derive inherent classes or hierarchical levels. It divides things around an average, according to their geometric, topological and/or semantic properties, into a few large ones in the head (those above the average) and many small ones (those below the average) in the tail, and recursively continues the division process for the large ones (or the head) until the notion of far more small things than large ones has been violated (c.f., Section 2 for a working example).

Natural cities refer to naturally and automatically derived human settlements, or human activities in general on the earth's surface, based on a meaningful cutoff averaged from a massive amount of units extracted from massive geographic information. For example, we build up a huge triangulated irregular network (TIN – a digital data structure commonly used for the representation of a surface) consisting of one-day Tweets locations indicated by GPS coordinates around the world. It is obvious that with the TIN there are far more short edges than long ones. The average length of the edges splits all the edges into two parts: a minority of long edges (longer than the average) in the head, and a majority of short edges (shorter than the average) in the tail of the rank-size plot (Zipf 1949). Aggregate all short edges to create thousands of natural cities around the world. The natural cities emerge from a collective decision of diverse, independent, and heterogeneous TIN edges, thus manifesting some wisdom of crowds (Surowiecki 2004). Interestingly, the natural cities demonstrate striking fractal structure and nonlinear dynamics (Jiang and Miao 2015). While conventional cities imposed by authorities from the top down are of great use for administation and mangement, natural cities defined from the bottom up are of more use for studying the underlying structure and dynamics. Natural cities are not constrained to individual countries, but are universally defined and delineated for the entire world with support of big data. Because of the unversality, natural cities defined at very fine spatial and temporal scales provide a useful means for scientific research.

This paper attempts to develop an argument that in the big data era head/tail breaks can become an efficient and effective visualization tool for illustrating structure and dynamics of natural cities. The fundamental logic of this argument is as such. A large number of natural cities as a whole can be classified into different hierarchical levels or classes. Instead of showing all the classes or the whole, we can deliberately drop out some low classes, yet without distorting the underlying scaling pattern of the whole. This is because the remaining classes as a sub-whole are self-similar to the whole. This logic applies to the time dimension as well, i.e., instead of showing all evolving patterns along a time line, we deliberately choose a part that reflects the whole. Head/tail breaks provides a simple instrument that helps us see fractals in nature and society, i.e., through examining whether there are far more small things than large ones, or more precisely whether the scaling pattern recurs multiple times with Ht-index being at least 3 (Jiang and Yin 2014). Conventionally, we must compute the fractal dimension to determine whether a set or pattern is fractal (Mandelbrot 1982). Fractal dimension (D) is rigorously defined, referring to the ratio of the change of details (N) to that of measuring scale (r), $D = \log(N)/\log(r)$. Following the rigorous definition, fractals are found to appear in a variety of phenomena such as mountains, trees, clouds, rivers, cities, streets, architectures, the Internet, the World Wide Web, social media, and even the paintings of Jackson Pollock (e.g., Batty and Longley 1994, Eglash 1999, Taylor 2006). Now with head/tail breaks, not only experts, but also the general public can simply judge the ubiquity of fractals relying on our intuitions, i.e., a set or pattern is fractal if the scaling pattern of far more small things than large ones recurs multiple times.

The remainder of this paper is structured as follows. Section 2 introduces head/tail breaks and discusses how it leads to a new definition of fractals using the Sierpinski carpet and Mandelbrot set as working examples. Section 3 reports several case studies applied to visualization of natural cities derived from POI and social media data. Section 4 adds some further discussions on head/tail breaks to meet challenges from big data. Finally Section 5 concludes the paper, and points to future work.

Head/tail breaks leading to a new definition of fractals

Head/tail breaks is largely motivated by heavy-tailed distributions such as power law, lognormal, and exponential distributions (c.f. Section 4 for a discussion) to derive inherent classes or hierarchical levels. The resulting number of classes is given by another term called Ht-index (Jiang and Yin 2014), as an alternative index to fractal dimension for characterizing the complexity of fractals. The higher the Ht-index, the more complex the fractals. Before illustrating its visualization capability, I shall briefly introduce head/tail breaks using the working example of the Sierpinski carpet.

The Sierpinski carpet, as a classic plane fractal, contains far more small squares than large ones, i.e., 1, 8, and 64 squares with respect to sizes 1/3, 1/9 and 1/27 given the carpet of one unit (Figure 1). The fractal dimension of the Sierpinski carpet can be calculated by $D = \log(8)/\log(3) = 1.893$, which indicates that every time the scale (r) is reduced three times, the number of squares (N) increases eight times. The calculation may look somewhat abstract and hard to grasp. Now let us take a simpler and easier way. There are far more small squares than large ones; at the smallest end there are 64 squares sized 1/27, at the largest end 1 square sized 1/3, and in the middle of the two ends 8 squares sized 1/9. If we create a scatterplot of these three points in an Excel sheet and fit them into a power function, one would observe $y = 0.125 \ x \wedge -1.893$ (see Figure 1). This is called the Richardson plot, showing the ratio of the change of details (N) to the change of scales (r). In the Richardson plot, three points are exactly on the distribution line, implying that the Sierpinski carpet is strict fractal, or alternatively, the parts are strictly self-similar to the whole. If we replaced the squares with city sizes, the points would be around rather than exactly on the distribution line. This is because city sizes are just statistically fractal rather than strictly fractal.

Now let us examine how head/tail breaks works for the Sierpinski carpet. There are a total of $1 + 8 + 64 = 73$ squares, and the average size of which is calculated by

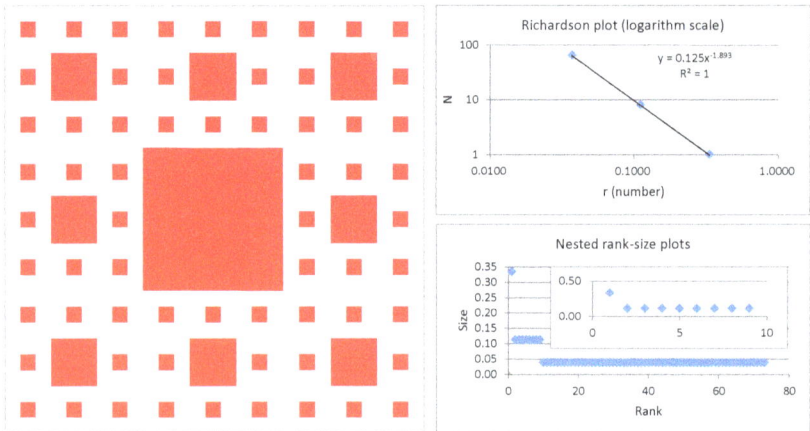

Figure 1: (Color online) Illustrstion of head/tail breaks and fractal dimension using the Sierpinski carpet.

Note: There are far more small squares than large ones for the Sierpinski carpet. The Richardson plot shows the fractal dimension, while the nested rank-size plots demonstrate the head/tail breaks process or the Ht-index.

# Squares	Mean	# head	# tail	% head	% tail
73	0.0492	9	64	12%	88%
9	0.1358	1	8	11%	89%

Table 1: The head/tail breaks for the Sierpinski squares.

$m_1 = (1/3 * 1 + 1/9 * 8 + 1/27 * 64) / (1 + 8 + 64) = 0.0492$. This first mean can split all the 73 squares into two unbalanced parts: a small portion of the large squares (nine squares larger than the mean) in the head, and a big portion of the small squares (64 squares smaller than the mean) in the tail. For the nine squares in the head, their average size is calculated by $m_2 = (1/3 * 1 + 1/9 * 8) / (1 + 8) = 0.1358$. This second mean can split all nine squares into two unbalanced parts: a small portion of the large squares (one square larger than the mean) in the head, and a large portion of the small squares (eight squares smaller than the mean) in the tail (Table 1). The above calculation indicates that the pattern of far more small squares than large ones recurs twice, and therefore Ht-index = 2 + 1 = 3. The recurring scaling pattern is also shown in the nested rank-size plots in Figure 1. The Ht-index is indeed 3 because there are only three scales: 1/3, 1/9, and 1/27. If we added 512 squares of the smaller size 1/81, the Ht-index would increase by one, but the fractal dimension would remain unchanged. From this, we see how Ht-index complements fractal dimension in capturing the complexity of fractals.

As the above example shows, head/tail breaks is quite simple and straight-forward, i.e., given that there are far more small things than large ones, split things into a few large and many small, and recursively continue the splitting for the large until the notion of far more small things than large ones is violated. Importantly, head/tail breaks leads to a relaxed definition of fractals: a set or pattern is fractal if the notion of far more small things than large ones recurs multiple times, Ht-index >= 3. This new definition based on the head/tail breaks is pretty intuitive, and may help refine our eyes or improve our intuitions for fractals. As remarked by Mandelbrot (1982), the most important instrument of thought is the eye rather than mathematical formula. With the new definition, anyone with little mathematical knowledge can easily rely on his/her intuitions to determine whether something is fractal. Now let us examine whether our intuitions have been improved with reference to the Mandelbrot set in Figure 2.

It is well known that the Mandelbrot set, probably the most complex shape known to man, comes from the amazingly simple equation: $z = z ^ 2 + c$. Despite the simplicity, I have decided not to consider the underlying mathematics; interested readers can refer to the literature for more details (e.g., Mandelbrot 2004). Instead, let us rely on our intuitions by concentrating on the Mandelbrot set shape and an infinite number of convoluted Julia sets shapes it generated, some of which are shown in Figure 2. What the stunning images of these shapes have in common is the ubiquity of far more small things than large ones. The Mandelbrot set can be zoomed into deeply to find similar patterns

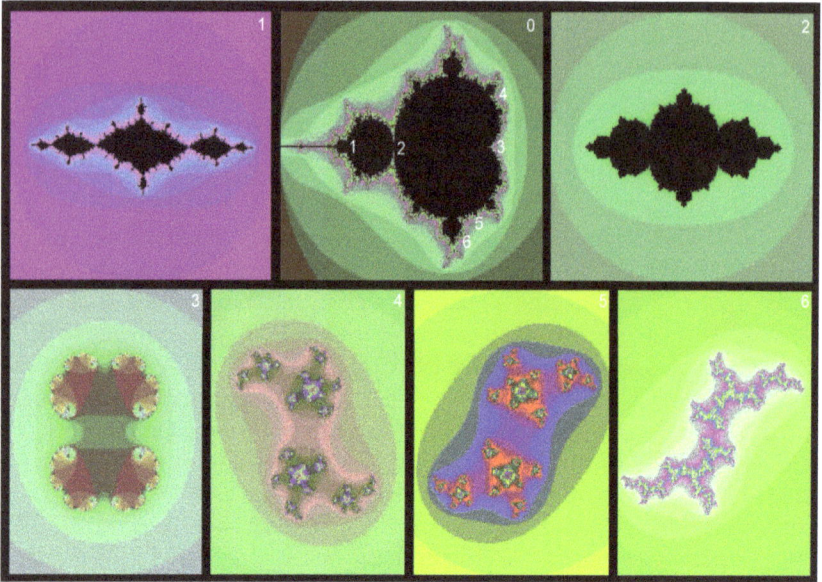

Figure 2: (Color online) Ubiquity of far more small things than large ones in both the Mandelbrot set (Panel 0) and the Julia sets (Panels 1 to 6).
Note: There are an infinite number of bulbs tangent to the main cardioid of the Mandelbrot set. The Julia sets in Panels 1 and 2 are generated from within the bulbs (black in Panel 0), whereas the Julia sets in Panels 3 to 6 are generated from outside of the bulbs (color in Panel 0). The figures in Panel 0 indicate the approximate locations where the Julia sets are generated.

again and again infinitely, so the Mandelbrot set can be said to be "big data". The Mandelbrot set (Panel 0) contains far more small bulbs than large ones, as do the two related Julia sets (Panel 1-2) generated from within the bulbs (black in Panel 0) of the Mandelbrot set. The Julia sets generated from outside the bulbs (color in Panel 0) of the Mandelbrot set have some dramatically different shapes and colorful images (Panel 3-6), which clearly evoke a sense or intuition that there are far more small structures than large ones. The images also look beautiful. Note that it is essentially not the colors but the underlying fine structures (or recurring pattern of far more small things than large ones) that make the patterns beautiful (Alexander 2002); see Section 4 for a further discussion.

Visualization of city structure and dynamics

When the social scientist Jacob L. Moreno (1934) first studied such human relationships as likes and dislikes, his dream was to map them for a whole city or nation. It now appears that what he dreamed of has been fully realized, not only for a whole nation, but for the entire world, with millions or billions of people connected through social media such as Facebook and Twitter. The New York Times praised Moreno's work as a new human geography (Jones 1933), because the map metaphor was used for portraying the acquired human relationships, with nodes for individuals, links for relationships between the individuals, red lines for liking, black lines for disliking, triangles for boys, and circles for girls. This semiology, together with the methods of data collection and data analysis, were typical social science methods in the age of data scarcity, or the so-called small data era. Nowadays, we have entered an unprecedented big data era, in which we are overwhelmed by crowdsourcing data, accumulated in social media and contributed by individuals (Goodchild 2007, Kwak et al. 2010, Gao and Liu 2014). In addition, advanced geospatial technologies have already produced a large amount of geographic information such as satellite images (National Research Council 2003). Big data requires new ways of thinking (Jiang 2015b) in order to better understand the underlying social and geographic structure and how the structure evolves over time. In this connection, visualization offers a powerful means to reach the better understanding.

Natural cities derived from POI

Points of interest (POI) are spread across countries, particularly within cities, represent interesting locations or facilities such as churches, schools, shops, and pubs. As a wiki-like collaboration to create a free editable map of the world, OpenStreetMap (Bennett 2010) has integrated millions of POI, including basic categories such as automotive, eating and drinking, government and public services, health care, and leisure. In this study, I took approximately 2 million POIs for the three European countries: France, Germany, and the United Kingdom from CloudMade (http://download.cloudmade.com/). Following the same procedure of extracting natural cities introduced in the previous work (Jiang and Miao 2014), we built a huge TIN for each country, and then derived natural cities for further scaling analysis. Table 2 presents the basic statistics about the derived natural cities. France, for example, has 280,117 POI, of which 254,008 unique points were used to generate a huge TIN with 835,009 edges. There are far more short edges than long edges, so the distribution is clearly L-shaped. I applied the head/tail division rule (Jiang and Liu 2012) into the massive number of edges, which resulted in two unbalanced parts: those above the mean in the head, and those below the mean in the tail. All those edges in the tail were aggregated, leading to the 9,391 natural cities. Figure 3 shows the resulting natural cities (Panel 1), together with those for the other two countries (Panels 2 and 3). Germany is the densest country in terms of both POI and natural cities, followed by the UK.

	France	Germany	UK
POI	280,117	1,299,638	505,051
Unique POI	254,008	977,357	462,424
TINEdge	835,009	3,238,695	1,511,023
Natural cities	9,391	48,830	16,814
Ht-index/hierarchy	6	7	6
Hierarchy shown	4	4	3

Table 2: Basic statistics about the natural cities derived from POI.

There are far more small natural cities than large ones in terms of the numbers of POI they contain. To effectively visualize the underlying scaling hierarchy of the natural cities, I applied the head/tail breaks to computing the Ht-index that is shown in Table 2. France, Germany, and the UK, respectively, have six, seven, and six hierarchical levels or classes. If all the classes were displayed by different sizes of red dots, no matter how small they are, the patterns would not be recognizable like hairballs. Instead, I chose the top four or three classes (Panels 4, 5, and 6 of Figure 3), which reflect the same scaling patterns of all the classes in the sense that the pattern of far more small things than large ones is retained. The fact that the top classes reflect the whole is the true power of head/tail breaks. In other words, the top classes retain the same scaling pattern of far more small things than large ones of all the classes.

Natural cities from Tweets locations

Like POI, social media users' locations can be aggregated to form individual natural cities. Unlike POI, Twitter users' geolocations contain very precise time information, up to minutes or seconds. In this way, we can slice the Tweets location data minute by minute, hour by hour, in order to track how the natural cities evolve. The derivation of the natural cities followed the same procedure in the previous work (Jiang and Miao 2015), and was based on the fact that there are far more low-density areas than high-density ones. That is, we generated a huge TIN for the unique locations of Tweets and then split the TIN edges into two unbalanced parts: those above the average in the head, and those below the average in the tail. Eventually, those edges in the tail are aggregated into the thousands of natural cities. The procedure is a simple application of the head/tail breaks, or that of the head/tail division rule (Jiang and Liu 2012). Let us consider the four snapshots to examine the underlying fractal structure and nonlinear dynamics of the natural cities (Figure 4). The evolution of the natural cities shows little difference from that of the Koch flake: the former being statistically self-similar, and the latter being strictly self-similar. Accordingly, I

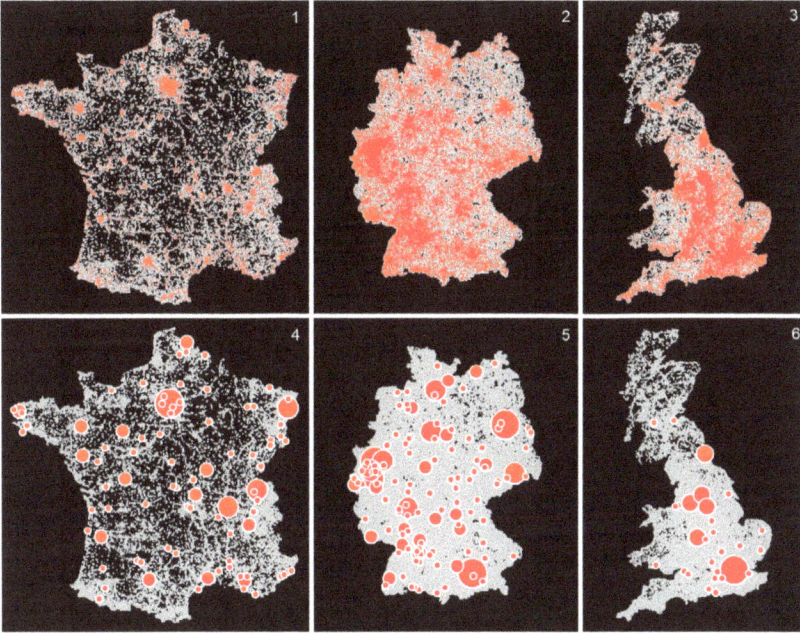

Figure 3: (Color online) The natural cities derived from POI of France (Panels 1 and 4), Germany (Panels 2 and 5), and the UK (Panels 3 and 6).
Note: The red patches indicate the natural cities or their boundaries, whereas the red dots indicate classified city sizes in terms of the number of POI. As mentioned in this paper, only a few top classes based on the head/tail breaks are shown for visual clarity. The grey background is the points of interest. The map scales are 1:15M.

claimed that social media could act as a good proxy for studying the evolution of real cities, in order to understand how they are generated and evolve through local and global interactions from the bottom up. This insight could fundamentally change the ways we studied cities in the small data era of the past.

The scaling patterns appear at different levels of geographic space. This is the true sense of ubiquity of fractal geographic features. It appears at a country level, a regional level, and a city level. Figure 5 presents an illustration of the ubiquity of scaling patterns. The large number of natural cities derived from Tweets locations are classified into six classes. I display only the top 4 classes for visual clarity, yet they reflect the pattern of the whole set (Panel 0). The enlarged regions of Chicago and New York (Panels 1 and 2) clearly show that there are far more small cities than large ones. At the city level, I computed the connectivity of individual streets, and the connectivity (or the number of other streets intersected) clearly shows a heavy-tailed distribution. All the streets are

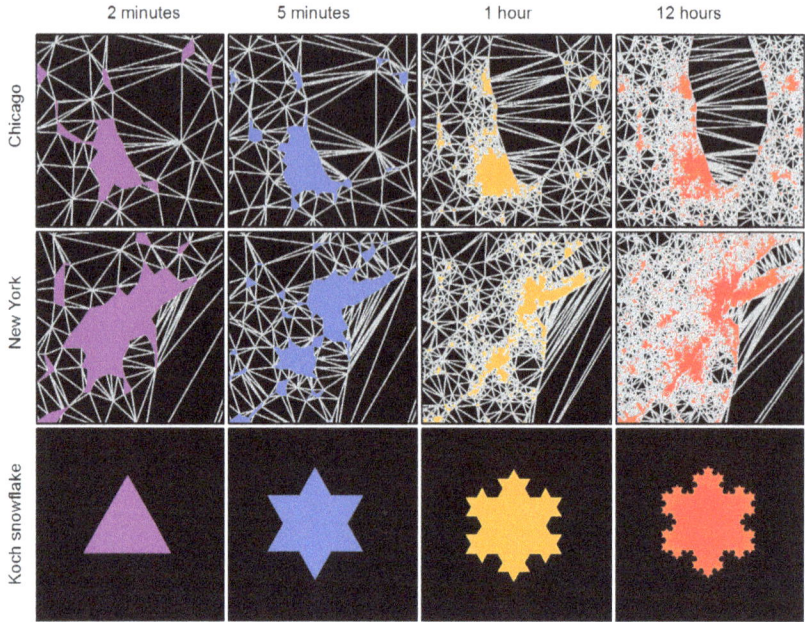

Figure 4: (Color online) The evolution of the natural cities on the background of TIN versus iteration of Koch flake.

Note: A few large pieces become more fragmented, whereas many small pieces are continuously added. Eventually there are far more small cities than large ones. This way of evolution looks very much like that of Koch flake. The major difference between the natural cities and Koch flake is the former being statistically self-similar, and the latter strictly self-similar. The map scales are 1:8.4M.

therefore classified and visualized based on the head/tail breaks. I invite the reader to compare Figures 2 and 5: the former being purely mathematical, and the latter geographic; the former being infinite, and the latter finite; the country as a whole equivalent to the Mandelbrot set as a whole, whereas the cities as parts equivalent to the Julia sets as parts; a city as a whole equivalent to the Mandelbrot set a whole, whereas the streets as parts equivalent to the Julia sets as parts. From the comparison, we see a nested or cascading structure for both the Mandelbrot set and the geographic space, and importantly the shared recurring scaling pattern of far more small things than large ones.

Further discussions on head/tail breaks

Head/tail breaks offers a new, less strict way of looking at our surrounding phenomena, in particular societal and organization phenomena. A phenomenon

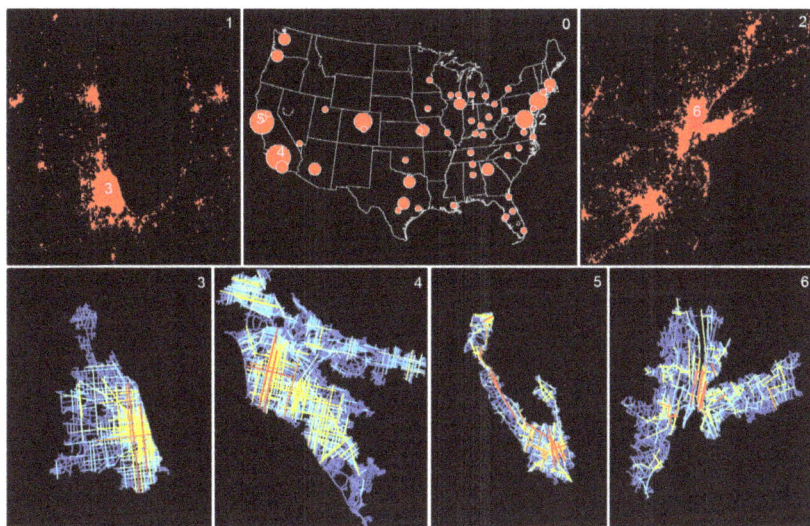

Figure 5: (Color online) Ubiquity of scaling patterns at different levels using the USA (mainland) as an example.
Note: The largest 55 natural cities in the top four classes at the country level (Panel 0); there are far more small cities than large ones at the regional level (Panels 1 and 2), and far more less-connected streets than well-connected ones at the city level (Panels 3 to 6), with blue being the least connected and red the most connected. The map scales for the USA (Panel 0), the regions (Panels 1, and 2), and the cities (Panels 3-6) are respectively 1:60M, 1:6M, and 1:2M.

or structure is fractal if there are far more small things than large ones in it. The notion of far more small things than large ones is not just in terms of geometric properties, but for topological and semantic properties as well, i.e., far more unpopular things than popular ones, or far more meaningless things than meaningful ones. Consequently, many societal and organizational phenomena are fractal, because they tend to be divided in an unbalanced way, which is known as the 80/20 principle (Koch 1998). Head/tail breaks, in particular the nested rank-size plots, provides a new interpretation of self-similarity. Conventionally, self-similarity refers to the property that the whole has the same shape as one or more of its parts (Mandelbrot 1982). Now the self-similarity can be interpreted by the repeated presence of far more small things than large ones, or alternatively, the repeated appearance of the small head and long tail division. It is the self-similarity that makes visualization of city structure and dynamics possible.

Head/tail breaks applies to data with a heavy-tailed distribution. The heavy-tailed distribution includes power laws as well as lognormal and exponential distributions. Strictly speaking, the exponential distribution is excluded from the heavy-tailed distribution, because its tail is quite short. I included it in the heavy-tailed distribution family because for most real-world data, there is a minimum threshold above which data are claimed to be power laws, lognormal or exponential distributions (Newman 2005). However, while conducting the head/tail breaks for the data, we consider all the data values including those below the minimum threshold. Thus, the data with an exponential distribution including those below the minimum would be heavy-tailed, and can therefore be broken multiple times using the head/tail breaks. It is also widely recognized that the bigger the data, the more likely they are heavy-tailed (Jiang 2015d).

Fractal structure, or the recurring scaling pattern of far more small things than large ones, possesses a new kind of beauty that positively impacts human well-being (Jiang and Sui 2014). The new kind of beauty, initially discovered and defined by Christopher Alexander (2002), differs fundamentally from conventional wisdom about aesthetics, being personal and subjective. The fractal beauty exists in deep structure, being objective and universal in nature. In other words, it is not the surface colors but the deep fractal structure that makes fractals beautiful. This deep structure is a kind of order that exists not only in nature but also in what we build and make (Alexander 2002), not only in science but also in humanities and social sciences. The beauty, the order revealed by head/tail breaks, or fractal geometry in general, cuts across multiple sciences and disciplines, bridging the two cultures (Snow 1959) to form the third culture. I believe that the visualization examples of city structure and dynamics shown in the paper embody some spirits of the third culture.

Many natural and societal phenomena demonstrate fractal structure and nonlinear dynamics (Mandelbrot and Hudson 2004). To better understand the complexity of social structure and dynamics, we must rely on a range of complexity modeling tools such as fractal geometry, chaos theory, and agent-based simulations (Miller and Page 2007) rather than conventional linear methods such as Euclidean geometry and Gaussian statistics. We must harness the large amounts of data accumulated on social media and the Internet for mining individual and collective behaviors. As a timely response to the challenges arising from big data, the emerging field of computational social science (Lazer et al. 2009, Watts 2007) has been fundamentally transforming the conventional social sciences into a data- and computational-intensive science. Unlike computational sciences in the twenty century, computational social science is a product of the twenty-first century, and it should be correctly interpreted as data-intensive computational social science in the big data era. This is the same for computational geography, which appeared first in the 1990s (Openshaw 1998), should be characterized as computational- and data-intensive in the twenty-first century (Jiang 2013b). In the big data era, cartography faces the same challenge of how to efficiently and effectively visualize the large amounts of crowdsourcing geographic information. I believe that recognition of the fractal nature of maps and mapping (Jiang 2015c) offers a way to meet the challenge.

Conclusion

This paper has developed an argument that head/tail breaks can be used for visualization of structure and dynamics of natural cities in the big data era. To support the argument, I have developed several case studies applied to natural cities derived from crowdsourcing data. This paper has also discussed how head/tail breaks leads to a new definition of fractals, helping improve our intuitions for seeing fractals in nature and society. Throughout the paper, we have seen both mathematical fractals (such as the Sierpinski carpet, Mandelbrot set, and Koch flakes) and geographic features (such as the natural cities and streets) share the same recurring scaling of far more small things than large ones. The scaling property is what drives the development of head/tail breaks. It is the scaling property that makes head/tail breaks an efficient and effective visualization tool for revealing city structure and dynamics. The power of head/tail breaks lies in its simplicity: split things around an average into a few large and many small, respectively in the head and the tail of the nested rank-size plots, and recursively continue the splitting process in the head until the condition of far more small things than large ones is violated. The simple head/tail breaks and its induced Ht-index can help even the general public to see a variety of fractals in science, art, and society.

The notion of natural cities, as a product of the big data era, provides a powerful tool to study human activities on the earth's surface, and enables us to develop new insights into geographic information harvested from crowdsourcing data. Compared with conventional real cities that are imposed by authorities from the top down, natural cities are defined from the bottom up, and from individual people and their interactions. Unlike real cities, natural cities can be naturally and objectively derived and delineated from big data such as VGI and social media data. This makes natural cities universally available for the entire world, i.e. all the natural cities in the world rather than those in some countries. In this regard, big data is probably not so much about bigness, but rather completeness. Natural cities may fundamentally change the ways cities were studied.

Acknowledgement

This chapter is a short version of the paper (Jiang 2015a), from which I have kept only VGI and social media related content for the book. Thanks to the initial publisher Elsevier for the permission of reprint.

References

Alexander. 2002. *The Nature of Order: An essay on the art of building and the nature of the universe: Book 1. The phenomenon of life*, Center for Environmental Structure: Berkeley, CA.

Batty, M., & Longley, P. 1994. *Fractal Cities: A geometry of form and function*, Academic Press: London.

Bennett, J. 2010. *OpenStreetMap: Be your own cartographer*, PCKT Publishing: Birmingham.

Eglash, R. 1999. *African Fractals: Modern Computing and Indigenous Design*, Rutgers University Press: New Jersey.

Gao, H., & Liu, H. 2014. Data Analysis on Location-Based Social Networks. In: Chin, A., & Zhang, D. (Eds.) *Mobile Social Networking: An innovative approach*. Springer, pp. 165–194.

Goodchild, M. F. 2007. Citizens as sensors: The world of volunteered geography, *GeoJournal, 69*(4): 211–221.

Jiang, B. 2013a. Head/tail breaks: A new classification scheme for data with a heavy-tailed distribution, *The Professional Geographer, 65*(3): 482–494.

Jiang, B. 2013b. Volunteered geographic information and computational geography: New perspectives. In: Sui, D., Elwood, S., & Goodchild, M. (Eds.) *Crowdsourcing Geographic Knowledge: Volunteered Geographic Information (VGI) in Theory and Practice*. Springer: Berlin, pp. 125–138.

Jiang, B. 2015a. Head/tail breaks for visualization of city structure and dynamics, *Cities, 43:* 69–77.

Jiang, B. 2015b. Geospatial analysis requires a different way of thinking: The problem of spatial heterogeneity, *GeoJournal, 80*(1): 1–13.

Jiang, B. 2015c. The fractal nature of maps and mapping, *International Journal of Geographical Information Science, 29*(1): 159–174.

Jiang, B. 2015d. Big data is not just a new type, but a new paradigm. Available at: https://www.researchgate.net/publication/283017967_Big_Data_Is_not_just_a_New_Type_but_a_New_Paradigm.

Jiang, B., & Liu, X. 2012. Scaling of geographic space from the perspective of city and field blocks and using volunteered geographic information, *International Journal of Geographical Information Science, 26*(2): 215–229.

Jiang, B., & Miao, Y. 2015. The evolution of natural cities from the perspective of location-based social media, *The Professional Geographer, 67*(2): 295–306.

Jiang, B., & Sui, D. 2014. A new kind of beauty out of the underlying scaling of geographic space, *The Professional Geographer, 66*(4): 676–686.

Jiang, B., & Yin, J. 2014. Ht-index for quantifying the fractal or scaling structure of geographic features, *Annals of the Association of American Geographers, 104*(3): 530–541.

Jones, D. 1933. Emotions mapped by new geography: Charts seek to portray the psychological currents of human relationships, *The New York Times* (Published on April 3, 1933).

Koch, R. 1998. *The 80/20 Principle: The secret of achieving more with less*, DOU-BLEDAY: New York.

Kwak, H., Lee, C., Park, H., & Moon, S. 2010. What is Twitter, a social network or a news media? In: *Proceedings of the 19th International World Wide Web Conference*, 2010.

Lazer, D., Pentland, A., Adamic, L., Aral, S., Barabási, A.-L., Brewer, D., Christakis, N., Contractor, N., Fowler, J., Gutmann, M., Jebara, T., King, G., Macy, M., Roy, D., & Van Alstyne, M. 2009. Computational social science. *Science, 323*: 721–724.

Mandelbrot, B. 1982. *The Fractal Geometry of Nature*. W. H. Freeman and Co.: New York.

Mandelbrot, B. B. 2004. *Fractals and Chaos: The Mandelbrot set and beyond*, Springer: New York.

Mandelbrot, B. B., & Hudson R. L. 2004. *The (Mis)Behavior of Markets: A fractal view of risk, ruin and reward*, Basic Books: New York.

Mayer-Schonberger, V., & Cukier, K. 2013. *Big Data: A revolution that will transform how we live, work, and think.* Eamon Dolan/Houghton Mifflin Harcourt: New York.

Miller, J. H., & Page, S. E. 2007. *Complex Adaptive Systems: An introduction to computational models of social life.* Princeton University Press: Princeton.

Moreno, J. L. 1934. *Who Shall Survive? A New Approach to the Problem of Human Interrelations*, Nervous and Mental Disease Publishing Co.: Washington D. C.

National Research Council. 2003. *IT Roadmap to a Geospatial Future*, The National Academies Press: Washington, D.C.

Newman, M. E. J. 2005. Power laws, Pareto distributions and Zipf's law, *Contemporary Physics, 46*(5): 323–351.

Openshaw, S. 1998. Towards a more computationally minded scientific human geography, *Environment and Planning A, 30*: 317–332.

Snow, C. P. 1959. *The Two Cultures and The Scientific Revolution*, Cambridge University Press: New York.

Surowiecki, J. 2004. *The Wisdom of Crowds: Why the Many Are Smarter than the Few*, ABACUS: London.

Taylor, R. P. 2006. *Chaos, Fractals, Nature: A New Look at Jackson Pollock*, Fractals Research: Eugene, USA.

Watts, D. J. 2007. A twenty-first century science, *Nature, 445*(x): 489.

Zipf, G. K. 1949. *Human Behavior and the Principles of Least Effort*, Addison Wesley: Cambridge, MA.

CHAPTER 14

Querying VGI by semantic enrichment

Rob Lemmens*, Gilles Falquet†, Stefano De Sabbata‡,
Bin Jiang§ and Benedicte Bucher¶

*University of Twente, The Netherlands, r.l.g.lemmens@utwente.nl
†University of Geneva, Switzerland, Gilles.Falquet@unige.ch
‡University of Leicester, UK, s.desabbata@le.ac.uk
§University of Gävle, Sweden, bin.jiang@hig.se
¶IGN, France, benedicte.bucher@ign.fr

Abstract

Volunteered geographic information (VGI) plays an increasing role in current geodata provision. At the same time, due to its lack of structure, it is hard to use as meaningful input in software applications. In this chapter, we embark upon the unstructured character of VGI and on ways to enrich the structure in order to make it suitable for information retrieval. We describe the characteristics of semantic enrichment and explain how folksonomies and ontologies play a role. We believe that they represent different levels of formality in a semantic reference space and determine the richness of the information retrieval.

Keywords

VGI, Query, Semantic enrichment, Folksonomy, Ontology

How to cite this book chapter:
Lemmens, R, Falquet, G, De Sabbata, S, Jiang, B and Bucher, B. 2016. Querying VGI by semantic enrichment. In: Capineri, C, Haklay, M, Huang, H, Antoniou, V, Kettunen, J, Ostermann, F and Purves, R. (eds.) *European Handbook of Crowdsourced Geographic Information*, Pp. 185–194. London: Ubiquity Press. DOI: http://dx.doi.org/10.5334/bax.n. License: CC-BY 4.0.

Introduction

Recent developments in personal computing, GPS and Web 2.0 technologies are enabling a wide web audience to actively contribute to geo-information through the internet. Information obtained in this way – commonly referred to as volunteered geographic information (VGI) – is often difficult to query due to several reasons.

The complexity of querying is rooted in the informal, unstructured, heterogeneous nature of VGI, which is often published without a description of its context. Those issues are inherent to the process by which VGI is produced, i.e. by individuals who are in most cases not concerned with the query process. This chapter investigates how the process of querying VGI can be improved by semantically enriching it during its production and after it is published. The enrichment connects VGI to well-known concepts which are captured in both informal structures (folksonomies) and formal structures (ontologies). A folksonomy represents a particular domain through a set of user-generated tags/topics of domain-related information, whereas an ontology constitutes a domain more rigorously through the representations of logical relationships between concepts used in that domain. In this research we differentiate the semantic enrichment along the line of informal-formal conceptualization, i.e. evaluating conceptual bases ranging between folksonomy and ontology, supporting the enrichment of VGI.

The main point we want to stress is that VGI implies further degrees of freedom and expression for the users, which can enable new, different narratives in collecting, describing, and representing geographic information. At the same time, this intrinsic diversity requires the creation of 'interfaces' between VGI datasets and any algorithm aiming to analyze them, in order to translate the folksonomy (representing the vocabulary used in the VGI) into the structure used by a query algorithm. This is a challenge in terms of 1) the ad-hoc work necessary to deal with the data and 2) the errors, misinterpretation, and information loss in the translation.

We pose the following main research question and set the scene for its discussion, but do not claim to answer it yet fully: how does varying the level at which a top-down ontology is applied to a bottom-up folksonomy change the understanding of underlying data, and thus the ability of querying VGI?

The goal is to query VGI sources such as Tweets, commented photos and news items about the named features they contain. Typical queries are

- what is the location of a feature named X?
- what is the footprint of a feature named X?
- what are the features located at or near P?
- what are the features with type T?

In some cases this involves the harvesting of implicit geographic information (see also Kessler et al. 2009) and in other cases such information cannot be directly extracted from the VGI itself and it needs to enriched with more formally structured information obtained from related sources, such as Wikipedia, OpenStreetMap, Geonames, etc. (see Smart et al. 2010).

Terminology of semi-structured data

VGI may appear as structured, semi-structured and unstructured data. Kitchin (2014) defines structured data as data 'that can be easily organized, stored and transferred in a defined data model', thus encompassing all data that can be represented and dealt with using relational databases and other technologies or representational models such as object-oriented languages or description logics. As a result, this kind of data can be straightforwardly processed through algorithms and visualized using graphs and maps. By contrast, semi-structured data don't have a, rigid, regular, or complete predefined data model/schema as required by traditional databases (Abiteboul 1997), though having 'a reasonably consistent set of fields' (Kitchin 2014). These include content that could barely be coded in a relational database, while still being characterized by irregular and flexible structures. Abiteboul (1997) provides a clear explanation of how HTML pages are a good example of semi-structured data, due to their lack of uniformity, and ample use of plain text. Finally, data is defined as unstructured if it has no structure that can be identifiable as common for the whole dataset, despite each element of the same dataset might have its own internal structure, which is not shared by any other element.

On such basis, most VGI content would be classified as semi-structured data, as few datasets could be straightforwardly dealt with in a relational database. Instead VGI datasets commonly use loose data definitions and categorizations, which are flexible and constantly edited by the same users, as well as more suitable to describe large quantities of vague information. A good example of loose categorization of geographic data can be found amongst the OpenStreetMap (OSM) map features — which include over eight thousand different user-defined kind of shops.

Most VGI content would also fit in Kitchin's (2014) definition of 'captured' data', that is data that has been directly captured through some device with the specific intention of capture the data. However, it might be argued that geo-tagged information, such as photos, entail 'exhaust' geographic data (implicitly included geodata) in the form of GPS coordinates in the image header— that is, as byproduct of capturing the photo, but not as main outcome of the process.

Characterizing the heterogeneity of VGI

VGI is very heterogeneous and diverse, due to three major reasons.

First, geographic features are very heterogeneous, since there are far more small geographic features than large ones. Using a more scientific terminology, geographic features are fractal or scaling (Jiang & Yin 2014), and they are best characterized by some heavy tailed (Zipf 1949) rather than Gaussian-like distributions. There are, for example, far more small street blocks than large ones. The small street blocks can be named as city blocks in cities, while the big street blocks are called field blocks in the countryside. The small street blocks constitute cities or natural cities to be more precise, whereas the large street blocks collectively form the countryside.

The heterogeneity of OSM data can be examined from various aspects such as element sizes, the number of edits, and the number of users for each element (Ma et al. 2015). For example, the element size ranges from 3 up to 5,000,000, the number of edits for each element can go up from 1 to 2,000. It is the heterogeneity that makes VGI unique and powerful in comparison to authoritative geographic information. It is the heterogeneity that makes VGI differ fundamentally from small data. It is the heterogeneity that makes researching VGI interesting and exciting. We should go beyond small data thinking such as Gaussian distributions and Euclidean geometry, and towards big data thinking such as heavy tailed distributions and fractal geometry.

Second, VGI can be produced through different methods and technologies, implying different levels of structural rigidity, ranging from menu entries to free text entries. This has important implications for semantic querying (see Section 6). The same VGI dataset may contain structured, semi-structured and unstructured data. For instance, the geometric part of OSM or the data/ time metadata of Twitter are structured, as there is a fixed schema for them; additional information about geographic features in OSM consists of semi-structured sets of tag-value pairs; and the content of some fields are unstructured texts. In OSM, tags can be freely created by the user and the way people assign a geometry to a feature is not always consistent throughout the project (with different scales typically (Touya & Brando 2013)).

Third, VGI contributors may come from very diverse geographic, cultural, and technical backgrounds, and thus might be accustomed with different terminologies, or have different narratives. Some VGI is produced with a shared conceptualization that can be a set of tags or a category graph (like OSM or DBpedia), yet the production of data with this conceptualization in mind is done differently depending on contributors (Brando & Bucher 2010). An example is mapping of crimes, where people can interpret the levels of violence differently. Besides, sets of tags evolve over time. Hence, if data have not been tagged with a specific tag, it might just be due to the fact that that tag did not exist when they were produced.

The heterogeneity of geographic features, modes of production, and contributors' background are all contributing to the fact that the quality of VGI is often disputed and that even the quality itself is heterogeneous.

Folksonomies and Ontologies for querying VGI

Writing an algorithm to perform a task on a given data source, or querying this source, can be better accomplished if the meaning of each element of the source is well defined. In traditional structured sources this meaning is conveyed by a database schema or a datatype definition expressed in a database or programming language. In VGI the situation is different because 1) the schema, if it exists, may not be sufficient, due to various interpretations by the users, and 2) many VGI sources are only semi-structured, without any centrally defined schema. Therefore it is necessary to rely on some semantic resource to represent the meanings of the data elements.

There are several types of such semantic resources, ranging from the most informal (folksonomies or glossaries) to the most formal (formal logical ontologies). These resources, generally known as knowledge organization systems, can be characterized along two axes: 1) the structure complexity of the underlying data (tags, classes, hierarchical relations, etc.) and 2) the formalism used to express concept definitions. Figure 1 presents a classification, along these axes, of the most frequently used knowledge organization systems.

Semantic enrichment of VGI

Semantic enrichment refers to the process of making information more meaningful by adding explicit structure, metadata, definitions, etc. Explicit means that the result is queryable.

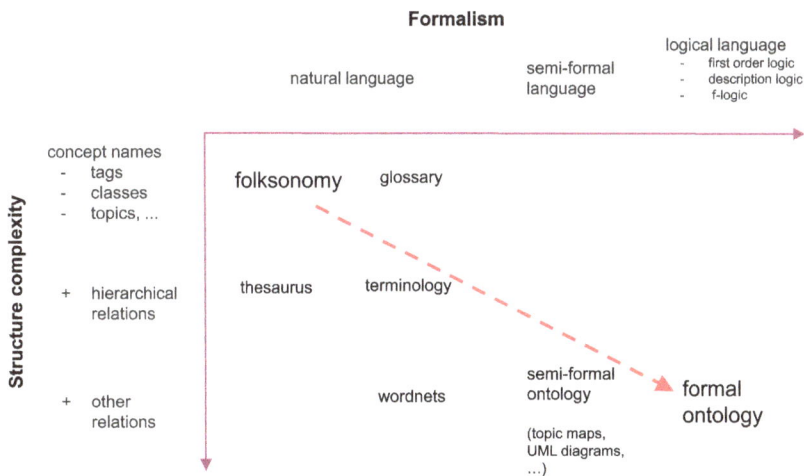

Figure 1: Informal and formal semantic reference space.

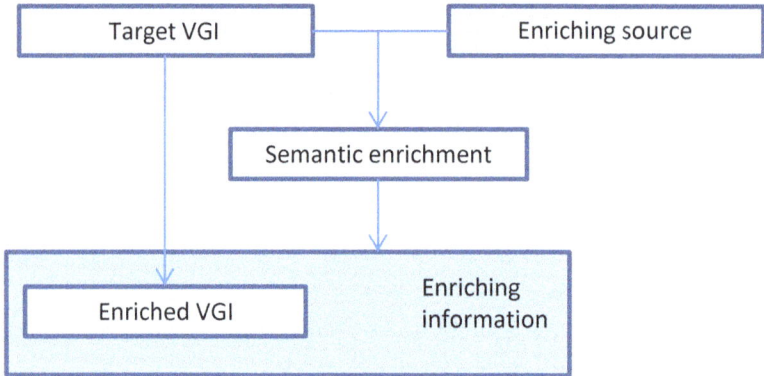

Figure 2: Semantic enrichment process.

The result of the enriching process obviously relies on the syntax and semantics of the target information and the enriching information source and the way in which the enrichment is performed (see Figure 2). We highlight three aspects which are crucial for a successful usage of the enriching process: 1) The semantics of the enriching source, 2) The semantics of the enriching information and 3) The syntax of the enriching information.

Targets can be free text in which case grammar rules provide the enriching information source and semantic enrichment is done through natural language processing (NLP) (see for example Peñas & Hovy 2010). In some cases the enriching information is intrinsically held in the target itself, such as relationships between items in a photo databases, such as Flickr. In such cases co-occurrence and data mining methods can be used to make these knowledge explicit (Deng et al. 2009).

In other cases the enriching information source is constituted by other sources, such as

1) (for exhaust-like data): web resources, sensors, gazetteer information (see Graham & De Sabbata 2015), etc.
2) (for 'captured data'): shared 'guidelines', each capturer's skills and intention, tasks assignment between several capturers, and the capturers' abilities to work together.

The second category can be captured by context models and provenance (background on how the information was produced) as reported in (Abel et al. 2012). The enriching information appears itself in different forms, for example as an ontology (see Lacasta et al. 2012).

In geospatial applications the semantic aspects of space put an extra constraint on semantic enrichment. Ballatore et al. (2011) combine a semantically-rich and spatially-poor ontology (DBpedia) with a spatially-rich and semantically-poor VGI dataset (OpenStreetMap) to facilitate spatial knowledge discovery. As geo-information is so often constructed through multi-function workflows, provenance plays an important role in understanding the created geo-information. In addition, in VGI projects we think it is relevant to capture what people intended to do with the VGI at hand.

The result of the semantic enrichment can range from an ontology to more light-weight schema elements. Such enriching information can exist as separate entities relating to the target information or can be embedded within the target as metadata. As such they provide a more meaningful view on the target data and the basis for more meaningful queries, as described in the next section.

Towards semantic queries

In this section we show the different uses of semantic enrichment when querying a VGI source. For each structure level (structured, semi-structured, unstructured) we study what can be done with and without semantic enrichment: what are the problems and limitations with 'direct' queries on the VGI source and how different types of enrichment can help.

The main problem that arises when querying the structured part of VGI lies certainly in the differences in terms of quality and semantics that occur in the attribute values. E.g. a time value may be expressed in local time or in UTC, a length with different units of measurement, etc. These variations may ultimately render query results very imprecise or even meaningless.

The semantic enrichment of structured data may connect structural elements (table and attribute names) or data elements (attribute values) to semantic entities in some knowledge organization system.

The semantic enrichment of structural elements can be exploited by meta-level queries that help build correct queries. For instance: Find the tables and attributes that hold information about employment rates. This is particularly useful when the database schema is large and complex. Any type of knowledge organization system can be used for this purpose. In a geographical context, enrichment can be done for example by making geographic properties explicit, e.g. that a bridge is part of a road, 'built-up area' is an aggregate of building features, etc.

If the enrichment is done with a formal logical ontology, it becomes possible to express deductive queries (such as in the programming language Datalog) that can produce results not computable with standard SQL queries.

The semantic enrichment at the data level consists in associating attribute values to descriptors that make them meaningful (units, scale, accuracy, etc). These descriptions can then be used to augment the queries with selection criteria or transformation functions to produce higher quality results (e.g. select only those data that have a sufficient accuracy and a given unit of measurement).

Since VGI sources essentially link data to geographic entities, a natural enrichment consists in annotating the data elements to entities in some geographic knowledge source, such as Geonames or a geographic ontology. This will allow for semantic queries that combine geographic knowledge and other data.

Semi-structured VGI presents additional types of problems. Since the schema is generally not controlled, users can create multiple structures to represent the same real world phenomenon. For instance, the DBpedia database has at least five different properties to describe a person's birthplace, even though these names are obtained from supposedly structured 'infoboxes' of Wikipedia. This leads to complex queries in which all the possible property and value names must appear, e.g. {?x ex:placeOfBirth ?p} union {?x ex:birthplace ?p} union {?x birthPlace ?p} in a SPARQL query. The problem is of course worse with sources in which users regularly define new attribute names and values, as is the case in OpenStreetMap. In some situations it may become almost impossible to express consistent and complete queries.

For querying semi-structured VGI, the role of the semantic enrichment is essentially to describe and unify (or differentiate) the multiple naming schemes produced by the users. If the names used in the VGI source are associated to corresponding entities in an ontology, then the ontology's vocabulary can be used to express 'unified' queries that can be automatically rewritten into the VGI's vocabulary to produce a (complex) query on the VGI source.

In the case of OpenStreetMap, many tags may designate the same concept and a single tag may designate different concepts in different contexts. For instance, the semantic query roadType *motorway* will return features (roads) tagged with roadCategory motorway, roadCategory highway, roadCategory turnpike, category, turnpike, type motorway, etc. And in case the formal language models more relationship types, it will also indicate related features such as bridges, traffic lights, etc.

The abstraction level of the ontology used for the enrichment will determine the level of (semantic) detail of the queries. A high level ontology will unify many different names of the VGI into a single high level concept, while more precise (domain specific) ontologies will enable queries that are closer to the VGI's level of granularity. A similar remark applies to the geographic axis. The geographic precision of the results will depend on the scale and precision the geographic ontology that is used to enrich the VGI source. Moreover, if the enrichment structure possesses a rich semantic structure, with subclass relations or more sophisticated axioms, it will support a more expressive query rewriting. For instance, a query about Artists could be rewritten as a query about its subclasses Painters, Sculptors, Musicians, etc. if the Artist concept is not directly represented in the VGI source. In more than one case one should combine several ontologies to support the queries (see for example Lemmens & Kessler 2014).

In semantic enrichment a basic effort consists of associating texts with the concepts they deal with, this is generally accomplished with techniques such as word sense disambiguation and named entity recognition (in particular geographic entity recognition). In this effort, one has to implement methods that are able to deal with the vagueness of information. With this kind of enrichment, semantic queries can answer questions such as 'find the data elements (texts) about the concept X'. Higher levels of enrichment consist in extracting precise information (facts) from texts. This amounts to transform unstructured sources into (semi-)structured ones, which is still an open research challenge.

Conclusions and recommendations

The semi-structured nature of VGI causes without doubt problems in the querying of its contents. We have presented several ways of imposing structure in order to facilitate more meaningful queries. Even the most basic queries, which go beyond text search, rely on some kind of structure. Whether the right degree of structure can be created depends on the success of the semantic enrichment process. In case of VGI, there are a variety of options, for which some of them rely on the reference to geodatabases. Semantic enrichment is basically constituted by linking the VGI to ontological concepts and their relationships.

We believe that some of the enriching information sources need curation as they are often ambiguous themselves. The level of enrichment needed, for which the semantic reference space is positioned between folksonomy and formal ontology, depends on the type of queries and needs to be further investigated with practical use cases.

References

Abel, F., Hauff, C., Houben, G., Tao, K., & Stronkman, R. 2012. Twitcident: fighting fire with information from social web streams. In: WWW 2012 Companion, pp. 305–308. DOI: http://doi.org/10.1145/2187980.2188035

Abiteboul, S. 1997. Querying Semi-Structured Data. In: *Proceedings of the 6th International Conference on Database Theory*. London, UK, Springer-Verlag, pp. 1–18. Retrieved from: http://dl.acm.org/citation.cfm?id=645502.656103.

Ballatore, A., & Bertolotto, M. 2011. Semantically enriching VGI in support of implicit feedback analysis. In: *Lecture Notes in Computer Science* (including Subseries Lecture Notes in Artificial Intelligence and Lecture Notes in Bioinformatics). 6574 LNCS, pp. 78–93. DOI: http://doi.org/10.1007/978-3-642-19173-2_8

Brando, C., & Bucher, B. 2010. Quality in User Generated Spatial Content. In: *A matter of specifications, 13th International Conference on Geographic Information Science (AGILE'10)*, 10–14 May, Guimarães, Portugal.

Deng, D.-P., Chuang, T.-R., & Lemmens, R. 2009. Conceptualization of place via spatial clustering and co-occurrence analysis. In: *Proceedings of the 2009*

International Workshop on Location Based Social Networks, pp. 49–56. DOI: http://doi.org/10.1145/1629890.1629902

Graham, M., & De Sabbata, S. 2015. Mapping Information Wealth and Poverty: The Geography of Gazetteers. *Environment and Planning A*: Forthcoming. Available at SSRN: http://ssrn.com/abstract=2587746.

Jiang, B., & Yin, J. 2014. Ht-index for quantifying the fractal or scaling structure of geographic features. *Annals of the Association of American Geographers, 104*(3): pp. 530–541.

Kessler, C., Janowicz, K., & Bishr, M. 2009. An Agenda for the Next Generation Gazetteer: Geographic Information Contribution and Retrieval. In: *Proceedings of the 17th ACM SIGSPATIAL International Conference on Advances in Geographic Information Systems*. Seattle, WA: ACM. Retrieved from: http://dl.acm.org/citation.cfm?id=1653771.

Kitchin, R. 2014. *The data revolution: Big data, open data, data infrastructures and their consequences*. Sage, London.

Lacasta, J., Nogueras-Iso, J., Falquet, G., Teller, J., & Zarazaga-Soria, F. J. 2013. Design and evaluation of a semantic enrichment process for bibliographic databases. *Data and Knowledge Engineering, 88*: 94–107. DOI: http://doi.org/10.1016/j.datak.2013.10.001

Lemmens, R., & Kessler, C. 2014. Geo-Information Visualizations of Linked Data. In: Huerta, Schade, & Granell (Eds.) *Connecting a Digital Europe through Location and Place. Proceedings of the AGILE'2014 International Conference on Geographic Information Science*. Castellón, June, 3–6, 2014. ISBN: 978-90-816960-4-3.

Ma, D., Sandberg, M., & Jiang ,B. 2005. Characterizing the heterogeneity of the OpenStreetMap data and community. *ISPRS International Journal of Geo-Information, 4*(2): 535–550.

Peñas, A., & Hovy, E. 2010. Semantic enrichment of text with background knowledge. In: *Proceedings of the NAACL HLT 2010 First International Workshop on Formalisms and Methodology for Learning by Reading*, (June), 15–23. Retrieved from: http://dl.acm.org/citation.cfm?id=1866775.1866778.

Smart, P. D., Jones, C. B., & Twaroch, F. A. 2010. Multi-Source Toponym Data Integration and Mediation for a Meta-Gazetteer Service. In: *Proceedings of GIScience 2010*. Zurich, Switzerland. Lecture Notes In Computer Science 6292, pp. 234–248.

Touya, G., & Brando, C. 2013. Detecting Level-of-Detail Inconsistencies in Volunteered Geographic Information Data Sets. *Cartographica: The International Journal for Geographic Information and Geovisualization, v48*(2): 134–143.

Zipf, G. K. 1949. *Human Behavior and the Principles of Least Effort*. Addison Wesley: Cambridge, MA.

CHAPTER 15

Extracting Location Information from Crowd-sourced Social Network Data

Pinar Karagoz*, Halit Oguztuzun, Ruket Cakici,
Ozer Ozdikis, Kezban Dilek Onal and Meryem Sagcan

Middle East Technical University, Computer Eng. Dept., Ankara, Turkey,
*karagoz@ceng.metu.edu.tr

Abstract

With millions of users worldwide, crowd-sourced social media data provide a valuable data source for events happening around the world. More specifically, microblogs, which are social networks that enforce short text messages, have a high popularity due to their availability as a mobile application and the practicality of short messages. Estimating the location of the events detected by following posts in microblogs have been the motivation of numerous recent studies. Extracting the location information and estimating the event location is a challenging task to maintain satisfactory situation awareness, especially for emergency cases such as fire or traffic accidents. Today, Twitter is among the most popular microblogging platforms, and there are recent research efforts aimed at detection of novel events online by following the Tweets. In order to analyze events, researchers generally focus on spatio-temporal features of the posts. Temporal features denote the time and ordering of posts, whereas spatial features are useful for location extraction or estimation. In this work, we present an overview on the process for toponym recognition and location estimation from microblogs.

How to cite this book chapter:
Karagoz, P, Oguztuzun, H, Cakici, R, Ozdikis, O, Onal, K D and Sagcan, M. 2016.
    Extracting Location Information from Crowd-sourced Social Network Data. In:
    Capineri, C, Haklay, M, Huang, H, Antoniou, V, Kettunen, J, Ostermann, F and
    Purves, R (eds.) *European Handbook of Crowdsourced Geographic Information*,
    pp. 195–204. London: Ubiquity Press. DOI: http://dx.doi.org/10.5334/bax.o.
    License: CC-BY 4.0.

Keywords

Location Estimation, Event Detection, Semantic Vector Expansion, Named-Entity Recognition (NER), Toponym Recognition

Introduction

Social media platforms constitute a rich and up-to-date information resource with their rapidly increasing number of users. With its popular features, including the text message limit and easy access on mobile environments, microblogging platforms in particular are intensively used by a large number of users. Among such platforms, Twitter appears as the most popular service with a high number of users. Since most of the users allow public access to posts and user profile information, it provides rich data for research areas including text analysis, text mining, information extraction or social analysis.

Event extraction and location estimation from crowd-sourced data in social networks are important for learning about the events happening. These techniques have applications in various domains involving being aware of what is happening in the vicinity and rapid access to emergency information. As a potential use case, for instance, on the basis of the Twitter posts of users witnessing a major accident in which a chain of cars are involved, it may be possible to estimate the event location, open alternative routes and provide first aid. Similarly, for management of disasters such as earthquakes or floods, accessing such information fast is valuable.

An important task to be fulfilled for location estimation is extracting evidence about location, especially recognizing location names, i.e. toponyms, within a social network message. This problem is a specific sub-problem under named entity recognition (NER) task, which is a well-known natural language processing (NLP) problem aimed at extracting certain types of entities including people, organizations, dates, locations etc. from text. NER techniques proposed in the literature mostly work on long and formal texts, such as newspaper articles. In long and formal texts, literature provides established NER solutions. On the other hand, short and informal texts obtained in microblogs pose important challenges. Short text limits the contextual information that can be obtained from the whole text, whereas informal language makes it difficult to recognize the words. In this paper, we particularly concentrate on toponym recognition in microblog messages under these challenges.

Once toponyms are recognized in microblog posts, they provide useful and rich input for location estimation tasks. However, this step includes several challenges as well. One important challenge is that, in addition to recognized toponyms, there may be other clues to the location of the event, such as the GPS annotation of the message provided by the mobile device. It is necessary to make use of these clues in a complementary manner. Another basic challenge is the existence of several contradictory toponyms or clues, such as having location names that point to different coordinates. It is necessary to devise a mechanism in order to weigh how well conflicting clues contribute to the location estimation.

It is important to note that, although the proposed techniques in the literature are generally demonstrated on Twitter posts, i.e. Tweets, they can be applied on other social media data, especially those having informal use of language.

In this work, we describe these steps in an overview. The organization of this paper is as follows. In the next section, toponym recognition in informal text is described. In Section 3, we describe the location estimation problem and the suggested solutions in the literature, and conclude the paper with a summary and overview in Section 4.

Toponym Recognition

As *toponym*, referring to location names, is a type of named entity, for toponym recognition, generally, techniques for Named Entity Recognition (NER) are used. However, the proposed solutions conventionally work on formal texts. Recently, with the increasing amount of potentially rich data from social networks, NER and toponym recognition in web resources has attracted attention. However, solutions on formal texts rely on very basic features such as capitalization of the first letter of a token, existence of an apostrophe character within the token or existence of the token in the gazetteer. Informal and non-standardized language use in social networks poses a challenge in applying the same techniques in this area. Therefore new NER and toponym recognition techniques are being proposed.

In the literature of toponym recognition, the proposed techniques can be categorized as:

• gazetteer-based,
• rule-based, and
• machine learning based approaches.

The first step in all these techniques is tokenizing the messages into words. In addition, *morphological analysis* and Part-of-Speech (POS) tagging may be employed as preprocessing steps if morphemes/suffixes and POS tags are utilized in the toponym recognition process. Morphological analysis enables decomposition of a word into its affixes and the stem. POS tag of a word and suffixes in a word are effective features used in NER systems (StanfordNLP), (Seker 2012). For example, for the word '*happily*', a morphological analyzer for English will show that the stem of the word is '*happy*' and having the suffix '*-ly*'. An English POS tagger will annotate this word as an *adverb*. For stemming and POS tagging, language specific morphological analysis tools are needed. For English, one of the well-known tools is the NLP library of Stanford NLP Group

(StanfordNLP). For example, for Turkish texts, the morphological analysis tool Zemberek (Zemberek 2015) is commonly used.

Another important preprocessing step is *normalization*. Due to the informal use of language, text may include spelling errors or unusual abbreviations. In normalization, such problems are fixed before applying the toponym recognition technique. In Turkish, one of the available tools for normalization is being developed by ITU NLP group (Eryigit 2014). This tool fixes some of the spelling errors and performs capitalization for some proper nouns. Current solutions for normalization mostly include rule-based corrections capturing previously known informal language patterns, such as repetition of characters for emphasizing emotion. For example, *'Soooooo coooooooool !!!!'* is such a message that can be frequently used in microblogs. Normalization process should be able to convert the words to *'So'* and *'cool'*.

Gazetteer-based Approach

In this approach, a predefined list of location names is used as the gazetteer. The recognition process basically relies on checking whether a given token is in the gazetteer. The content and the granularity of the list depend on the context of the toponym recognition application. For general-purpose solutions, the list may contain country, city, town or Point of Interest (POI) names. For a specific geographical region, this list may be more detailed, including hospitals, banks, pharmacies etc. depending on the context of the application. For general-purpose gazetteers, OpenStreetMap (OpenStreetMap 2015) and Wikipedia (Wikipedia 2015) are commonly used resources.

In this approach, toponym recognition is based on checking whether text includes any toponym from the gazetteer. To this end, very simply, each token in the text is looked up in the gazetteer. The ones that are found are marked as toponyms. For example, for the post *'Enjoying good weather in Bebek Park, in Istanbul'*, with a general-purpose gazetteer that includes POIs as well as city names, it is possible to recognize the toponyms *Bebek Park* and *Istanbul*. On the other hand, with a more limited gazetteer of city names, only *Istanbul* will be recognized. In this approach, normalization is more crucial, since it is based on matching between the token at hand and the toponym in the gazetteer.

Rule-based Approach

In this approach, certain patterns are defined in the form of the rules in order to recognize toponyms. For instance, if the word *street* follows a token, the token is considered to be a toponym referring to a street name (such as *'Oak Street'*). Some other patterns rely on the morphology or the POS tag of the word. One basic pattern for English is that if a token is preceded by a pronoun, it is likely to be a toponym (such as *'in Istanbul'*). However, such rules overestimate the toponyms, leading to false positive recognition for the phrases such as 'on the table'. Therefore, they may have high recall performance, however precision performance, generally, is not high. A recent study conducted on Turkish Tweets has shown that POS tag/morphology based rules achieved 30% precision, whereas with the gazetteer-based approach 67% precision is obtained (Onal 2014).

Machine Learning based Approach

In the machine learning based approach, supervised learning is the most common sub-approach employed for toponym recognition. Given a set of texts in which toponyms are annotated, a toponym model is constructed. More specifically, Conditional Random Field (CRF) is the most commonly used supervised learning technique providing satisfactory recognition ratios for formal texts. The advantage of CRF is that it is possible to capture contextual information in the model, such as having words in the neighborhood of a toponym. For Turkish texts, in Seker (2012), a CRF-based NER solution is presented. However, the success of this approach on informal texts heavily relies on the normalization step.

In a more recent study (Sagcan 2014), a CRF-based solution that focuses on toponym recognition in microblogging messages is proposed. The main motivation behind the work is performing toponym recognition without using any gazetteer and with less preprocessing effort for normalization. The main architecture of this approach is given in Figure 1.

As seen in the figure, the architecture is composed of two parts. In the first part, data preparation is performed. The second part contains CRF-based learning modules.

In the data preparation phase, initially, conventional tokenization and morphological analysis operations are applied. Afterwards, two simple normalization steps are applied. In the first normalization step, repeated characters are eliminated. For instance, 'çooooook güzeel' (çok güzel (Turkish) = very good (English)), 'gooooool' (gol (Turkish) = goal (English)) are commonly seen misspellings in Turkish Tweets. Such character repetitions occur in all languages in Twitter to denote an emphasis or exaggeration of emotion. The second normalization step is for the cases in which the English alphabet is used instead of the Turkish alphabet characters. Most of the misspellings in Turkish Tweets originate from replacement of Turkish characters with diacritics (ç, g, ı, ö, ş, ü) with non-accentuated characters (c, g, i, o, s, u). Such usage is also very common in other non-English microblogging posts. This step can be customized according to the language alphabet under interest. In the literature, the proposed normalization step involves intensive cleaning and pre-processing on the text. In Sagcan (2014), only two simple normalization steps are applied and the other informal language problems are expected to be resolved during the learning phase.

In the second part of the architecture, for CRF-based learning, an annotated training data set is prepared in order to be used for model construction. In this

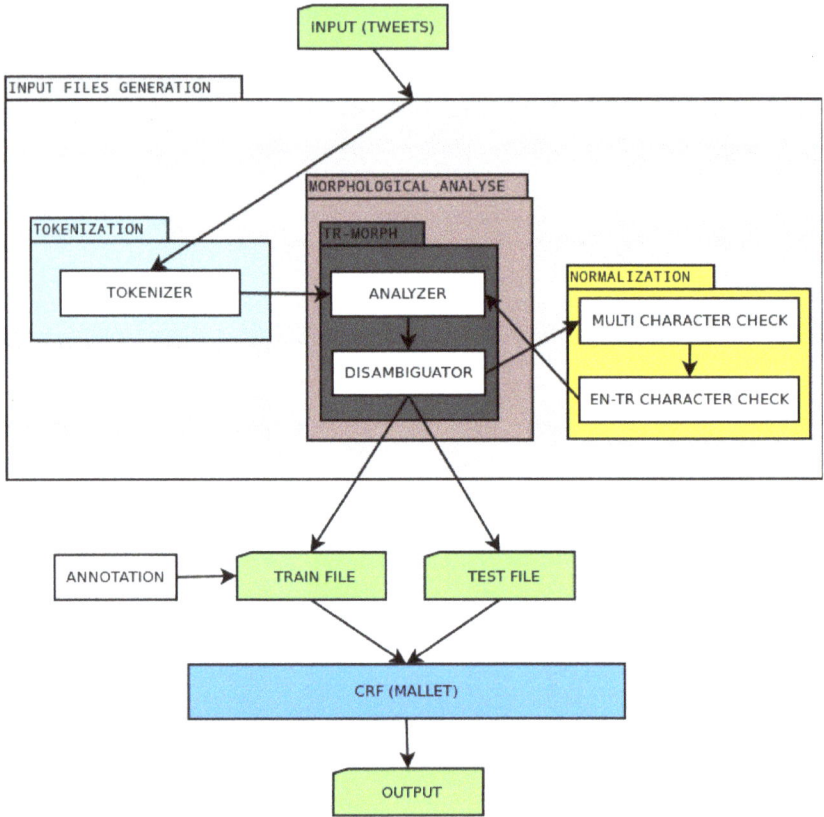

Figure 1: The architecture for CRF-based Toponym Recognition (Sagcan 2014).

step, the important issue is which attributes to use in the term vector representing the text message. Within the study, toponym recognition performance under several attribute compositions is analyzed. Under the best attribute composition, the precision performance is reported as 88%.

Location Estimation

Studies on location estimation are commonly related with event detection efforts. Recently, numerous studies have been conducted to detect real-world events by collecting public Tweets in Twitter. These include methods to identify various types of events including earthquakes (Sakaki 2010, 2013), disasters and crises (Yin 2012), and accidents (Rui 2012). The location estimation techniques proposed in these studies use different data and clues extracted from Tweets. In this section, we revise the most basic location estimation techniques according to the data used.

Geo-Annotation based Approach

In most of these event detection studies, GPS metadata of geo-tagged Tweets are the main resources for estimating the locations of events. A very simple procedure is followed. Once the location is estimated, location information about the detected events is presented on the map by plotting the geo-tagged Tweets about the event.

User Profile based Approach

In several other studies, the textual content in the location attribute of user profiles has been utilized in the place of GPS data for non-geo-tagged Tweets in, for example, Sakaki (2010, 2013) and Yin (2012). However, an in-depth study of user profiles show that since this attribute is a free-text field limited to 30 characters, it may contain multiple location names or even non-existing locations. According to the observation of Rui (2012), only 12% of users specified a location in their profile. Hence the authors tried to predict the user locations by analyzing their previous Tweets and locations of their friends.

Toponym based Approach

Due to the scarcity of geo-tagged Tweets and location information in user pro-files, in Middleton et al. (2014), researchers focused on analyzing the Tweet content for location references by implementing a geo-parser. Location estimation studies concentrating on Tweet content make use of the toponym recognition techniques as discussed in Section 2. For example, in Sankaranarayanan et al. (2009), the authors used a gazetteer for toponym recognition and described a heuristic for toponym resolution in Tweets. Within the scope of event detection, some of the studies concentrate on estimating the location by using the whole cluster of Tweets that correspond to an event. The most frequently mentioned toponym in Tweet contents and user profiles is designated as the event location.

Evidence Combination based Approach

In Ozdikis (2013) a Bayesian approach is employed that uses evidences from several resources for location estimation. To this aim, three resources are used:

- Toponyms within the Tweet text
- GPS annotations on Tweets
- Toponyms in user profiles

As a basic difference from previous studies, for location estimation, Dempster-Shafer Theory (DST) (Dempster 1967), which can combine evidence from various resources in a single model, is used. DST is a generalization of Bayesian inference technique. DST is a technique under cases in which evidences are limited or they provide conflicting data. In addition, estimation result can be given in a belief interval such as [X%, Y%] instead of a single probability value.

In social networks, the number of postings sent from populated regions may distort the result, since the average number of messages posted from such locations are already higher than the other places and this leads to a bias towards such places. Hence, in Ozdikis et al. (2013) a normalization operation is devised and applied on evidence probabilities to prevent this bias.

As the evidence from the toponyms within the Tweet texts, the authors initially employed a gazetteer-based approach, however, other toponym recognition techniques such as machine learning-based approaches can be incorporated into the system.

In Ozdikis (2013), experiments are conducted on a set of Tweets posted about minor two earthquakes in Turkey, which occurred on May 17, 2013 near the city of Mugla, and on July 30, 2013 on an island near the city of Canakkale. They were not strong quakes, and did not cause any damage, but they were felt by people in these regions and triggered a reaction in Twitter traffic. For the first event, the belief interval found for Mugla is [0.77, 0.97], which can be considered as a confident decision. For the second event, the highest belief interval is found for Canakkale, as [0.24, 0.42]. The next highest belief intervals are [0.16, 0.34] for Edirne and [0.14, 0.31] for Tekirdag. Although Istanbul is a metropole from which many messages are posted, the belief interval as the location of the event is calculated only as [0.03, 0.21]. Hence, the results show the applicability of the method for location estimation.

Conclusion

Social networks, especially microblogs, contain valuable data including geographical footprints of users. A message becomes geo-annotated in terms of latitude and longitude if the user posts it using a GPS-enabled device and allows the sharing of this geographic information. In addition, users tell their location in their social network profile. Moreover, users may talk about places in their messages. Such attributes of Tweets are valuable resources for spatial analysis. There are various studies on location estimation that use these attributes. Most of them rely on a single attribute. However, techniques that combine several attributes in a single model for location estimation have more potential for accurate estimation.

Each of these attributes cause different challenges. In geo-annotated messages, GPS coordinates provide precise geographic position in terms of latitude and longitude. However, the number of geo-tagged messages is very limited. In addition, GPS coordinate and the location of the event mentioned in the Tweet may not be the same.

Similar issues appear for location attribute of user profiles. The number of profiles including valid location information is very limited. In addition, having valid location information in the profile does not guarantee that the user is actually at that location at the time of message posting, or the user is mentioning that location in the Tweet.

Challenges with the Tweet content and location attributes of user profiles mostly concern text processing and toponym recognition. The quality of content is not as good as that in news articles due to misspellings, extraordinary writing conventions and abbreviations. Therefore, state-of-the-art NLP parsers do not perform as accurately on informal social media message texts. The current efforts mostly rely on applying a normalization process before toponym recognition. The research trend is towards minimizing the normalization effort and proving a more general adaptive approach.

Toponym recognition and location estimation on social media research in the literature are mostly applied on Twitter posts. This is due to the fact that Twitter has a high number of users and most of the data in Twitter is publicly available through its application-programming interface. However the described techniques are also applicable to other crowd-sourced social media.

As a future work, several other research dimensions can be pursed for toponym recognition and location estimation. For toponym recognition, various other features can be analyzed. Hashtags appear to be valuable resources. In addition, links or photographs attached in the messages can be further investigated. Meta features or tags of the pictures and videos, too, may provide toponyms. For location estimation, in addition to combination of features from a single data source, evidence from several complementary data resources, such as Twitter and Foursquare, may be combined for increasing the precision of the location prediction.

References

Dempster, A. 1967. *Upper and Lower Probabilities Induced by a Multi-valued Mapping. Annals of Mathematical Statistics, 38*: 325–339.

Eryigit, G. 2014. ITU Turkish NLP web service. In: *Proceedings of the Demonstrations at the 14th Conference of the European Chapter of the Association for Computational Linguistics (EACL).* Gothenburg, Sweden: Association for Computational Linguistics, April 2014.

Middleton, S. E., Middleton, L., & Modafferi, S. 2014. Real-Time Cri- sis Mapping of Natural Disasters Using Social Media. *Intelligent Systems, IEEE, 29*(2): 9–17.

Onal, K. D., Karagoz, P., & Cakici, R. 2014 (April). Turkce Twitter Gonderilerinde Lokasyon Tanima (Toponym Recognition on Turkish Tweets), SIU 2014, Trabzon, Turkey, pp. 1758–1761.

OpenStreetMap. Avaialble at: https://www.openstreetmap.org (Last accessed 15.2.2015).

Ozdikis, O., Oguztuzun, H., & Karagoz, P. 2013 (November). Evidential Location Estimation for Events Detected in Twitter. In: *Proceedings of ACM SIG Spatial/GIS Workshop on Geographic Information Retrieval (GIR)*. Orlando, USA, pp. 9–16.

Rui, L., Lei, K. H., Khadiwala, R., & Chang, K. C. C. 2012. TEDAS: A Twitter-based Event Detection and Analysis System." In: *Proc. 28th Int'l Conference on Data Engineering (ICDE '12)*, pp. 1273–1276.

Sakaki, T., Okazaki, M., & Matsuo, Y. 2010. Earthquake Shakes Twitter Users: Real-time Event Detection by Social Sensors. In Proc. of the 19th Int'l Conference on World Wide Web (WWW '10), pp. 851-860, 2010.

Sakaki, T., Okazaki, M., & Matsuo, Y. 2013. Tweet Analysis for Real-Time Event Detection and Earthquake Reporting System Development. *IEEE Transactions on Knowledge and Data Engineering, 25*(4): 919–931.

Sankaranarayanan, J., Samet, H., Teitler, B. E., Lieberman, M. D., & Sperling, J. 2009. TwitterStand: News in Tweets. In: *Proceedings of the 17th ACM SIG-SPATIAL International Conference on Advances in Geographic Information Systems*, November 04–06, 2009. Seattle, Washington.

Seker, G. A., & Eryigit, G. 2012. Initial explorations on using CRFs for Turkish named entity recognition. *COLING*: 2459–2474.

Sagcan, M. A. 2014. Hybrid Method for Toponym Recognition on Informal Turkish Texts (M.Sc. Thesis), METU Computer Eng. Dept.

Stanford NLP Library. 2015. Available at: http://nlp.stanford.edu/software/index.shtml.

Wikipedia. Available at: http://www.wikipedia.org (Last accessed 15.2.2015).

Yin, J., Lampert, A., Cameron, M., Robinson, B., & Power, R. 2012. Using Social Media to Enhance Emergency Situation Awareness. *Intelligent Systems, IEEE, 27*(6): 52–59.

Zemberek. Available at: https://github.com/ahmetaa/zemberek-nlp-distributions (Last accessed 15.2.2015).

CHAPTER 16

Spatial and Temporal Sentiment Analysis of Twitter data

Zhiwen Song and Jianhong (Cecilia) Xia*

Department of Spatial Sciences, Curtin University

*c.xia@curtin.edu.au

Abstract

The public have used Twitter world wide for expressing opinions. This study focuses on spatio-temporal variation of georeferenced Tweets' sentiment polarity, with a view to understanding how opinions evolve on Twitter over space and time and across communities of users. More specifically, the question this study tested is whether sentiment polarity on Twitter exhibits specific time–location patterns. The aim of the study is to investigate the spatial and temporal distribution of georeferenced Twitter sentiment polarity within the area of 1 km buffer around the Curtin Bentley campus boundary in Perth, Western Australia. Tweets posted in campus were assigned into six spatial zones and four time zones. A sentiment analysis was then conducted for each zone using the sentiment analyser tool in the Starlight Visual Information System software. The Feature Manipulation Engine was employed to convert non-spatial files into spatial and temporal feature class. The spatial and temporal distribution of Twitter sentiment polarity patterns over space and time was mapped using Geographic Information Systems (GIS). Some interesting results were identified. For example, the highest percentage of positive Tweets occurred in the social science area, while science and engineering and dormitory areas had the highest percentage of negative postings. The number of negative Tweets

How to cite this book chapter:

Song, Z and Xia, J. 2016. Spatial and Temporal Sentiment Analysis of Twitter data. In: Capineri, C, Haklay, M, Huang, H, Antoniou, V, Kettunen, J, Ostermann, F and Purves, R. (eds.) *European Handbook of Crowdsourced Geographic Information*, Pp. 205–221. London: Ubiquity Press. DOI: http://dx.doi.org/10.5334/bax.p. License: CC-BY 4.0.

increases in the library and science and engineering areas as the end of the semester approaches, reaching a peak around an exam period, while the percentage of negative Tweets drops at the end of the semester in the entertainment and sport and dormitory area. This study will provide some insights into understanding students and staff's sentiment variation on Twitter, which could be useful for university teaching and learning management.

Keywords

spatial and temporal analysis; sentiment analysis; Twitter; georeference

Introduction

Twitter as one of vital platforms for people to publically express their opinions and feelings about events and their private lives, has attracted enormous attention with millions of followers (Li et al. 2013). Numerous studies have been conducted on opinion mining and sentiment analysis on Twitter (Pang & Lee 2008; Poria et al. 2014; Taboada et al. 2011; Liu 2012). The sentiment classification methods have been developed from simple text mining to advanced symbol and feature recognition (Liu 2012), from a pure sentiment analysis to a sentiment and subjective analysis (Pang & Lee 2004), from machine learning or lexicon-based approaches to more advanced hybrid methods (Serrano-Guerrero et al. 2015) and from sentiment orientation with only two directions (e.g. positive and negative) coarse measurement scale to a fine grained classification (Fink et al. 2011). However, limited sentiment analysis research has been conducted from a spatial and temporal perspective. This study tested a research question of whether sentiment polarity on Twitter exhibits specific time–location patterns.

The aim of this paper is to investigate the spatial and temporal distribution of georeferenced Twitter sentiment polarity within Bentley campus, Curtin University in Perth Western Australia. The campus was divided into six zones – science and engineering buildings, social science buildings, library, lecture theatre, dormitory, entertainment and parking areas and four periods of time – beginning of the semester, middle of the semester, end of semester and after examination to investigate how Twitter sentiments vary across different zones and time periods. The Starlight Visual Information System[1] was used to conduct a sentiment analysis. The Feature Manipulation Engine (FME[2]) was employed to convert non-spatial files into spatial and temporal feature class.

---

[1] http://starlight.pnnl.gov.
[2] http://www.safe.com.

The sentiment polarity patterns across six spatial zones and four temporal zones were mapped using the ArcGIS[3] software.

The paper is structured as follows. Section 2 presents the research context including a review of the relevant research literature associated with methods for a Twitter sentiment analysis. The research method in Section 3 describes the overall approach taken in this research. Section 4 presents the results, and the paper concludes with a discussion of key findings and implications of our observations on different Twitter topics in Section 5.

Related work

Sentiment analysis is the Natural Language Processing work, which involves opinion detection and classification of attitudes in texts (Balahur et al. 2014). Numerous studies have been conducted for automatically detecting opinions and emotions. This section summarised these studies into two categories.

Sentiment classification trends

The early studies of sentiment classification are mostly based on text mining techniques. Opinion was classified into positive/negative or positive/negative/ neutral. It can be simple two, five or even eleven point scale depending on the complexity of a task (Taboada et al. 2011; Pang & Lee 2008; Pang & Lee 2004; Whitelaw et al. 2005). Human language tends to be subjective. The same sentence in different tones or contexts could in different emotional states. It creates a great challenge to identify the affective state or intended emotional communication (Sarvabhotla et al. 2011; Pang & Lee 2004; Wilson et al. 2005). Therefore, subjectivity analysis can go beyond simple category of positive, negative or neutral (Liu 2012). Some studies focus on detecting ironic and sarcastic content of texts. However, there is a huge debate of how to formally define irony and sarcasm, which add another dimension of subjectivity analysis (Reyes & Rosso 2012). Except extracting polarity of a given text, more fine-grained methods have been developed to detect emotion or opinions from symbols, such as emoticons (e.g. '☺', '☹') (Read 2005; Go et al. 2009) and visual features or images (Liu 2012). This progression of the studies has taken the sentiment analysis research to a new level.

Classification methods

In order to perform different sentiment classification tasks, various sentiment algorithms were developed (Medhat et al. 2014; Serrano-Guerrero et al. 2015).

---

[3] http://www.esri.com/software/arcgis.

Medhat et al. (2014) grouped the SA into two categories: machine learning and lexicon-based approaches (see Figure 1). Generally, machine-learning methods were used to automatically discover sentiment polarity pattern rules in large data in order to learn opinions or emotions of given texts or features. A variety of algorithms have been developed (Ye et al. 2009; Rushdi Saleh et al. 2011). Most algorithms fall into the category of supervised machine learning. For example, Rushdi Saleh et al. (2011) applied Support Vector Machines (SVM) to detect whether the opinion expressed in a document is positive or negative about a given topic using several weighting schemes. Ye et al. (2009) compared Naïve Bayes, SVM and the character based N-gram model for analysing the sentiment of travel blogs for seven popular travel destinations in US and Europe. The SVM and N-gram approach was found to outperform the Naïve Bayes approach with accuracies reaching to at least 80%. Balahur (2013) developed an unsupervised method especially for a Twitter data sentiment analysis using the SVM. The major contribution of the study is to employ methods in normalising Tweet language, including higher order n-grams to spot modifications in sentiment polarity articulated and selecting features using simple heuristics.

Lexicon-based approaches focus on measuring subjectivity and opinions in texts using Semantic orientation (SO) (Osgood et al. 1957), which capture orientations of opinions (positive or negative) and strengths or degrees of orientation (Taboada et al. 2011). Sentiment lexicons are the key for this type

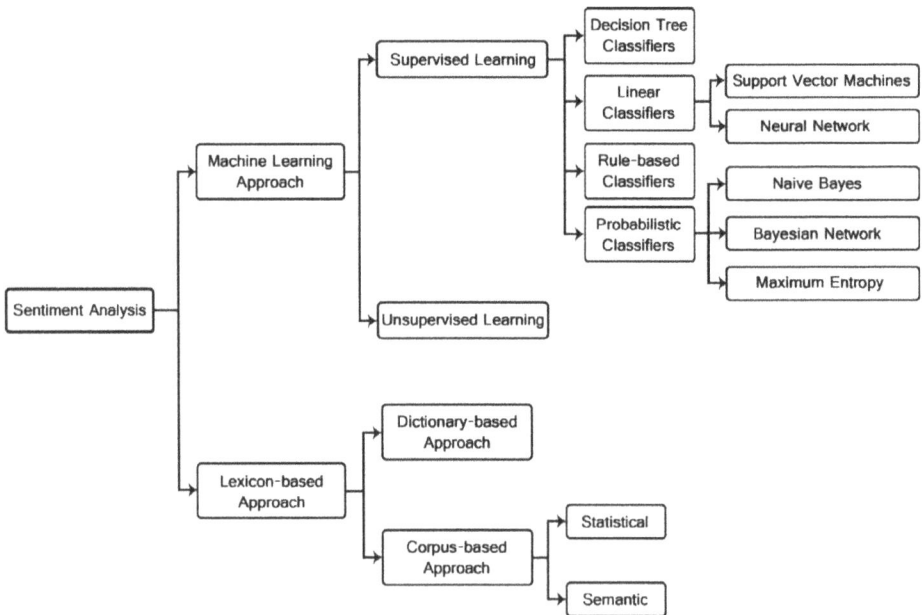

Figure 1: Sentiment classification techniques (Source: Medhat et al. 2014).

of methods. For example, Paltoglou and Thelwall (2012) proposed a lexicon-based approach to identify whether a text conveys negative or positive attitudes and to estimate the level of emotional intensity of a text in social media and microblogging environments. They added extensive linguistically function-alities (negation/capitalization detection, intensifier/diminisher detection and emoticon/exclamation detection) to the traditional classifiers such as Support Vector Machines (SVM), Maximum Entropy classifier and Naive Bayes clas-sifiers. Khan et al. (2014) utilised a hybrid system framework, which contains unsupervised learning algorithms and a dictionary-based method named Twitter Opinion Mining Framework (TOM). This method applied a variety of techniques for Twitter analysis and classification including a hybrid scheme of Enhanced Emoticon Classifier (EEC), SentiWordNet Classifier (SWNC) and an improved polarity classifier (IPC) using a list of positive/negative words. The findings reveal that the proposed algorithm resolved previous technical issues and increased the classification accuracy, effectively reduced the number of classified neutrals. Dacres et al. (2013) conducted a topic analysis and sen-timental analysis in understanding the contents of Tweets and trends posted over a 10-day period using machine learning and natural language processing techniques. The researchers examined and compared the commenly-used Data Science Toolkit's text2sentiment4, which is based on different methods, such as sentiment lexicon (Nielsen 2011), the lexicon-based but data-driven hybrid SentiStrength (Thelwall et al. 2012), and Charrerbox's Sentimental API (Purver & Battersby 2012). In Dacres et al. (2013) analysis, best result was achieved by the machine learning method (Charrerbox's Sentimental API) with 84% accu-racy. In our study, we have also adopted a machine learning method using Sub-space Transformation (TRUST) engine in Starlight for vector space modelling and supervised learning for a sentiment analysis.

Methods

Study area and data collection methods

Study Area

The developed content analysis methods were implemented using Tweets within the area of 1 km buffer around Curtin Bentley campus boundary (see Figure 2). Curtin University is one of the largest universities in Australia. It has more 60,000 students enrolled each year (OFFICE OF STRATEGY AND PLANNING 2014). The Bentley campus, as the main campus of Curtin University, is located about six km southeast from the Perth CBD. It covers 116 hectares with a variety of facilities, such as a library, lecture theatres, teaching rooms, cafés, dormitories and parking areas (Curtin University 2015).

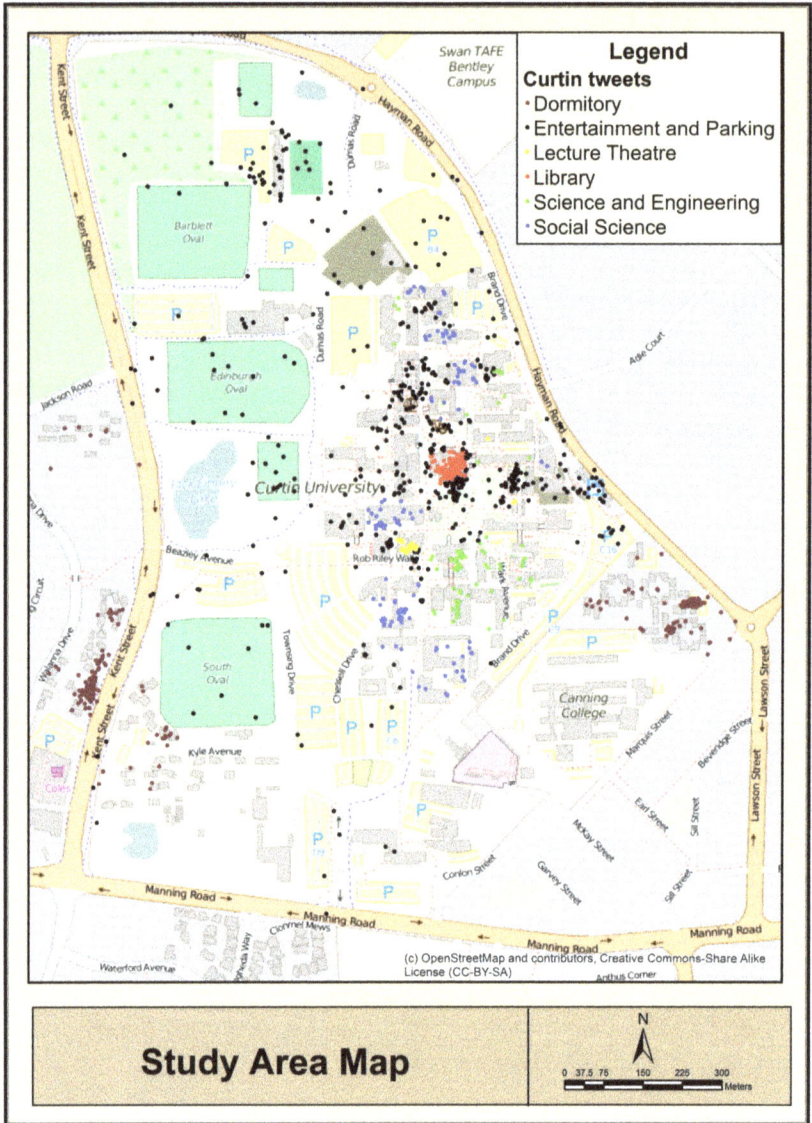

Figure 2: Curtin University Area Map (Map data © OpenStreetMap contributors).

Data Gathering And Pre-processing

Because the objective of this paper is to understand spatial and temporal patterns of georeferenced Tweets' emotional polarity, we only used geotagged Tweets posted between 12 May 2014 and 5 Jan 2015 within the area of 1 km buffer around Curtin Bentley campus boundary as JSON files via Twitter API and converted to XML format using the Ruby built-in JSON library. More than 5,000 Tweets were downloaded from Twitter API during the study period but around one third of them were located outside the Curtin University boundary. After removing the ones outside Curtin University boundary, there were a total of 3,172 Tweets gathered. These XML files of Tweets consist of both geo-location information and attribute information such as text of Tweets, time created at and geolocation, which are directly relevant to our research. Tweets of non-English languages were removed for this study in order to simplify the analysis, leaving 3,097 Tweets in the dataset. In our study, location associated with each Tweet is in the format of (x, y) coordinates, which were automatically captured using built-in Global Positioning System (GPS) receivers in mobile or tablet applications, such as a smart phone. This is the exact location where a user Tweeted. The accuracy of location recorded by GPS is usually around a few meters. However, if the location captured by triangulation outside a cellular network, the accuracy can range between 30-3000 m, subjecting to the cell distribution (Li et al., 2013). These geotagged Twitter data were imported into a geodatabase using GIS software, such as, FME and ArcGIS, as point features, Curtin Buildings and boundary geography files were extracted from Open Street Map[4] and converted into the same geodatabase using the ArcGIS software.

Sentiment analysis methods

There are three levels of sentimental analysis –1) document level sentiment classification; 2) sentence level sentiment classification; and 3) aspect level sentiment analysis (Liu 2012). This research chose the sentence level sentiment classification. The sentence level sentiment classification is suitable because of its assumption that each sentence contains only one entity or one aspect of entity in many cases (Liu 2012). Each Tweet has a limitation of 140 characters in order to ensure that information posted on Twitter is straight forward to the theme. As a result, it is more suitable to perform a sentence level sentiment analysis on Tweets. This study carried out a sentiment analysis by using the Starlight's Sentiment Analyzer function in Starlight Data Engineer (SDE), which adopted the Boeing Text Representation using Subspace Transformation (TRUST) engine for vector space modelling and text summarisation (Simoff et al. 2008). The input for Sentiment Analyzer is XML files. Sentiment Analyzer analyses individual words of a Tweet and calculates a score of sentiment orientation. It returns statistics from the sentiment analysis, such as, sentimentTotal, sentimentDiff, sentimentScore, wordCount, sentimentNegative, sentimentPositive. In this study, sentimentDiff is the sentiment orientation of Tweets. SentimentDiff is the result of sentimentPositive subtracting sentiment-Negative (Liu 2012). If sentimentDiff is positive number, it means this text hold

---

[4] http://www.openstreetmap.org

a positive attitude. If sentimentDiff is negative, it means the text expresses a negative opinion. These simple statistics returned from the Sentiment Analyzer are defined as the followings:

- SentimentTotal: the sum of sentimentPositive and sentimentNegative.
- SentimentDiff: sentimentPositive subtract sentimentNegative.
- SentimentScore:sentimentDiff divided by WordCount.
- WordCount: the total number of words in the text.
- SentimentNegative: each negative term in the text represent by 1. SentimentNegative is the total number of negative words in the text. For example, text 'such a bad day! Everything is wrong'. 'Bad' and 'wrong' makes the sentimentNegative to be 2.
- SentimentPositive: each positive term in the text represents by 1. SentimentPositive is the total number of positive words in the text. For instance, sentence 'study hard and the review is quite efficient', 'hard' and 'efficient' makes the sentimentPositive score to be 2.

From the analysis above, the range of sentiment orientation scores were derived from this study from 5 to −6. The sentiment category is shown in Table 1. The scale we used does not consider severity of individual words, but their frequency.

Sentiment orientation scores ranging from 4 to 5 represent Tweets expressing very positive sentiment and sentiment orientation scores ranging from 1 to 3 mean Tweets holding positive sentiments. Sentiment orientation score 0 means Tweets do not express any opinions. Sentiment orientation scores from −4 to −6 represent Tweets holding very negative sentiment while sentiment orientation scores ranging from −1 to −3 mean negative attitudes.

Spatial and temporal comparison of Twitter sentiment polarity patterns

We divided Curtin Bentley University campus into six spatial zones and four time periods (see Table 2). The spatial distribution of georeferenced Tweets

Sentiment Category	Sentiment orientation scores
Very Positive	4–5
Positive	1–3
Natural	0
Negative	(−1)–(−3)
Very Negative	(−4)–(−6)

Table 1: Sentiment category.

Study Period Category	Time Span
Beginning of the semester	28 July 2014 – 7 September 2014
Middle of the semester	6 May 2014 – 18 May 2014 and 8 September 2014 – 19 October 2014
End of the semester	19 May 2014 – 27 June 2014 and 20 October 2014 – 28 November 2014
After examination	28 June 2014 – 27 July 2014 and 29 November 2014 – 5 January 2015

Table 2: Study period category.

among six zones was derived using ArcGIS software. In addition, the ArcGIS software was used to assign all Tweets with time span information and exported to an excel spread sheet for further analyses.

Research Hypothesis

- *Spatial difference*
  Tweets posted in different campus locations are various in sentiments. For example, Tweets posted in Entertainment and parking area may be more positive while Tweets posted in study areas may be more negative.
- *Temporal difference*
  Adnan et al. (2012) conducted research on student stress levels at the beginning and the end of the semester and found the stress level of university students varies throughout a semester. Generally speaking, students felt more stressful at the end of the semester than at the beginning of the semester. Based on this study, we proposed a hypothesis that more positive Tweets are posted at the beginning of semester and after examination compared with the ones posted at the middle of the semester and at the end of the semester.

Results

Overall distribution of the Tweets by sentiments

Three thousand and ninety-seven Tweets were loaded into the Starlight Data Engineer for a sentiment analysis and output Tweets were assigned sentiment polarity. Then output Tweets from Starlight were further processed in FME to be converted into feature classes. Table 3 illustrates the overall distribution of the Tweets by sentiments. Around 45% of Tweets contain neutral opinions, such as 'I'm at Curtin University'. Besides, the number of Tweets holding positive

Sentiments	Score range	Number of Tweets
Very positive	4–6	32
Positive	1–3	1,091
Neutral	0	1,383
Negative	(−1)–(−3)	577
Very Negative	(−4)–(−6)	14

Table 3: Overall distributions.

opinions is nearly as twice as Tweets of negative opinions. Fourteen Tweets contain very negative feelings while thirty-two Tweets have very positive opinions.

Spatial distribution of sentiment polarity patterns

As it is showed in Figure 3 and Table 4, most Tweets were posted in the entertainment and parking and dormitory area. Tweets posted in the science and engineering area do not have very negative opinions, while Tweets posted in lecture theatre and library do not contain very positive opinions. Tweets posted in social science areas have the highest percentage of positive Tweets (42.1%), while Tweets gathered in the library area have the largest percentage of negative opinions (21.8% of total Tweets in the area). About 20.7% of total Tweets posted in science and engineering area hold negative opinions, while 18% of Tweets in social science are negative. Therefore the descriptive data potentially indicates that students or staff in the science and engineering areas could feel slightly more negative compared with students or staff in the social science areas.

Temporal distribution of sentiment polarity patterns

Sentiment orientation at different study time zones is showed in Table 5. It is interesting to note that the largest percentage of negative feeling actually occur at the beginning of semester, which is 21.6%. This is different from our hypothesis. The percentage of negative sentiments decreases from the beginning of the semester zone to after exam zone, reaching to smallest number of negative opinions (0.08) after examination. The percentage of very negative Tweets is roughly the same throughout all four study zones. After examination time zone holds the largest percentage of positive Tweets, which is align with the research hypothesis.

The library area contains a large cluster of Tweets (See Figure 4), which shows an interesting temporal pattern. At the beginning of the semester, still more positive Tweets occurred than negative ones in the library area. However, as a semester goes, more negative Tweets were posted in the library area. The percentage of positive Tweets in each temporal zone decreased over time gradually. We also summarised the number of negative opinions on Twitter across five

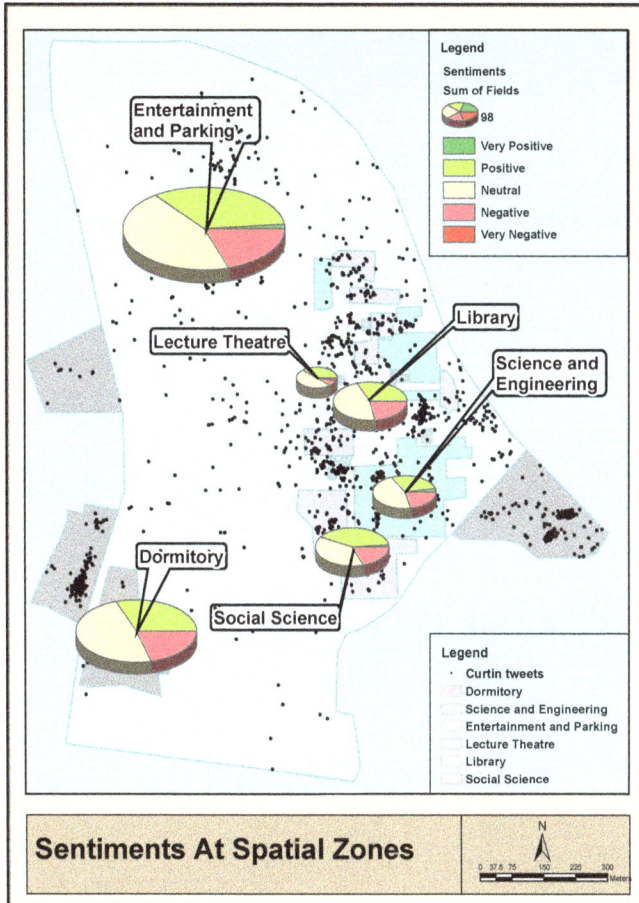

Figure 3: Sentiments at spatial zones.

Spatial zones	Very Positive	Positive	Neutral	Negative	Very Negative	Total
Science and Engineering	5 (2.7%)	59 (31.4%)	85 (45.2%)	39 (20.7%)	0 (0%)	188
Social Science	3 (1.1%)	110 (42.1%)	99 (37.9)	47 (18%)	2 (0.8%)	261
Library	0 (0%)	85 (32.6%)	118 (45.2%)	57 (21.8%)	1 (0.4%)	261
Lecture Theatre	0 (0%)	22 (40.0%)	27 (49.1%)	4 (7.3%)	2 (3.6%)	55
Dormitory	5 (1.6%)	260 (32.1%)	387 (47.8%)	156 (19.3%)	2 (0.2%)	810
Entertainment and Parking	19 (1.2%)	555 (36.4%)	667 (43.8%)	275 (18.1%)	7 (0.5%)	1523

Table 4: Sentiments for spatial zones.
*The percentage of very positive Tweets in the Science and Engineering area.

TimeZones	Very Positive	Positive	Neutral	Negative	Very Negative	Total
Beginning of semester	12 (1.5%)	275 (35.6%)	316 (40.9%)	167 (21.6%)	3 (0.4%)	773
Middle of semester	11 (0.9%)	407 (33.6%)	542 (44.8%)	244 (20.2%)	6 (0.5%)	1210
End of semester	7 (0.8%)	334 (36.0%)	431 (46.5%)	151 (16.3%)	4 (0.4%)	927
After exam	2 (1.1%)	76 (40.4%)	94 (50.0%)	15 (8.0%)	1 (0.5%)	188

Table 5: Sentiment polarity of temporal zones.

Time Zones	Very Positive	Positive	Neutral	Negative	Very Negative	Total
Beginning of the semester	0	12 (35%*)	15 (44%)	7 (21%)	0	34
Middle of the semester	0	34 (33%)	47 (46%)	22 (21%)	0	103
End of the semester	0	38 (31%)	56 (46%)	28 (23%)	1 (1%)	123
After examination	0	1 (100%)	0	0	0	1

Table 6: Sentiment polarity of temporal zones in the library area.
*The percentage of positive Tweets in the beginning of semester

Time Zones	Dormitory	Science and Engineering	Entertainment and Parking	Lecture Theatre	Library	Social Science	Total
Beginning the of semester	49	11	86	1	7	13	167
Middle of the semester	72	13	116	3	22	19	245
End of the semester	31	14	63	0	28	15	151
After examination	4	1	10	0	0	0	15

Table 7: The number of negative opinions on Twitter over space and time.

spatial zones and four temporal zones (see Table 7). Interestingly, except the science and engineering area, the number of negative opinions dropped at the end of semester in the other three areas. Certain temporal patterns of Tweeter polarity only occurred at specific locations. This may indicate that Twitter sentiments are time-location specific and they might depend on the activities conducted at certain locations.

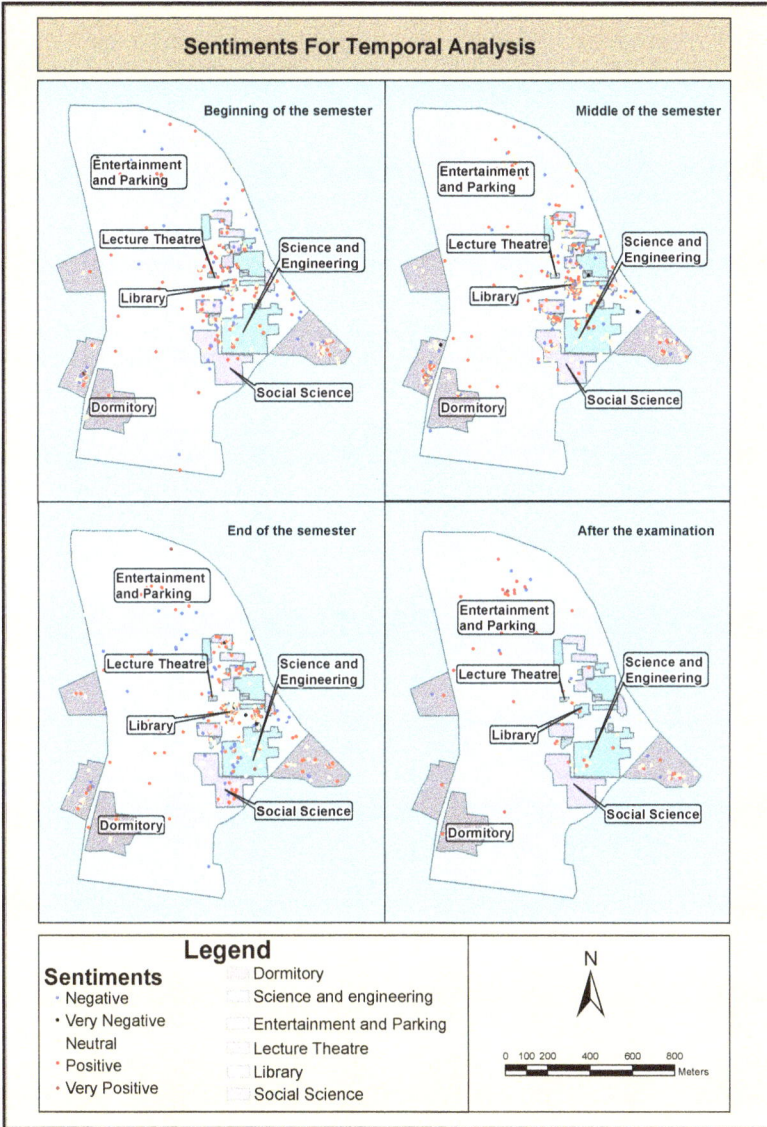

Figure 4: The distribution of sentiment polarity over time.

Concluding remarks

This study presents methods for analysing spatial and temporal patterns of Twitter sentiment polarity at Curtin University. By using this case study, we hope the results can provide some insights into understanding the spatial and temporal variation of students and staff's sentiment polarity. We separated Tweets into five different spatial zones and four time zones and tested a research question of whether sentiment polarity on Twitter exhibits specific time-location patterns.

Interestingly, although the number of positive Tweets posted in the entertainment and parking area were larger than the ones posted in the study areas, the highest percentage of positive Tweets were found in social science with over 40%. Besides, science and engineering and dormitory areas had the highest percentage of negative postings (both over 19%). It indicates that the sample of students in social science students tended to post twitter messages containing a higher ratio of positive-to-negative words than the sample of students in science and engineering in Curtin university. In addition, a trend of increasing number of negative Tweets was identified in library and science and engineering areas as a semester goes from the beginning to the end, reaching a peak around the exam period. While for the entertainment, sport and dormitory area, the percentage of negative Tweets dropped at the end of semester. This could mean that negative feelings might be associated with exam induced stress and study workload (Adnan et al. 2012).

The spatial and temporal sentiment analysis used geotagged Twitter data, which allow a sentiment polarity analysis at a fine-grained level. This method can be applied in many areas, such as polls (Wang et al. 2012a), consumer opinions concerning brands (Jansen et al. 2009), stock market performance (Bollen et al. 2011), crime prediction (Wang et al. 2012b) and tourism information (Shimada et al. 2011). However, there are a few limitations to be considered when making conclusions from this study. For example, outputs from the sentiment analysis in Starlight Data engineer are not perfectly accurate. For example, emotion tokens cannot be analysed in the sentiment analysis. Some emotion tokens, such as '☺' and '☹', actually express very obvious attitudes. But our methods in the sentiment analysis cannot process them. Besides, our sentiment analysis methods cannot handle sarcasm properly. For example, a Tweet of 'Well done. You have forgotten your umbrella' expresses a negative feeling, but the sentiment analysis misclassified it as positive. Some Tweets do not have sentiment words but they imply some emotions, which the sentiment analysis could misclassify them. For instance, 'Sleeping pattern is sooo screwed up, no more 4am' appears to be a negative sentiment, but the sentiment analysis classified it as neutral. In the future, this study will adopt more advanced algorithms, such as methods developed by Liu (2012), Katz et al. (2015) and Poria et al. (2014) for the sentiment analysis. In addition, in this study, we tested our hypotheses in only one university. We will collect Twitter data of major universities in Western Australia and conduct a comparison study in the future. In addition, we will develop a better measure of sentiment polarity, which will take both severity and frequency of individual words into account in the future.

Acknowledgements

The authors are very grateful to Curtin University who provides a research grant to support this study. Constructive comments from the editor and anonymous reviewers have significantly improved this paper. We would also like to thank Dr Wei Liu to share Twitter data with the research team, The views expressed in this article are those of the authors, and do not necessarily reflect the policies of any organisation.

References

Adnan, N., Murat, Z. H., Abdul Kadir, R. S. S., & Noradibah Hj Mohamad, Y. 2012. University students stress level and brainwave balancing index: Comparison between early and end of study semester. Research and Development (SCOReD), 2012 IEEE Student Conference, 5–6 Dec., 2012. Pulau Pinang, pp. 42–47.

Balahur, A. 2013. Sentiment analysis in social media texts. Citeseer.

Balahur, A., Mihalcea, R., & Montoyo, A. 2014. Computational approaches to subjectivity and sentiment analysis: Present and envisaged methods and applications. *Computer Speech & Language, 28*: 1–6.

Bollen, J., Mao, H., & Zeng, X. 2011. Twitter mood predicts the stock market. *Journal of Computational Science, 2*: 1–8.

Curtin University. 2015. *About* [Online]. Available at: http://about.curtin.edu.au/campus-locations/directions/ [Accessed March 26 2015].

Dacres, S., Haddadi, H., & Purver, M. 2013. Topic and Sentiment Analysis on OSNs: a Case Study of Advertising Strategies on Twitter.

Fink, C. R., Chou, D. S., Kopecky, J. J., & Llorens, A. J. 2011. Coarse- and Fine-Grained Sentiment Analysis of Social Media Text. *Johns Hopkins APL Technical Digest, 30*: 22–30.

Go, A., Bhayani, R., & Huang, L. 2009. Twitter sentiment classification using distant supervision. *CS224N Project Report, Stanford*: 1–12.

Jansen, B. J., Zhang, M., Sobel, K., & Chowdury, A. 2009. Twitter power: Tweets as electronic word of mouth. *Journal of the American Society for Information Science and Technology, 60*: 2169–2188.

Katz, G., Ofek, N., & Shapira, B. 2015. ConSent: Context-based sentiment analysis. *Knowledge-Based Systems*.

Khan, F. H., Bashir, S., & Qamar, U. 2014. TOM: Twitter opinion mining framework using hybrid classification scheme. *Decision Support Systems*, 57: 245–257.

Li, L., Goodchild, M. F., & Xu, B. 2013. Spatial, temporal, and socioeconomic patterns in the use of Twitter and Flickr. *cartography and geographic information science, 40*: 61–77.

Liu, B. 2012. Sentiment Analysis and Opinion Mining. San Rafael: Morgan & Claypool Publishers.

Medhat, W., Hassan, A., & Korashy, H. 2014. Sentiment analysis algorithms and applications: A survey. *Ain Shams Engineering Journal, 5*: 1093–1113.

Nielsen, F. Å. 2011. A new ANEW: Evaluation of a word list for sentiment analysis in microblogs.

Office of Strategy and Planning. 2014. *Curtin University Student Statistics 2009–2013* [Online]. Available at: https://planning.curtin.edu.au/stats/students2009-2013.cfm [Accessed March 26 2015].

Osgood, C. E., Suci, G., & Tannenbaum, P. 1957. *The measurement of meaning, 81.*

Paltoglou, G., & Thelwall, M. 2012. Twitter, MySpace, Digg: Unsupervised sentiment analysis in social media. *ACM Transactions on Intelligent Systems and Technology, 3.*

Pang, B., & Lee, L. 2004. A sentimental education: Sentiment analysis using subjectivity summarization based on minimum cuts. In: *Proceedings of the 42nd annual meeting on Association for Computational Linguistics*, 2004. Association for Computational Linguistics, p. 271.

Pang, B., & Lee, L. 2008. Opinion mining and sentiment analysis. *Foundations and trends in information retrieval, 2*: 1–135.

Poria, S., Gelbukh, A., Cambria, E., Hussain, A., & Huang, G.-B. 2014. EmoSenticSpace: A novel framework for affective common-sense reasoning. *Knowledge-Based Systems, 69*: 108–123.

Read, J. 2005. Using emoticons to reduce dependency in machine learning techniques for sentiment classification. In: *Proceedings of the ACL Student Research Workshop*, 2005. Association for Computational Linguistics, pp. 43–48.

Reyes, A., & Rosso, P. 2012. Making objective decisions from subjective data: Detecting irony in customer reviews. *Decision Support Systems, 53*: 754–760.

Rushdi Saleh, M., Martín-Valdivia, M. T., Montejo-Ráez, A., & Ureña-López, L. A. 2011. Experiments with SVM to classify opinions in different domains. *Expert Systems with Applications, 38*: 14799–14804.

Sarvabhotla, K., Pingali, P., & Varma, V. 2011. Sentiment classification: a lexical similarity based approach for extracting subjectivity in documents. *Information Retrieval, 14*: 337–353.

Serrano-Guerrero, J., Olivas, J. A., Romero, F. P., & Herrera-Viedma, E. 2015. Sentiment analysis: A review and comparative analysis of web services. *Information Sciences, 311*: 18–38.

Shimada, K., Inoue, S., Maeda, H., & Endo, T. 2011. Analyzing tourism information on twitter for a local city. In: *Software and Network Engineering (SSNE)*, 2011. First ACIS International Symposium on, 2011. IEEE, pp. 61–66.

Simoff, S., Böhlen, M. H., & Mazeika, A. 2008. *Visual data mining: theory, techniques and tools for visual analytics*, Springer Science & Business Media.

Taboada, M., Brooke, J., Tofiloski, M., Voll, K., & Stede, M. 2011. Lexicon-based methods for sentiment analysis. *Computational linguistics, 37*: 267–307.

Thelwall, M., Buckley, K., & Paltoglou, G. (2012). Sentiment strength detection for the social web. *Journal of the American Society for Information Science and Technology, 63*: 163–173.

Wang, H., Can, D., Kazemzadeh, A., Bar, F., & Narayanan, S. 2012a. A system for real-time twitter sentiment analysis of 2012 us presidential election cycle. In: *Proceedings of the ACL 2012 System Demonstrations.* Association for Computational Linguistics, pp. 115–120.

Wang, X., Gerber, M. S., & Brown, D. E. 2012b. Automatic crime prediction using events extracted from twitter posts. In: *Social Computing, Behavioral-Cultural Modeling and Prediction.* Springer.

Whitelaw, C., Garg, N., & Argamon, S. 2005. Using appraisal groups for sentiment analysis. In: *Proceedings of the 14th ACM international conference on Information and knowledge management*, 2005. ACM, pp. 625–631.

Wilson, T., Wiebe, J., & Hoffmann, P. 2005. Recognizing contextual polarity in phrase-level sentiment analysis. In: *Proceedings of the conference on human language technology and empirical methods in natural language processing*, 2005. Association for Computational Linguistics, pp. 347–354.

Ye, Q., Zhang, Z., & Law, R. 2009. Sentiment classification of online reviews to travel destinations by supervised machine learning approaches. *Expert Systems with Applications, 36*: 6527–6535.

CHAPTER 17

Social Networks VGI: Twitter Sentiment Analysis of Social Hotspots

Dario Stojanovski*, Ivan Chorbev, Ivica Dimitrovski and Gjorgji Madjarov

Faculty of Computer Science and Engineering, Ss. Cyril and Methodius University, Rugjer Boshkovikj 16, 1000 Skopje, Macedonia
*stojanovski.dario@gmail.com

Abstract

The enormous amount of data generated on social media provides vast quantities of geo-referenced data. Volunteered Geographic Information (VGI) originating from social networks has produced new challenges for research and has opened opportunities for a wide range of use cases. Smartphones with built-in GPS sensors enabled users to easily share their location and with the growing number of such devices available, VGI data is expanding at a rapid rate. Twitter is one of the most popular microblogging services. It's a social network that enables access to the data that is being created on the platform. It also allows for real-time retrieval of data from a given geographic area.

In this paper we give an overview of a system for detecting and identifying social hotspots from Twitter stream data and applying sentiment analysis on the data. Utilizing the Twitter Streaming Application Programming Interface (API), we collected a significant number of Tweets from New York and we evaluated the quality of the retrieved data. In this paper, we outline advantages and disadvantages of using various clustering algorithms over the data for this purpose, namely hierarchical agglomerative clustering and DBSCAN. We also elaborate on techniques for identifying social hotspots from spatially localized

---

How to cite this book chapter:
Stojanovski, D, Chorbev, I, Dimitrovski, I and Madjarov, G. 2016. Social Networks VGI: Twitter Sentiment Analysis of Social Hotspots. In: Capineri, C, Haklay, M, Huang, H, Antoniou, V, Kettunen, J, Ostermann, F and Purves, R. (eds.) *European Handbook of Crowdsourced Geographic Information*, Pp. 223–235. London: Ubiquity Press. DOI: http://dx.doi.org/10.5334/bax.q. License: CC-BY 4.0.

clusters. Finally, we present a deep learning approach to sentiment analysis used to determine the attitude of users participating in the identified social hotspots.

Keywords

social hotspots, sentiment analysis, Twitter, VGI, visualization, geo-clustering

Introduction

Volunteered Geographic Information (VGI) as coined by Goodchild (2007) is defined as the harnessing of tools to create, assemble and disseminate geographic data provided voluntary by individuals. VGI received a lot of traction in recent years with the continuing evolution of Web technologies that provide for easier user participation in the creation of such data with users assuming a more active role in the creation of VGI data (Sui 2011). This idea stimulated the popularity of a number of services such as Wikimapia, OpenStreetMap and many others. Wikimapia is based on the same concept as Wikipedia and it contains over 23.8 million objects. Flickr on the other hand, a system for hosting images allows users to upload geo-tagged photos on the platform with the corresponding latitude and longitude pair that are associated with the picture. Alternatively, people can also act as sensors and just provide their location through the use of various social network services such as Facebook, Twitter, Foursquare etc.

Taking advantage of VGI or otherwise known as User Generated Spatial Content (UGSC) leads to a vast number of applications ranging from public health, early disaster warning and crisis management to various types of analytics useful to marketing agencies and companies.

Social networks and microblogging platforms have attracted the attention of users on a huge scale over the recent years. Platforms such as Facebook, Instagram, Foursquare and many others generate massive amounts of data. These services also allow users to geo-locate the information they share on these networks. This lead to popularization of Social Media Geographic Information (SMGI) which refers to the geo-referencing of multimedia data extracted by social media applications. With the spread of mobile devices equipped with GPS sensors, it become even more accessible for users to share their location in order to provide more context to the content they are sharing. As of 2015, Twitter, the most popular microblogging platform of all, has over 300 million monthly active users and generates over 500 million messages on a daily basis[1]. One key advantage of Twitter is its real-time component, positioning the plat-

[1] https://about.twitter.com/company

form as one of the most up to date data source, as witnessed by its ability to break news before other sources.

Bloggers in the Twitter community use the platform to express their views and ideas on different topics, share thoughts on their daily activities, celebrity gossip etc. Although only a small percentage of Tweets (1.2%) (Dredze 2013) contain exact location, the sheer volume of messages generated every day, makes Twitter a gold mine for mining VGI data. Since users Tweet about events around them in real-time, we could tap into this information stream and identify social hotspots as they are emerging. Furthermore, applying sentiment analysis on the content shared related to a social hotspot, can provide additional insight about the place or event.

Sentiment analysis on social media has wide applications as it can be used to provide feedback for the reaction products and services receive, the public opinion towards different candidates during political elections etc. The presented work, focuses on analysis of sentiment related to social hotspots.

Related Work

A lot of research has been conducted to explore the various applications of volunteered geographic data from social media (Sui 2011). Companies have also showed interest in the area along with the field of sentiment analysis because of its potential to provide valuable insight into people's reaction regarding related products and the distribution over geographical areas (Liu 2014).

Dredze et al. (2013) explored the application of Twitter geo-located data to public health. They developed a system that infers structured location information from Tweets and showed how this information can be used for influenza tracking. However, their approach only detects location on city level and it's not able to detect finer grained locations. In the work of Li (2013), he addressed the issues of extracting local information and discovering communities of interest in local social media. Bosch et al. (2013) propose a visual analytics approach to facilitate sensemaking of geo-located microblog posts by enabling analysts to create automatic methods for extracting messages and by applying those methods when monitoring topics of interest.

Twitter as a source of geo-data has been used in various domains, many of which focus on event detection from Twitter data. Abdelhaq et al. (2013) presented a framework for detection of localized events in real-time from Twitter streams and tracking their evolution over time. The proposed system uses both geo-located and non-geo-located Tweets to identify event describing words, but only geo-located Tweets are used to determine the spatial distribution of such words. Spatial and temporal characteristics of keywords are continuously extracted to identify meaningful candidates for event descriptions. The system selects words that show bursty frequency in the current time frame and have local spatial distribution. Keywords are then clustered by their spatial signatures and clusters are scored to show how likely is that they represent a localized event. However, this approach does not perform very well in situations when there are multiple geo-terms within the same text.

Kisilevich et al. (2010) developed a new version of the DBSCAN clustering algorithm for analysis of places and events using a collection of geo-tagged photos. The assumption is that a high photo activity in a specific area is indicative of an interesting place or an ongoing event. The proposed algorithm addresses an issue that is specific to identifying social hotspots, which occurs when a significant portion of the samples in a cluster originates from a single user. In our work, we also utilize the DBSCAN clustering algorithm and tackle this issue in the cluster detection phase where features are generated indicative of the number of users in a cluster. Evaluation of the approach is done on Flickr images from Washington, D.C.

Walther et al. (2013) propose a system that detects geo-spatial events from the Twitter stream. The proposed system create clusters or EventCandidates in a manner similar to the DBSCAN clustering algorithm. Clusters are further analyzed to detect if any overlaps have occurred, both spatially and temporally. Several hand-crafted features were developed that the authors consider to be indicative of an actual event and these features are extracted from each cluster. Finally, an evaluation whether a cluster represents an actual event or not is made using a machine learning approach. Our system builds on the work of Walther et al. (2013). Additionally, we extend the system by applying sentiment analysis on the identified social hotspots.

Data Retrieval

Retrieving Twitter messages is available through the Twitter Application Program Interface (API) which offers a variety of REST endpoints. Nonetheless, in order to continuously collect Tweets from a certain geographic location, generating repeated REST calls is infeasible due to Twitter rate limits. As a result, we must utilize the Twitter Streaming API that gives low latency access to Twitter's global stream of data. The stream can return Tweets originating from a set of users or messages that contain certain keywords. However, the possibility of supplying the Streaming API with a filter specified by a set of spatial bounding boxes defined by latitude and longitude pairs is of interest in our work. This filter provides real-time access to all messages originating from a given geographic area.

A Tweet and its location are available through the Twitter public streams if the user explicitly consents to sharing the location in the post. Users can enable locations on their devices and provide exact coordinates obtained from a GPS sensor of the device they used to post the message. Additionally, Twitter enables for manual embedding of places in messages. Places on Twitter have specific IDs, defined by a bounding box and can be of several different types (city, POI, country). For a Tweet with an embedded place to be returned, the bounding boxes of the place and the filter applied in the stream must intersect. One drawback of retrieving Tweets with places is that the user location may not be the same as the one mentioned in the Tweet. A user can post a Tweet containing a mention of a place while Tweeting from somewhere else. Furthermore, many places refer to larger areas, cities or even countries which may feed the stream with Tweets that are outside of the defined bounding box. As a result, it is necessary to additionally filter the incoming data. MongoDB and its document model is probably the most suitable database system for storing social media posts. In addition, it supports temporal and geo-spatial indexes which is essential to the task at hand as we are dealing with geographical and temporal data.

In future work, it would be valuable to explore enriching the data with Twitter messages that are not explicitly geo-tagged. As for geo-tagged Tweets, the percentage of geo-tagged ones with exact coordinates goes as high as 1.2% (Dredze 2013), while 1.3% contain a Place object. In order to increase the utilization of the geo-referenced information generated on Twitter, one must look beyond explicitly volunteered geo-data. Users often include references to geographical information in the content they post on the Web, without tying it to specific coordinates. Location recognition in social media is a challenging problem, even more so in Twitter due to the 140 character limitation and the abundance of abbreviations, informal language or terms used only in the Twitter community.

In order to enrich dataset with VGI, one must first define a set of keywords to track using the Streaming API, as the number of keywords that can be fed to the API is limited to 400. In order to get Tweets most related to social hotspots, the stream should be fed with keywords relevant to social hotspots and with words related to the area that is being monitored. Liu (2014) developed an extensive system to disambiguate and identify locations mentioned in text and for estimating user location out of their activity. The approach relies on sequential learning methods to automatically learn the relations between parts of locations where the classifiers are fed with hand-crafted features.

Dataset

The system that is showcased in the remainder of the paper is based on Twitter data from New York between February 22 and April 16 2015. We set the bounding box to the following longitude and latitude pairs: $(-74, 40), (-73, 41)$. New York City is chosen because it's one of the most active cities Twitter-wise. Also, we assume that the majority of the messages will be in English. New York's big population and its dense social places structure pose both difficulties and advantages for mining geo-data. On one hand, the abundance of social places suggests that a relatively high number of Tweets will be related to social places.

On the other hand, the dense structure poses challenges to precise clustering of Tweets.

For the above mentioned period, we collected 4,125,542 Twitter messages, generated from a total of 226,114 distinct users. Out of these, 3,274,724 messages contained exact coordinates, while the others had a place entity only, which were attached manually. However, we observed that among these Tweets that also had place entities attached, a significant portion was outside of the defined bounding box. Upon filtering, the dataset was reduced to 2,350,739 messages.

In the set collected, we observed that place entities generally are related to greater geographical areas. For example, places such as 'Manhattan', 'New York' or 'Brooklyn' appear very often, as opposed to points of interests. Such places are insignificant to our analysis and have to be filtered out because they refer to a very broad area. Only 9,119 Tweets with embedded POIs that are within the defined bounding box were retrieved, while 269 messages have POIs outside of the bounding box.

System Overview

The presented system in this work monitors Twitter streams for a defined geographic area and identifies social hotspots as they are emerging. Figure 1

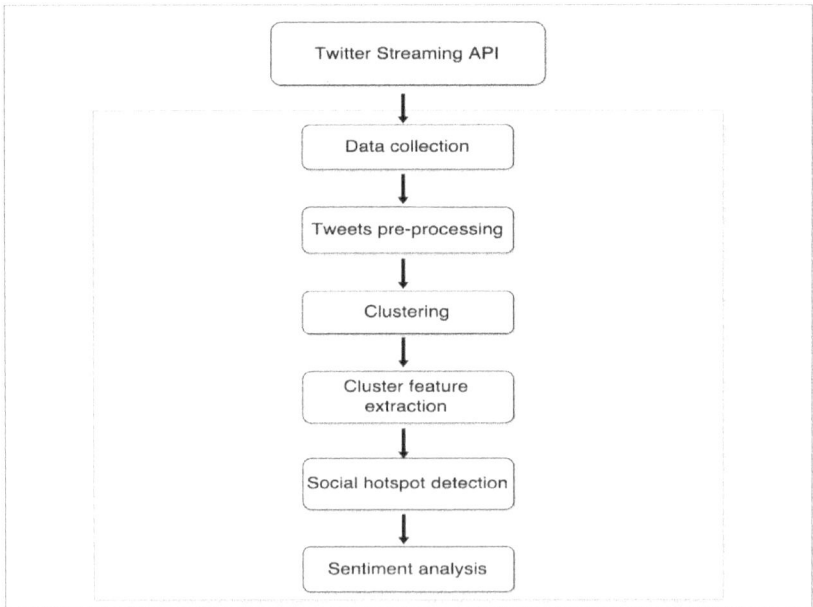

Figure 1: System architecture.

depicts an overview of the system architecture. The system can be broken down to the following key components:

- Data retrieval – connects to the Twitter Streaming API and collects messages from a certain geographic area.
- Tweet pre-processing – cleans Tweets from noisy tokens and characters.
- Cluster generation – creates clusters or social hotspots candidates from Tweets retrieved in a limited time period.
- Cluster feature extraction – extracts relevant features from Tweets in a social hotspot candidate
- Social hotspot detection – analyses clusters and evaluates using machine learning whether they represent a social hotspot or not.
- Sentiment analysis – analyses the sentiment of the Tweets in the social hotspot clusters.

Social Hotspot Detection

A social hotspot is a geographic POI which attracts the attention of many people in a limited period of time. In the context of Twitter, detecting social hotspots requires locating places with highly concentrated activity. In order to identify social hotspots from Twitter streams, clusters of geographically close Tweets must be created. There are several ways of generating such clusters, few of which are elaborated in the remainder of this work.

Clustering is an unsupervised machine learning method where samples from a given set of data are grouped based on features describing each sample. For purposes of the system described here, only the geographical component of the Tweets is taken into consideration for the clustering phase. Clustering algorithms compute distance between samples from the dataset. Since we are dealing with geo-data in relatively confined areas, the most appropriate metric is Euclidean distance.

Algorithms such as the K-means algorithm that require the number of clusters to be predefined are not appropriate for the specific problem, because the number of clusters or social hotspots candidates cannot be anticipated. One way of overcoming this issue is by using hierarchical agglomerative clustering. Each observation or Tweet starts-off as single cluster. Iteratively, pairs of clusters are merged together as the algorithm moves up the hierarchy. The result is a dendogram from which the threshold value can be observed which is used to cut the dendogram and to prevent it from building into the complete hierarchy. In order to get sufficiently localized clusters the threshold has to be set to a very small value. However, the complexity of hierarchical clustering ranges from $\Theta(N^2)$ to $\Theta(N^3)$, depending on the selected linkage criteria. Another deficiency is that it requires the pairwise distances of all the observations in the set. Computing pairwise distances has a $\Theta(N^2)$ memory complexity, which can be infeasible if the number of input data is huge.

Another appropriate clustering technique is DBSCAN. It is a density based clustering algorithm, not limited to shapes of clusters and only relies on a neighborhood count parameter (minPts) and a neighborhood distance ε. The general algorithm classifies each sample as core points, reachable points and outliers as follows:

- Core points have at least minimum number (minPts) of other points within ε distance
- A point p is density reachable if there is a point q and a path $p_1 \dots p_n$ where $p_1 = p$ and $p_n = q$ and p_{i+1} is directly reachable from p_i and is a core point
- A point is an outlier if it is not reachable from any other point

Points with a density above the specified threshold are constructed as clusters. DBSCAN handles outliers well, and has been proven as very effective in processing very large databases. It is by far most suitable for the task at hand as it gives close control over what is considered a social hotspot candidate. ST-DBSCAN is also a density-based algorithm for clustering spatial-temporal data. Birant et al. (2007) first proposed this approach as an extension on the existing DBSCAN in relation to the identification of core and noise objects and adjacent clusters. ST-DBSCAN takes into consideration the non-spatial, spatial and temporal attributes of the data. This is especially important in this case as social hotspots are not fixed in time. The algorithm requires two additional parameters, the distance parameter for non-spatial attributes and Δ_ε which is used to prevent the discovering of combined clusters. Additional modification is that a region is dense if the minPts criteria is satisfied by both of the distance parameters. ST-DBSCAN is also efficient at handling noise points when there are clusters with different densities. In Figure 2, the figure depicting clusters generated using hierarchical clustering, only clusters containing at least minPts messages are presented. Different marker colors are used for better clarity. We observe that DBSCAN generates less clusters than using hierarchical clustering. The generated clusters need to be further analyzed in order to determine if the Twitter messages refer to an actual social hotspot or are just random non-related posts or conversations. For this, we borrow on the work of Walther et al. (2013).

They developed several features divided into two categories. We only present the ten most effective features.

- Unique posters – the total number of unique users.
- Common theme – calculates word overlap between different Tweets in the cluster.
- @ Ratio – the number of user mentions relative to the number of Tweets
- Unique coordinates – the total number of unique coordinates within a cluster
- Ratio of Foursquare posts – fraction of Tweets originating from Foursquare
- Tweets count – total number of Twitter messages

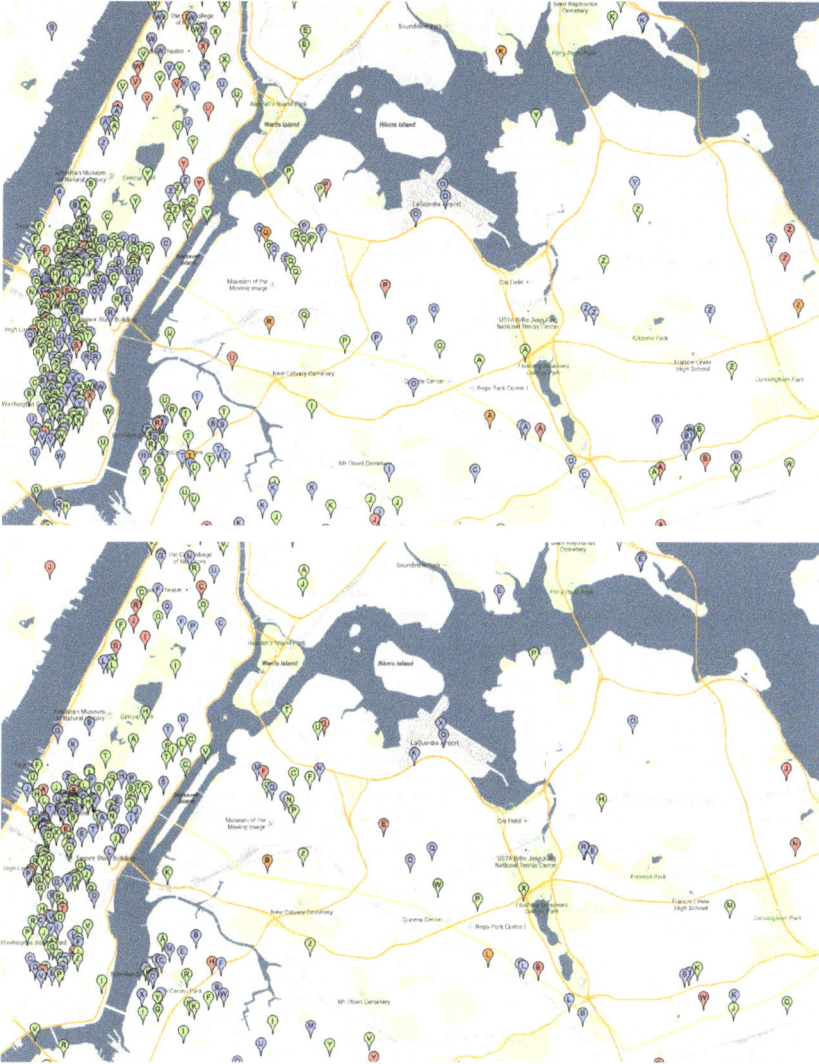

Figure 2: Clusters generated with hierarchical and DBSCAN clustering. Each marker represents a separate social hotspot candidate.

- Semantic Category – whether the cluster belongs to one of several predefined categories
- Subjectivity – indicates whether users share subjective posts or just various information
- Positive sentiment – indicates positive sentiment
- Ratio of unique posters – the number of unique users in relation to the number of Tweets

The effectiveness of the described features are experimentally evaluated in the work of Walther el al. (2013). They manually annotated 1000 clusters with binary labels in respect to whether they represent a real-world event or not. Textual features proved more significant than the Other group, but a combination of both feature categories provides for best performance. Walther et al. (2013) used three different machine learning algorithms, specifically Naive Bayes, Multilayer Perceptron and C4.5 decision trees. However, it would be beneficial to evaluate the effectiveness of Support Vector Machines and other machine learning approaches.

Sentiment Analysis

Sentiment analysis is the task of identifying human emotion in text. Social networks and media sparked interest in sentiment analysis amongst both academia and industry as users often share opinions and feeling on social networks (Pak 2010). Observing sentiment regarding social hotspots can provide valuable information about the popularity of a place or an event that is taken into consideration.

So far, Twitter sentiment analysis has heavily relied on hand-crafted features, which are both incomplete and too domain specific and depend on lexicons with sentiment polarity. Additionally, the process of manual feature generation is time-consuming and requires extensive domain knowledge. Deep learning techniques on the other hand, automate the feature generation and are more robust and flexible when applied to various domains. Convolutional Neural Networks (CNN) have been shown to achieve state-of-the-art results in sentence classification and specifically in sentiment analysis (Kim 2014), (dos Santos 2014).

We have developed an architecture for sentiment analysis that uses a CNN with multiple filters with varying window sizes. The model is built on the work of (Kim 2014) where they report state-of-the-art performances on 4 out of 7 sentence classification tasks. It consists of one convolutional layer and a max-over-time pooling layer which outputs a fixed sized vector. This vector is then fed to a three layer feed-forward network with two non-linear layers and a soft-max output layer which gives the probability distribution over the sentiment classes. The architecture maps each token in a given Tweet to an appropriate word representation. The approach leverages large Twitter corpora for unsupervised learning of these word representations, which capture syntactic and semantic characteristics of words. Instead of doing the pre-training of word embeddings ourselves, we use available word vectors.[2] We continuously update word vectors by back-propagation during training time and by doing so we capture and encode sentiment information into the word embeddings. We train

---

[2] http://nlp.stanford.edu/projects/glove/

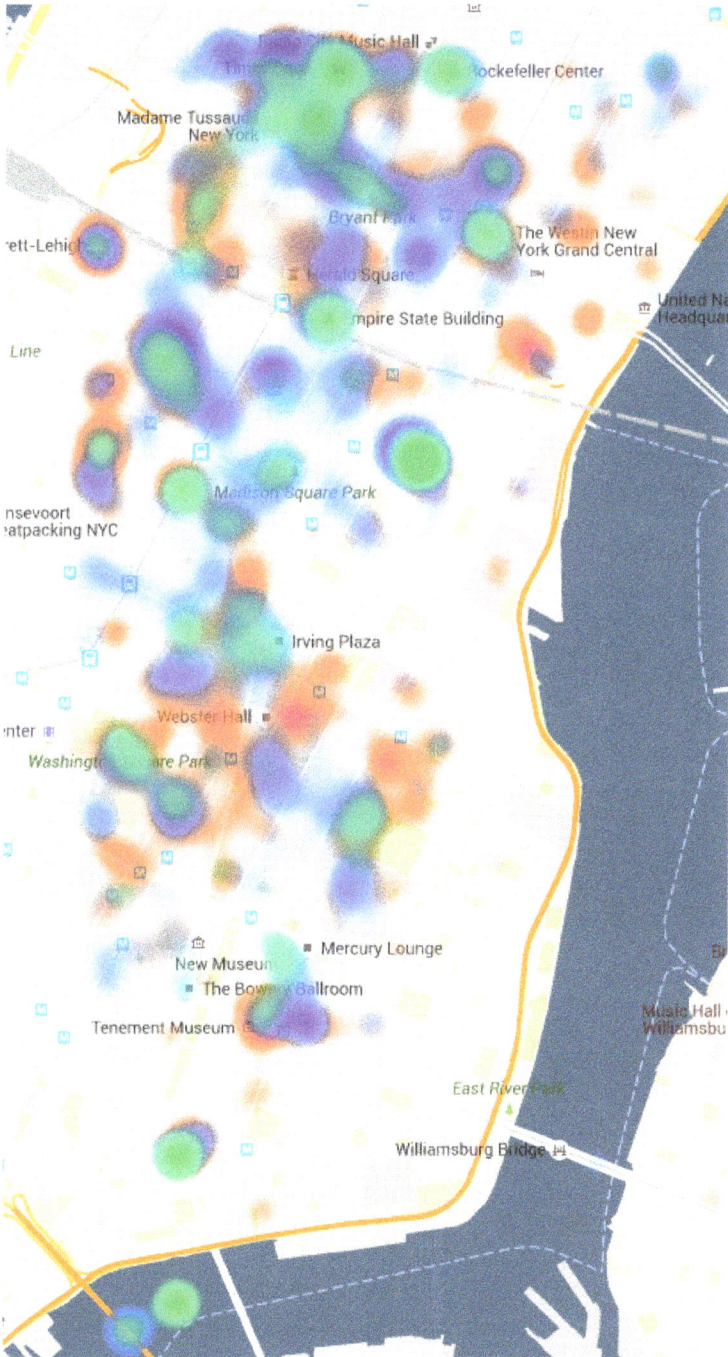

Figure 3: Sentiment heatmap.

the deep convolutional neural network with manually annotated Tweets provided by the Sentiment Analysis in Twitter task on SemEval-2015. The neural network in our system classifies Tweet sentiment into 3 classes, where the labels can be positive, negative or neutral. For future work, it would be interesting to use the proposed architecture for emotion identification which may provide an even deeper insight into the social hotspot popularity.

The system presents the overall sentiment of a social hotspot which can even be used for recommending points of interests to users as it can provide them with feedback for the popularity of a social hotspot in real-time. In Figure 3 a sentiment heatmap is depicted, where the red color represents negative, blue represent positive and green neutral Tweets.

Conclusion

In this paper, we have presented a system for detecting and identifying social hotspots from Twitter stream data, and applying sentiment analysis on the data. Utilizing the Twitter Streaming API, we have collected a significant number of Tweets from New York City and we have evaluated the quality of the retrieved data. Hierarchical and DBSCAN clustering algorithms have been analyzed for their usefulness in generating spatial clusters. We also elaborated on techniques for identifying social hotspots out of spatial clusters. Finally, we present an approach for sentiment analysis based on deep learning that is used to determine the attitude of users that participate in the hotspots.

Acknowledgments

We would like to acknowledge the support of the European Commission through the project MAESTRA – Learning from Massive, Incompletely annotated, and Structured Data (Grant number ICT-2013-612944).

References

Abdelhaq, H., Sengstock, C., & Gertz. 2013. Eventweet: Online localized event detection from twitter. In: *Proceedings of the VLDB Endowment (VLDB Endowment) 6*, pp. 1326–1329.

Birant, D., & Kut, A. 2007. ST-DBSCAN: An algorithm for clustering spatial—temporal data. *Data & Knowledge Engineering, 60*: 208–221.

Bosch, H., Thom, D., Heimerl, F., Puttmann, E., Koch, S., Kruger, R. Worner, M., & Ertl, T. 2013. Scatterblogs2: Real-time monitoring of microblog messages through user-guided filtering. *Visualization and Computer Graphics, IEEE Transactions on (IEEE), 19*: 2022–2031.

dos Santos, C., & Gatti, M. 2014. Deep Convolutional Neural Networks for Sentiment Analysis of Short Texts. In: *Proceedings of COLING 2014, the 25th International Conference on Computational Linguistics: Technical Papers*. Dublin City University and Association for Computational Linguistics, pp. 69–78.

Dredze, M., Paul, M. J., Bergsma, S., & Tran, H. 2013. Carmen: A twitter geolocation system with applications to public health. *AAAI Workshop on Expanding the Boundaries of Health Informatics Using AI (HIAI)*, pp. 20–24.

Goodchild, M. F. 2007. Citizens as sensors: the world of volunteered geography. *GeoJournal, 69*: 211–221.

Kim, Y. 2014. Convolutional Neural Networks for Sentence Classification. In: *Proceedings of the 2014 Conference on Empirical Methods in Natural Language Processing (EMNLP)*. Association for Computational Linguistics, pp. 1746–1751.

Kisilevich, S., Mansmann, F., & Keim, D. 2010. P-DBSCAN: a density based clustering algorithm for exploration and analysis of attractive areas using collections of geo-tagged photos. *Proceedings of the 1st international conference and exhibition on computing for geospatial research & application*. 38.

Li, R-Y. 2013. A Study of Volunteered Geographic Information and Social Media (University of Calgary).

Liu, J. 2014. *A Location-Aware Social Media Monitoring System*.

Pak, A., & Paroubek, P. 2010. *Twitter as a Corpus for Sentiment Analysis and Opinion Mining, LREC, 10*: 1320–1326.

Sui, D., & Goodchild, M. 2011. The convergence of GIS and social media: challenges for GIScience. *International Journal of Geographical Information Science, 25*: 1737–1748.

Walther, M., & Kaisser, M. 2013. Geo-spatial event detection in the twitter stream. *Advances in Information Retrieval*: 356–367.

CHAPTER 18

Research on social media feeds – A GIScience perspective

Enrico Steiger, Rene Westerholt and Alexander Zipf*

*zipf@uni-heidelberg.de

Introduction

During the last two decades, the role of internet users changed dramatically. While they were mostly passive content consumers before, they are now considered proactive data producers. This phenomenon is summarized by the term "Prosumer" (Ritzer & Jurgenson (2010) and gets facilitated through major technological advancements such as ubiquitous access to the mobile Internet and a widespread use of smartphones equipped with positioning and sensing capabilities. These outlined developments do not just happen recently, but trace back to the much older development around the so called "Web 2.0" (ITU 2014). In geospatial terms, these developments are well reflected by Mike Goodchild's popular definition of 'Citizens as Sensors' (Goodchild 2007), where ordinary people capture and disseminate "Volunteered Geographic Information (VGI)." Haklay further puts this development into broader context and rather coined the term "GeoWeb" (Haklay *et al.* 2008). OpenStreetMap (OSM) is probably the most prominent example of VGI.

Projects like OSM provide a well-defined data capturing protocol as well as a clear mission regarding their contributed contents. In contrast, data originating from online social networks (another source of VGI) is way more heterogeneous and diverse. At the same time, however, it may also provide high levels of semantic detail and is generated by a larger number of users. Consequently, it gained the interest of various research disciplines. These range from sociology

How to cite this book chapter:
Steiger, E, Westerholt, R and Zipf, A. 2016. Research on social media feeds – A GIScience perspective. In: Capineri, C, Haklay, M, Huang, H, Antoniou, V, Kettunen, J, Ostermann, F and Purves, R. (eds.) *European Handbook of Crowdsourced Geographic Information*, Pp. 237–254. London: Ubiquity Press. DOI: http://dx.doi.org/10.5334/bax.r. License: CC-BY 4.0.

toward linguistics, and of course geography and GIScience. The latter one is facilitated by the fact that a great deal of information contributed to social media is geotagged. Thus, the remainder of this chapter focuses on the spatial aspects of social media, and potential applications that can be derived from this kind of data.

Section A highlights the general potential of social media analysis for investigating social phenomena. We do that by outlining selected case studies from the exemplary field of human mobility analysis. These have demonstrated the usefulness of social media for investigating mobility patterns as well as human spatial behavior. Section B then provides an overview of several different application domains of social media analyses, with a particular focus on Twitter. Finally, Section C discusses some technical issues of established spatial analysis methods and their application to social media data. We conclude the chapter by summarizing its different parts. We further provide recommendations regarding future areas for GIScience research on social media data.

Utilization of social media data for investigating urban environments

The spatial and social structures of a city as well as the dynamic nature of human activities result in certain collective and individual human behavior patterns. Social media data can help to "sense" this type of information from urban environments in an in-situ manner. GIScience research thereby is focused on the overall question how corresponding spatiotemporal patterns from ubiquitous sensor networks and heterogeneous data streams can be explored, extracted, validated and aggregated. In turn, such information might enable us to sense everyday spatial processes and to gain knowledge about urban environments, especially with respect to collective human dynamics. The study of these issues has become one of the primary objectives of GIScience (Giannotti & Pedreschi 2008).

The information originating from social media messages (e.g., Tweets in case of Twitter) may contain spatial, temporal and semantic attributes. Considering these dimensions, social media can be considered as a (partial) proxy of real world happenings. However, space, time as well as semantics are influenced by each user's individual perception of the surrounding space. It is thus important to figure out ways to circumvent these issues for gaining trustworthy and objective information from these data sources.

The following short paragraphs outline case studies in which a range of GIScience researchers has drawn human mobility and urban study related knowledge from Twitter. We group these studies in accordance to their underlying research goals. The listed paragraphs thus provide the reader a quick overview of both the types of studies that have been conducted as well as methods and outcomes.

Mobility and social behavior. Studying the social dynamics of a city remains a challenging endeavor, which has recently been carried in a qualitative manner. Thus, social media might be a promising source of information in order to provide a better understanding of social dynamics within urban environments and resulted in various research efforts. Regarding the analysis of collective human mobility and activity patterns from social media, Cho et al. (2011) investigate social ties and their influence on human mobility patterns by comparing social media check-in data and cellphone location data. They found a stronger association of social network ties influencing long-distance travel than short range spatially and temporally periodic movements. Within the observed Twitter user pattern, Lee & Sumiya (2010) study user behavior by measuring geographic regularities and detecting geo-social events through identifying Regions of Interests (RoI). Another approach conducted by Noulas et al. (2011), Cranshaw et al. (2012) and Kafsi and Cramer (2015) is the identification of characteristic neighborhoods, collective movement patterns and social ties within certain user communities from Foursquare and other Social media data. In a similar approach for Twitter, Li *et al.* (2014) measure the spatial dispersion of users in a community and their trajectories. Hawelka et al. (2012) aim to further empirically validated the observed human behavior patterns and found a correlation between the conducted Twitter census and economic key figures. Furthermore, Li et al. (2013) explore spatiotemporal patterns of Twitter and Flickr data and investigated a relationship between socioeconomic characteristics of people who are generating social media posts in the US.

Mobility and underlying urban structures. The exploration of the relationships and the impact of urban structures on human mobility is an interesting study area for social media researcher. Wakamiya et al. (2011) investigate temporal patterns of crowd behavior over Japan by spatial partitioning Tweets in order to extract urban characteristics. On a smaller scale several studies investigate the connection with extracted urban activities from social media and their connection with the underlying urban structure. Kling et al. (2012) were able to detect spatiotemporal clusters of frequently occurring urban topics in New York. Furthermore, Ferrari et al. (2012) also work with georeferenced Tweets and a semantic probabilistic topic modeling approach to automatically extract urban patterns from location-based social networks. The study concluded that extracted urban motion patterns and identified hotspots in the city allow the inference of crowd behaviors that recur over time and space. A similar approach by using Foursquare data by Cheng et al. (2011) and Hasan et al. (2013) also resulted in the characterization of urban human mobility and activity patterns. Andrienko & Andrienko (2013) correlated the spatiotemporal clusters of keyword based filtered georeferenced Tweets of places where people Tweet with US population densities. The results have shown strong correlations between the observed Twitter distribution and census data, suggesting that social media is a reliable proxy for the inference of mobility patterns. One further application is to derive intra-urban events showing distinct mobility patterns over time. This spatiotemporal movement has proven to reflect typical mobility behavior in the underlying urban structures (Steiger, Westerholt, et al. 2015).

Mobility and human activities. Several studies infer individual and collective human daily activity patterns by analyzing crowdsourced information, such as taxi trip records (Liang et al. 2012), GPS traces (Azevedo & Bezerra 2009) (Jiang, Yin, & Zhao 2009) or mobile phone records (Candia & González 2008) (Gao 2014). Consequently, a large literature body also focus on studying human mobility and activity pattern from social media data. Krumm et al. (2011) estimate individual home locations of heavy Twitter user and apply machine learning algorithms to classify and predict individual travel behavior. Jin *et al.* (2014) developed a method to infer users' mobility patterns from check-ins in Foursquare. Coffey & Pozdnoukhov (2013) go one step further and semantically annotate mobility flow datasets with activity information and trip purposes from Tweets. Similarly, Wu *et al.* (2015) utilize social media to annotate the location history of mobile phone users for the characterization of certain social activities. Focusing on the content of Tweets, Grinberg *et al.* (2013) proposed a method to detect semantic patterns to infer clusters of users' real world activity. Gao (2014) developed a probabilistic approach to make place recommendations based on the users' geo-social circles, as extracted from Foursquare. In another study, the authors estimate spatiotemporal mobility flows from Twitter for the area of greater Los Angeles to infer origin- and destination trips (Gao et al. 2014). Results have shown similar pattern when comparing with community survey data. In a previous study we introduced a semantic and spatial analysis method (Steiger, Lauer, *et al.* 2014), through which we were able to extract geographic features from uncertain Twitter data and have shown that observed clusters correspond to landmarks, such as highly frequented squares and major transportation hubs. A further investigation revealed similar semantic layers that represent collective human mobility flows in co-occurrence with underlying social activity (Steiger, Ellersiek, *et al.* 2014) and could thus lead to new insights in characterizing urban mobility.

Future research recommendations

Further research needs to be conducted to assess the reliability of social media datasets. It also must be noted that the data collected from wireless devices are influenced by GPS/WIFI inaccuracy issues (Zandbergen and Barbeau 2011). Moreover, users can individually choose to share their precise location to a Tweet or just a general location information (such as a city or neighborhood). This resulting location uncertainty leads to imprecise location information of geotagged Tweets(Li *et al.* 2011).

Within the semantic attribute one must consider that the containing information may relate to events in the past, present or even future (Sengstock and

Gertz 2012). Principally the text corpora as such in social media posts are relatively sparse and vague. It may also be fairly ambiguous and hence observed phenomena may only be a weak indicator of a real world event. This uncertain semantic knowledge is a result of the fact that people using Twitter have individual motivations to post information and their main intention is to primarily serve their own communication needs. One further typical characteristic of social media is that users do not post equally distributed in geographic space and time leading to a heterogeneous dispersion of posts. Jatowt *et al.* (2015) further assess these varying temporal patterns and dynamics within social media. Furthermore, georeferenced social media posts only represent a small fraction of the overall available data. Not all user groups use all types of social media platforms similarly, which produces a potentially strong socio-demographic bias (Longley and Adnan 2015). Last, the application of spatial and semantic methods themselves creates uncertainties, since the distribution of specific geographic phenomena and their semantic complexities within Tweets are not known beforehand (Westerholt *et al.* 2015). Hence, it is important to compare and validate results with other acquired sensor data.

Conducting further research in this area however will be worthwhile, since study results may provide new additional insights into the complex human-sensor-city relationship at a much more fine-grained spatial and temporal level than before. New knowledge gained from this research will provide a better understanding of individual and collective human behavior within urban environments and may assist stakeholders and decision makers in their planning processes.

Application Domains of Social Media Analyses

Location-based social networks (LBSN) (Roick and Heuser 2013) offer a vast amount of voluntary content. The investigation of human activities in location-based social networks is one promising example of exploring spatial structures in order to infer underlying spatiotemporal patterns. Twitter for example is more and more recognized by numerous research domains. In particular it provides an opportunity for GIScience to understand geographic processes and spatial relationships comprised in social networks. Summarizing the current state of research concerning the application for spatiotemporal analyses, one outcome of a previously conducted systematic literature (Steiger, Albuquerque, *et al.* 2015) revealed that Twitter analyses are mainly focused on the spatiotemporal classification and detection of events. Principal investigated application domains are:

> ***Event Detection.*** To detect events, researchers are currently looking for spatial, temporal and semantic patterns within Twitter. In this respect people act as a social sensors for events (Yardi and Boyd 2010, Chae *et al.*

2012). *Disaster- and emergency management* as one event detection sub-field has been the primarily identified application in nearly a third of all reviewed studies (Sakaki *et al.* 2010, Murthy and Longwell 2013, Crooks *et al.* 2013). Further research has been conducted on utilizing Twitter in *traffic management*. This can be found in 14% of reviewed studies (Kosala and Adi 2012, Wakamiya and Lee 2012, Lenormand *et al.* 2014). Another area which seems to be quite popular is research on Twitter data for *disease/ health management* adding up to another 5% of the reviewed studies (Lampos and Cristianini 2010, Veloso and Ferraz 2011, Sofean and Smith 2012). A famous example is the derivation and prediction of information on infection sources and the spreading of an illness from Twitter messages (Culotta 2010, Collier *et al.* 2011). One prominent example is earthquake detection from Twitter data (Longueville *et al.* 2010, Zook *et al.* 2010). This has been successfully accomplished in a number of studies correlating results with official earthquake sensor data (Tapia *et al.* 2011, Thomson *et al.* 2012). Sakaki *et al.* (2010) have developed an algorithm that uses Twitter to calculate earthquakes' epicenters and the typhoons' trajectories. Moreover, situational information can be derived from location-related short messages to coordinate emergency responses (Vieweg *et al.* 2010). Also in the context of disease and health management similar outcomes have been derived. Tweets showing disease incidents have shown similar spatiotemporal distributions as those in with official reports. With these studies research has proven the trustworthiness and a high level of representativeness of Tweets throughout different application domains (Albuquerque *et al.* 2015).

Location Inference. Locations of users within social networks can be inferred or even predicted with the help of direct or indirect geolocation information derived from the provided metadata or from the semantic content (Kinsella *et al.* 2011, Hong *et al.* 2012, Hiruta *et al.* 2012). The geographic accuracy could be increased by extracting the textual information from the Tweet or from the metadata itself. For example, Lamprianidis and Pfoser (2011) have extracted locations and their names from Flickr pictures by clustering user-generated data points associated with geo-referenced pictures. Kelm *et al.* (2013) discusses various methods to extract place names from textual data from articles, posts or tags in geo-social networks, including place name gazetteer and statistical language modeling. Some methods follow an opposite approach and infer the location of a feature from implicit location information. Serdyukov *et al.* (2009) model the probability that a group of tags be assigned to a location. Similarly, (Gallagher *et al.* 2009) used location probability maps generated from tags for the same purpose. Van Laere *et al.* (2010) have pursued the same goal using k-medoids and Naive Bayes clustering methods. Some approaches focus on inferring a user's or a group of users' location. Cheng *et al.* (2010) have proposed a probabilistic method to determine users′

location from the content of their Twitter messages. Other authors have proposed to use the location of users´ friends to achieve the same goal (Backstrom *et al.* 2010). Stefanidis *et al.* (2013) have proposed a framework to harvest ambient geospatial information from social media feeds to locate social hotspots or to map social networks in a given geographical area. Ajao *et al.* (2015) summarize the broad range of available techniques applied to infer direct and indirect location from Twitter messages and social media users.

Geo-Social Network Analysis. Another important domain of research is social analysis which investigates relationships of individual users within a social network (Wu *et al.* 2011, Cranshaw *et al.* 2012). Geo-social network analysis seeks to identify the structure of social networks and their distribution in geographic space (Scellato and Mascolo 2010, Lee and Sumiya 2010). Social ties may feature distinct spatial distributions enabling spatiotemporal analyses. These distributions can help finding collective social activities and ultimately understanding geographical processes. A subfield of geo-social network analysis are *sentiment and emotion analysis* (Wang *et al.* 2012, Quercia *et al.* 2012). This field of research also offers a great potential for GIScience in the context of extracting contextual emotional information within urban and rural environments. One promising further field of research within social analysis which should be mentioned is urban planning and management which also could benefit from the rich data found in location based social networks such as Twitter. In the context of disaster management, several studies aim to infer the social dimensions within certain geo-located communities in twitter during disaster events (Conover *et al.* 2013, Bakillah *et al.* 2014).

Future research recommendations

Social Media data for research has proven to be a valuable source, as it not only comes for free, but also features a high spatiotemporal resolution. This kind of data especially enables possibilities to find spatial patterns and events which can help validating existing information sources. One identified main research gap is the exploration of human spatial behavior (Miller & Goodchild 2014) in order to gain knowledge about the underlying geographic processes and dynamics. Furthermore, the current research foci allow to transfer established methods from various disciplines (e.g. Computer- and Information Science, Social Science etc.) into other disciplines and enhancing new applications. As one example, more use of computer linguistic approaches to leverage knowledge from textual information, combined with methods for spatiotemporal analysis from computational sciences could lead to new insights within specific geographic application domains, such as disaster management or human mobility analysis.

Spatial Analysis of Social Media Feeds – Challenges and Approaches

The primary goal of spatial analysis is to explore structures within spatial data. This typically involves tasks like finding clusters on a map or figuring out distributional characteristics of data. One theoretical field underlying spatial analysis is spatial statistics. This field provides the basic principles that are underlying many spatial analysis problems. Key to this field is identifying spatial correlations, and thus hints on systematic patterns in geographic data (Fischer & Getis 2010). Respective methods and techniques are thus useful tools for gaining geographic insight into social media data.

The spatial analysis of social media data is typically conducted in an exploratory manner. This is due to lacking knowledge about potential underlying spatial processes, and thus about social media messages and their dispersal in geographic space in general. Useful tools on that regard are the K-Function (Ripley 1976) (purely geometric) and the mark correlation function (Stoyan & Stoyan 1994) (attribute values), both originating from spatial point pattern analysis. These methods allow identifying significant geometric clustering and regularity within stochastic point patterns. When the geometry is fixed (or rather treated as such) spatial autocorrelation statistics like Moran's I (Moran 1950, Cliff & Ord 1973) and hot spot statistics like Getis-Ord's G statistics (Getis & Ord 1992, Ord & Getis 1995) are suitable alternatives. These assess the degree of randomness within georeferenced attributes associated to units on a fixed geographic layout. In fact, many of the latter are essentially identical to different variants of the mark correlation function (see, e.g., Shimatani 2002). Thereby, Moran's I tests for correlations between neighbored observations across space, while G separates between extremal values (i.e., high and low).

As mentioned earlier, thorough spatial knowledge about social media datasets is typically lacking. Consequently, analysts oftentimes proceed with a trial-and-error approach when parameterizing the methods mentioned above. It is common practice to apply these techniques to different scales. The goal then is to sort out that scale at which patterning seems to be most pronounced. However, the techniques mentioned so far were designed long before the appearance of social media and similar kinds of user-generated data. The idea of the following two sections is thus to briefly reflect differences between social media and more traditional data, and to give some recommendations with respect to the spatial analysis of these.

Potential issues and pitfalls

The issues presented in the following are likely to occur when analyzing social media feeds with established methods from spatial analysis. It is important to note that social media feeds provide a mixture of indications from different real-world (and also some solely virtual) phenomena. This is due to the autonomous manner in which the data is being collected. Users can contribute any type of content from any place at any time. Such a mixture might be beneficial in terms of the wealth of contained information about the users' everyday lives. However, it also imputes some critical problems when it comes to spatial analysis. Probably the most trivial yet critical among these is the mere mixture of information as such. Any attribute which is derived from social media is highly likely to include information from several different real-world phenomena. Analyzing social media therefore comes at the risk of drawing conclusions about a mixture population that might not exist in reality. In most circumstances this is not desirable, since it does not lead to reasonable insight about any real-world process. One way to overcome this problem would be an accurate a priori semantic separation. However, that is a non-trivial task on its own right given the colloquial language used in corresponding messages.

Another issue with social media data is the implicit subjectivity that is per se introduced by the notion of "humans as sensors" (Goodchild 2007). One implication from that concept is the diversity at which people perceive environments (see also Section A). Similar phenomena might lead to varying responses among different users. This inevitably leads to an increased difficulty in analyzing the semantics (i.e. the attribute value) of the observations; and thus to a potential misclassification of phenomena. The implication of that for spatial analysis is crucial: techniques such as measures of spatial autocorrelation or spatial regression techniques are based on both, spatial characteristics as well as the attribute values. Consequently, spatial analysis techniques might end up in spurious results when the analyst fails controlling such effects.

The analysis of social media can also lead to an artificial increase in the number of type I / type II errors. This problem is likely to occur whenever testing hypotheses about spatial patterns with social media datasets. One might be interested in assessing spatial heterogeneity by means of local statistics like local Moran's I (Anselin 1995) or G_i^* (Ord and Getis 1995). It is common sense that these methods lead to an increase in type I errors due to alpha error inflation (Nelson 2012). Thus, it is important to control the alpha level accordingly (e.g., through techniques such as False-Discovery-Rate (Benjamini & Hochberg 1995)). With social media datasets, however, phenomena operating at smaller scales than the adjusted analysis scale might be considered by accident; and inadvertently influence the analysis. This is due to the mixture described above which is leading to spatially overlapping representations of different phenomena. The result is an increased amount of spurious indications of significant spatial effects.

Another critical implication of the scale-mixture outlined above is a potential creation of wrong and misleading relationships across scale levels. Recall that observations from smaller scale levels are prone to inherently being included in analyses at larger scales due to potential geometric mixture. Effects from smaller scales are therefore likely to be propagated towards analyses at larger scales. Due to this effect, some results become impossible, e.g., in scenarios where one wants to assess spatial autocorrelation at some large scale that is influenced by highly autocorrelated observations from smaller scales. If there is spatial autocorrelation present at some small scale (e.g. one "heavy" Twitter user recurrently posting from a particular location), it will be carried through to all larger scales being observed in the same geographic neighborhood.

Further discussion of these and related problems (including some empirical results) can be found in Westerholt *et al.* (2015) (including a discussion of a multi-scale modification of the local G statistic) and Lovelace *et al.* (2016). The presented list of effects is of course not exhaustive. There might be many more effects, some of which are still about to be discovered. The subsequent section provides some hints and recommendations about how to precede with the spatial analysis of social media data.

Some recommendations

Spatial autocorrelation is the core principle underlying a great deal of spatial analysis methodology. Therefore, it is crucial to accurately assess this characteristic in order to design applicable methods, and for drawing reasonable geographic conclusions. This is not just important for exploratory tests on spatial clustering and heterogeneity, but also crucial for model-driven spatial regression scenarios such as Geographically Weighted Regression (GWR) (Fotheringham *et al.* 2003) and for assessing model misspecification (Cliff and Ord 1981). Unfortunately, in case of social media analysis, the assessment of spatial autocorrelation is strongly affected by the problems depicted in the previous section. Therefore, one recommendation in terms of future research is to work on appropriate adaptations of corresponding measures and techniques in order to account for multi-scale (or rather: "mixed-scale") and multi-categorical effects. As long as these are not available, one should carefully parameterize respective techniques. Another (aspatial) approach might be to decompose social media datasets a priori, probably based on some other characteristic such as the Tweets' semantics. The worst option of all, however, would be to neglect the specific spatial characteristics of social media data when conducting spatial analysis. That would lead to a wrong evaluation of spatial effects; and thus to wrong analysis results.

Another recommendation is related to one of the promising opportunities that come with social media datasets: their wealth of information. We can obtain an array of valuable and potentially interrelated properties from social media data. These include temporal, semantic and spatial information. Correspondingly, one should try to analyze all these dimensions simultaneously instead of considering them in a separated fashion. This might unveil a much deeper understanding of social phenomena that are reflected in such datasets. Recent research efforts like, e.g., Steiger *et al.* (2015) reflect this idea. However, it yet remains a challenge to find measures to incorporate these different kinds of information in joint methodology in a reasonable way.

Conclusion and an outlook on future work

We outlined some potential pitfalls when analyzing social media data spatially. These are caused by the inherent characteristics of the data, i.e., the way in which the data is collected and what such services are used for. Potential problems include geometric mixtures of differently scaled data; semantic mixtures that get blurred in joint attributes derived from the data; and (more generally) spurious assessments of spatial correlations and thus pattern in the data.

The previous paragraphs are clearly biased towards the concept of spatial autocorrelation. On the one hand this focus is due to the research focus of the authors. On the other hand this is due to the central role which spatial autocorrelation plays throughout the entire field of spatial analysis. However, there are of course other important characteristics and pitfalls that might also influence the spatial analysis of social media data. The observations come, for instance, with considerable uncertainties with respect to relevant dimensions: The text snippets are colloquial and oftentimes difficult to interpret (semantics), the time stamp is sometimes not in line with real-world happenings (temporal) and the geographic coordinates are prone to positioning inaccuracies (spatial). The intensities of all these uncertainties appear to be varying across different users, devices, regions, etc. All these uncertainties indeed have impact on the results of spatial analysis.

Future methodological research should focus on the specific spatial characteristics of social media data (that are not yet known to a full extent). For now, across all disciplines and domains, it is common sense to apply established standard methodology to social media data. Relatively little emphasis is put on purely methodological research on the background of the special characteristics of these datasets. Thus, there is still plenty of room for improvement. The discipline of GIScience could play a vital role in these developments. Beyond purely empirical research, the impact of the spatial disciplines has been quite small so far. However, given that many research questions around social media are distinctive spatial ones, we should put much more emphasis on specialized spatial analysis techniques for social media.

Conclusion

On the one hand, social media data offers an array of new perspectives regarding many research questions and applications. On the other hand, however, these datasets also come with a set of issues that need to be taken into account, in particular when it comes to spatial analysis. GIScience can contribute to the development of new spatial analysis methods for social media data. Current major issues from a GIScience perspective include:

- the need of spatial analysis methods to be adapted towards uncertain and unstructured data types from LBSN;
- the handling of geographic scale effects when analyzing social media data;
- the need for combining different methods across disciplinary boundaries (e.g. social network analysis, semantic analysis, spatiotemporal analysis), in order to better utilize all available information dimensions;
- the development of data fusion and information extraction methods that take several different data sources simultaneously into account.

This would support exploring latent patterns and sensing geographical processes from social media data in a more realistic manner. GIScience could thus contribute to answering these important geographic questions and may play a major role in the further exploration of social media data.

References

Ajao, O., Hong, J., & Liu, W., 2015. A survey of location inference techniques on Twitter. *Journal of Information Science, 1*: 1–10.

Albuquerque, J., de, Herfort, B., & Brenning, A. 2015. A geographic approach for combining social media and authoritative data towards identifying useful information for disaster management. *International Journal of Geographical Information Science*, pending (pending).

Andrienko, G., & Andrienko, N. 2013. Thematic Patterns in Georeferenced Tweets through Space-Time Visual Analytics. *Computing in Science & Engineering, 15*(3): 72–82.

Anselin, L., 1995. Local Indicators of Spatial Association-LISA. *Geographical Analysis, 27*(2): 93–115.

Azevedo, T., & Bezerra, R. 2009. An analysis of human mobility using real traces. In: *Wireless Communications and Networking Conference WCNC*, pp. 1–6.

Backstrom, L., Sun, E., & Marlow, C., 2010. Find me if you can: improving geographical prediction with social and spatial proximity. In: *Proceedings of the 19th international conference on World wide web*. ACM, pp. 61–70.

Bakillah, M., Li, R.-Y., & Liang, S. H. L. 2014. Geo-located community detection in Twitter with enhanced fast-greedy optimization of modularity: the case study of typhoon Haiyan. *International Journal of Geographical Information Science*.

Benjamini, Y., & Hochberg, Y. 1995. Controlling the false discovery rate: a practical and powerful approach to multiple testing. *Journal of the Royal Statistical Society. Series B (Methodological)*, 57 (1): 289–300.

Candia, J., & González, M. 2008. Uncovering individual and collective human dynamics from mobile phone records. *Journal of Physics A: Mathematical and Theoretical, 41*(22).

Chae, J., Thom, D., Bosch, H., Jang, Y., Maciejewski, R., Ebert, D. S., & Ertl, T. 2012. Spatiotemporal social media analytics for abnormal event detection and examination using seasonal-trend decomposition. In: *2012 IEEE Conference on Visual Analytics Science and Technology (VAST)*, pp. 143–152.

Cheng, Z., Caverlee, J., & Lee, K. 2010. You Are Where You Tweet : A Content-Based Approach to Geo-locating Twitter Users. In: *Proceedings of the 19th ACM international conference on Information and knowledge management*, pp. 759–768.

Cheng, Z., Caverlee, J., Lee, K., & Sui, D. Z. D. 2011. Exploring Millions of Footprints in Location Sharing Services. *ICWSM*: 81–88.

Cho, E., Myers, S. A., & Leskovec, J. 2011. Friendship and mobility. In: *Proceedings of the 17th ACM SIGKDD international conference on Knowledge discovery and data mining – KDD '11*. New York, NY: ACM, pp. 1082–1090.

Cliff, A., & Ord, J. 1981. Spatial processes: models & applications.

Coffey, C., & Pozdnoukhov, A., 2013. Temporal decomposition and semantic enrichment of mobility flows. In: *Proceedings of the 6th ACM SIGSPATIAL International Workshop on Location-Based Social Networks – LBSN '13*. ACM, New York, NY: ACM, pp. 34–43.

Collier, N., Son, N. T., & Nguyen, N. M. 2011. OMG U got flu? Analysis of shared health messages for bio-surveillance. *Journal of biomedical semantics, 2*(Suppl 5): S9.

Conover, M. D., Davis, C., Ferrara, E., McKelvey, K., Menczer, F., & Flammini, A. 2013. The geospatial characteristics of a social movement communication network. *PloS one, 8*(3): e55957.

Cranshaw, J., Schwartz, R., Hong, J., & Sadeh, N. 2012. The Livehoods Project: Utilizing Social Media to Understand the Dynamics of a City. In: *ICWSM*. AAAI.

Crooks, A., Croitoru, A., Stefanidis, A., & Radzikowski, J. 2013. #Earthquake: Twitter as a Distributed Sensor System. *Transactions in GIS, 17*(1): 124–147.

Culotta, A., 2010. Towards detecting influenza epidemics by analyzing Twitter messages. In: *Proceedings of the first workshop on social media analytics*. ACM, pp. 115-122.

Fischer, M. M., & Getis, A. 2010. Introduction. In: Fischer, M. M., & Getis, A. (Eds.) *Handbook of Applied Spatial Analysis*. Heidelberg: Springer, pp. 1–26.

Fotheringham, A., Brunsdon, C., & Charlton, M. 2003. *Geographically weighted regression: the analysis of spatially varying relationships*. John Wiley & Sons.

Gallagher, A., Joshi, D., Yu, J., & Luo, J. 2009. Geo-location inference from image content and user tags. *Computer Vision and Pattern Recognition Workshops, 2009*. CVPR Workshops 2009, IEEE, pp. 55–62.

Gao, S., 2014. Spatio-Temporal Analytics for Exploring Human Mobility Patterns and Urban Dynamics in the Mobile Age. *Spatial Cognition & Computation, 15*(2): 86–114.

Getis, A., & Ord, J. K. 1992. The analysis of spatial association by use of distance statistics. *Geographical Analysis, 24*(3): 189–206.

Goodchild, M. 2007. Citizens as sensors: the world of volunteered geography. *GeoJournal, 69*(4): 211–221.

Grinberg, N., Naaman, M., Shaw, B., & Lotan, G. 2013. Extracting Diurnal Patterns of Real World Activity from Social Media. *ICWSM.*

Haklay, M., Singleton, A., & Parker, C. 2008. Web mapping 2.0: The neogeography of the GeoWeb. *Geography Compass, 2*(6): 2011–2039.

Hasan, S., Zhan, X., & Ukkusuri, S. V. 2013. Understanding urban human activity and mobility patterns using large-scale location-based data from online social media. In: *Proceedings of the 2nd ACM SIGKDD International Workshop on Urban Computing – UrbComp '13.* ACM, New York, NY: ACM Press, p. 1.

Hiruta, S., Yonezawa, T., Jurmu, M., & Tokuda, H. 2012. Detection , Classification and Visualization of Place-triggerd Geotagged Tweets. In: *Proceedings of the 2012 ACM Conference on Ubiquitous Computing.* ACM.

Hong, L., Ahmed, A., Gurumurthy, S., Smola, A.J., & Tsioutsiouliklis, K. 2012. Discovering geographical topics in the twitter stream. In: *Proceedings of the 21st international conference on World Wide Web – WWW '12.* New York, USA: ACM, p. 769.

International Telecommunication Union (ITU). 2014. *Measuring the Information Society. Report.*

Jatowt, A., Antoine, É., Kawai, Y., & Akiyama, T. 2015. Mapping Temporal Horizons: Analysis of Collective Future and Past related Attention in Twitter, pp. 484–494.

Jiang, B., Yin, J., & Zhao, S. 2009. Characterizing the human mobility pattern in a large street network. *Physical Review E, 80*(2).

Jin, L., Long, X., Zhang, K., Lin, Y., & Joshi, J. 2014. Characterizing users' check-in activities using their scores in a location-based social network. *Multimedia Systems.*

Kafsi, M., & Cramer, H. 2015. Describing and Understanding Neighborhood Characteristics through Online Social Media. In: *Proceedings of the 24th International Conference on World Wide Web.* International World Wide Web Conferences Steering Committee, pp. 549–559.

Kelm, P., Murdock, V., & Schmiedeke, S. 2013. Georeferencing in social networks. *Social Media Retrieval.* Springer London, pp. 115–141.

Kinsella, S., Murdock, V., & Hare, N. O. 2011. 'I' m Eating a Sandwich in Glasgow: Modeling Locations with Tweets. In: *Proceedings of the 3rd international workshop on Search and mining user-generated contents.* ACM, pp. 61–68.

Kling, F., Kildare, C., & Pozdnoukhov, A. 2012. When a City Tells a Story: Urban Topic Analysis. *In: Proceedings of the 20th International Conference on Advances in Geographic Information Systems.* New York, USA: ACM, pp. 482–485.

Kosala, R., & Adi, E. 2012. Harvesting Real Time Traffic Information from Twitter. *Procedia Engineering, 50* (Icasce), pp. 1–11.

Krumm, J., Caruana, R., & Counts, S. 2011. Learning Likely Locations. In: *User Modeling, Adaptation, and Personalization*. Springer Berlin Heidelberg, pp. 64–76.

Van Laere, O., Schockaert, S., & Dhoedt, B. 2010. Towards automated georeferencing of Flickr photos. *In: Proceedings of the 6th Workshop on Geographic Information Retrieval – GIR '10*. New York, USA: ACM Press, p. 1.

Lampos, V., & Cristianini, N. 2010. Tracking the flu pandemic by monitoring the Social Web. *Cognitive Information Processing (CIP)*. 2010 2nd International Workshop on IEEE, pp. 411–416.

Lamprianidis, G., & Pfoser, D. 2011. Jeocrowd – Collaborative Searching of User-Generated Point Datasets. *ACM*.

Lee, R., & Sumiya, K. 2010. Measuring geographical regularities of crowd behaviors for Twitter-based geo-social event detection. In: *Proceedings of the 2nd ACM SIGSPATIAL International Workshop on Location Based Social Networks – LBSN '10*. New York, USA: ACM, p. 1.

Lenormand, M., Tugores, A., Colet, P., & Ramasco, J. J. 2014. Tweets on the road. *PloS one, 9*(8).

Li, C., Zhao, Z., Luo, J., Yin, L., & Zhou, Q. 2014. A spatial-temporal analysis of users' geographical patterns in social media: a case study on microblogs. *Database Systems for Advanced Applications*. Springer Berlin Heidelberg, pp. 296–307.

Li, L., Goodchild, M. F., & Xu, B. 2013. Spatial, temporal, and socioeconomic patterns in the use of Twitter and Flickr. *Cartography and Geographic Information Science, 40*(2): 61–77.

Li, W., Serdyukov, P., de Vries, A. P., Eickhoff, C., & Larson, M. 2011. The where in the Tweet. *In: Proceedings of the 20th ACM international conference on Information and knowledge management – CIKM '11*. New York, USA: ACM, p. 2473.

Liang, X., Zheng, X., Lv, W., Zhu, T., & Xu, K., 2012. The scaling of human mobility by taxis is exponential. *Physica A: Statistical Mechanics and its Applications, 391*(5): 2135–2144.

Longley, P. A., & Adnan, M. 2015. Geo-temporal Twitter demographics. *International Journal of Geographical Information Science*: 1–21.

Longueville, B., De Annoni, A., Schade, S., Ostlaender, N., & Whitmore, C. 2010. Digital earth's nervous system for crisis events: real-time sensor web enablement of volunteered geographic information. *International Journal of Digital Earth, 3*(3): 242–259.

Lovelace, R., Birkin, M., Cross, P., & Clarke, M. 2016. From big noise to big data: Toward the verification of large data sets for understanding regional retail flows. *Geographical Analysis, 48*(1): 59–81.

Miller, H J., & Michael, F. 2015. Goodchild. Data-driven geography. *GeoJournal, 80*(4): 449–461.

Murthy, D., & Longwell, S. a. 2013. Twitter and Disasters. *Information, Communication & Society, 16*(6): 837–855.

Nelson, T. 2012. Trends in spatial statistics. *The Professional Geographer, 64*(1): 83–94.

Noulas, A., Scellato, S., Mascolo, C., & Pontil, M. 2011. Exploiting Semantic Annotations for Clustering Geographic Areas and Users in Location-based Social Networks. In: *The Social Mobile Web 11*. AAAI.

Ord, J., & Getis, A. 1995. Local spatial autocorrelation statistics: distributional issues and an application. *Geographical analysis, 27*(4): 286–306.

Quercia, D., Capra, L., & Crowcroft, J. 2012. The Social World of Twitter: Topics, Geography , and Emotions. In: *ICWSM 12*. AAAI, pp. 298–305.

Ritzer, G., & Jurgenson, N. 2010. Production, Consumption, Prosumption The nature of capitalism in the age of the digital 'prosumer'. *Journal of Consumer Culture, 10*(1): 13–36.

Roick, O., & Heuser, S. 2013. Location Based Social Networks – Definition, Current State of the Art and Research Agenda. *Transactions in GIS, 17*(5): 763–784.

Sakaki, T., Okazaki, M., & Matsuo, Y. 2010. Earthquake shakes Twitter users: real-time event detection by social sensors. In: *Proceedings of the 19th international conference on World wide web*. ACM, New York, NY, pp. 851–860.

Scellato, S., & Mascolo, C. 2010. Distance matters: geo-social metrics for online social networks. In: *Proceedings of the 3rd conference on Online social networks*.

Sengstock, C., & Gertz, M. 2012. Latent geographic feature extraction from social media. In: *Proceedings of the 20th International Conference on Advances in Geographic Information Systems – SIGSPATIAL '12*. New York, New York, USA: ACM Press, p. 149.

Serdyukov, P., Murdock, V., & van Zwol, R. 2009. Placing flickr photos on a map. In: *Proceedings of the 32nd international ACM SIGIR conference on Research and development in information retrieval – SIGIR '09*. New York, New York, USA: ACM Press, p. 484.

Shimatani, K. 2002. Point processes for fine-scale spatial genetics and molecular ecology. *Biometrical Journal, 44*(3): 325–352.

Sofean, M., & Smith, M. 2012. A Real-Time Architecture for Detection of Diseases using Social Networks: Design , Implementation and Evaluation. In: *Proceedings of the 23rd ACM conference on Hypertext and social media*. ACM, pp. 309–310.

Stefanidis, A., Crooks, A., & Radzikowski, J. 2013. Harvesting ambient geospatial information from social media feeds. *GeoJournal, 78*(2): 319–338.

Steiger, E., De Albuquerque, J. P., & Zipf, A. 2015. An advanced systematic literature review on spatiotemporal analyses of Twitter data. *Transactions in GIS* (pending).

Steiger, E., Ellersiek, T., & Zipf, A. 2014. Explorative public transport flow analysis from uncertain social media data. In: *Third ACM SIGSPATIAL Interna-*

tional Workshop on Crowdsourced and Volunteered Geographic Information (GEOCROWD).

Steiger, E., Lauer, J., Ellersiek, T., & Zipf, A. 2014. Towards a framework for automatic geographic feature extraction from Twitter. In: *Eighth International Conference on Geographic Information Science.*

Steiger, E., Resch, B., & Zipf, A. 2015. Exploration of spatiotemporal and semantic clusters of Twitter data using unsupervised neural networks. *International Journal of Geographical Information Science,* X(X): xx-xx.

Steiger, E., Westerholt, R., Resch, B., & Zipf, A. 2015. Twitter as an indicator for whereabouts of people ? Correlating Twitter with UK census data. *Computers, Environment and Urban Systems, 54:* 255–265.

Stoyan, D., & Stoyan, H. 1994. Fractals, random shapes and point fields: methods of geomtetrical statistics. Hoboken: John Wiley & Sons.

Tapia, A. H., Bajpai, K., Jansen, B. J., & Yen, J., 2011. Seeking the Trustworthy Tweet : Can Microblogged Data Fit the Information Needs of Disaster Response and Humanitarian Relief Organizations. *ISCRAM*(May): 1–10.

Thomson, R., Ito, N., Suda, H., Lin, F., Liu, Y., Hayasaka, R., Isochi, R., & Wang, Z. 2012. Trusting Tweets: The Fukushima Disaster and Information Source Credibility on Twitter. *In: Proceedings of the 9th International ISCRAM Conference,* pp. 1–10.

Veloso, A., & Ferraz, F. 2011. Dengue surveillance based on a computational model of spatio-temporal locality of Twitter. In: *Proceedings of the 3rd International Web Science Conference.* ACM, 3.

Vieweg, S., Hughes, A., Starbird, K., & Palen, L. 2010. Microblogging during two natural hazards events: what twitter may contribute to situational awareness. In: *Proceedings of the SIGCHI conference on human factors in computing systems,* pp. 1079–1088.

Wakamiya, S., & Lee, R. 2012. Crowd-sourced Urban Life Monitoring: Urban Area Characterization based Crowd Behavioral Patterns from Twitter Categories and Subject Descriptors. In: *Proceedings of the 6th International Conference on Ubiquitous Information Management and Communication.* ACM, 26.

Wakamiya, S., Lee, R., & Sumiya, K. 2011. Crowd-based Urban Characterization: Extracting Crowd Behavioral Patterns in Urban Areas from Twitter. In: *Proceedings of the 3rd ACM SIGSPATIAL International Workshop on Location-Based Social Networks.* ACM, New York, NY: ACM, pp. 77–84.

Wang, H., Can, D., Kazemzadeh, A., Bar, F., & Narayanan, S., 2012. A System for Real-time Twitter Sentiment Analysis of 2012 U . S . Presidential Election Cycle. In: *Proceedings of the Association for Computational Linguistics 2012 System Demonstrations.* ACM, pp. 115–120.

Westerholt, R., Resch, B., & Zipf, A., 2015. A local scale-sensitive indicator of spatial autocorrelation for assessing high- and low-value clusters in multiscale datasets. *International Journal of Geographical Information Science* (pending).

Wu, F., Li, Z., Lee, W., Wang, H., & Huang, Z. 2015. Semantic Annotaion of Mobility Data using Social Media. *Proceedings of the 24th International Conference on World Wide Web. International World Wide Web Conferences Steering Committee,* pp. 1253–1263.

Wu, S., Hofman, J. M., Watts, D. J., & Mason, W. A. 2011. Who Says What to Whom on Twitter. In: *Proceedings of the 20th international conference on World wide web.* ACM, pp. 705–714.

Yardi, S., & Boyd, D. 2010. Tweeting from the Town Square: Measuring Geographic Local Networks. In: *ICWSM 10.* AAAI.

Zandbergen, P. a., & Barbeau, S. J. 2011. Positional Accuracy of Assisted GPS Data from High-Sensitivity GPS-enabled Mobile Phones. *Journal of Navigation, 64*(03): 381–399.

Zook, M., Graham, M., Shelton, T., & Gorman, S. 2010. Volunteered Geographic Information and Crowdsourcing Disaster Relief: A Case Study of the Haitian Earthquake. *World Medical & Health Policy, 2*(2): 6–32.

PART IV

VGI and crowdsourcing in environmental monitoring

.

CHAPTER 19

Changing role of citizens in national environmental monitoring

Juhani Kettunen*, Jari Silander, Matti Lindholm, Maiju Lehtiniemi, Outi Setälä and Seppo Kaitala

Finnish Environment Institute, Address: Mechelininkatu 34a, 00250 Helsinki, Finland
*juhani.kettunen@ymparisto.fi

Abstract

During the last few decades the role of citizens in environmental monitoring has changed remarkably in Finland. In this chapter, we briefly describe this change by using examples of both traditional and modern monitoring systems. According to our findings, there are at least four important drivers challenging traditional monitoring systems. First, the monitoring is undergoing a rapid process of globalisation and e.g. the systems that earlier focused on national problems are today controlled by European legislation or influenced by international problems, agreements and practices. Second, public obligations for monitoring have grown much more rapidly than economic resources and it requires the monitoring systems to have a new kind of ability to adapt to changes. Third, the migration of people from rural areas to towns has reduced the potential of a voluntary workforce. The forth driver is the aging of the volunteers. All drivers, without new monitoring strategies, challenge both the performance and geographical coverage of monitoring systems. We expect that a combination of new technologies, such as remote sensing, the Internet of Things and Big Data, can empower new groups of volunteers and increase the

How to cite this book chapter:
Kettunen, J, Silander, J, Lindholm, M, Lehtiniemi, M, Setälä, O and Kaitala, S. 2016.
　Changing role of citizens in national environmental monitoring. In: Capineri, C,
　Haklay, M, Huang, H, Antoniou, V, Kettunen, J, Ostermann, F and Purves, R. (eds.)
　European Handbook of Crowdsourced Geographic Information, Pp. 257–267.
　London: Ubiquity Press. DOI: http://dx.doi.org/10.5334/bax.s. License: CC-BY 4.0.

social impact and effectiveness of voluntary monitoring to fulfil our national and international obligations.

Keywords

Monitoring, Environmental, Drivers, Voluntary, Urbanisation, Globalisation

Introduction

Monitoring provides a sufficient level of information to support decision-making and comply with legal requirements. Volunteers are one of the greatest resources for enforcing environmental laws and regulations. Their role is ever-changing as there are more obligations than ever before and fewer resources. There is a growing interest to motivate volunteers and increase the efficiency and social impact of monitoring.

The European Union has rapidly expanded and many new directives have come into force influencing environmental monitoring. The EU legislation has had an impact on the national legislation of all Member States and the daily lives of people living in the EU. Finland joined the EU in 1995, and today directives are influencing monitoring and their data dissemination practices. New monitoring projects, such as beach littering, have been established, and global obligations require the implementation of the INSPIRE Directive, which is based on the infrastructures for spatial information.

In the past ten years we have seen an immense increase in the number and volume of citizen science projects. The Internet, sensor technology and smart phones have made it easy to record observations with stamps on position and time, and the communication with data has become quick and easy. However, the participation of non-scientists in scientific research and data collection is not a new phenomenon. Well into the 19th century it was possible for non-professional scientists to contribute remarkably to scientific research (Haklay 2015). Significant scientific figures, such as Charles Darwin or Robert Boyle, may be considered, by today's standards, to have been non-professional scientists (Shapin 1994).

Urbanisation began relatively late in Finland and the process has been more rapid than in other European countries (Heikkilä 2003). The migration of people from rural areas to towns has reduced the potential of a voluntary workforce. Aging is another challenge, with the number of people over 60 being over four times higher than it was in 1900 (Official Statistics of Finland 2016). A whole new set of options have been used to solve challenges. The hydrological monitoring service is currently using new technologies and has increased marketing effort to recruit new observers. Modern monitoring projects, such as the algal watch, the Lake&Seawiki and jellyfish, have utilised new mobile sensor technologies as well as social networking and processes to mitigate the impacts of drivers. Different monitoring projects seem to have different dominating drivers, as shown in the following table.

	Globalisation	Grown obligations	Urbanisation	Ageing
Hydrology	0	0	−	−
Birds	+	+	0	−
Game animals	0	+	−	−
Algal watch	+	+	0	0
Lake&Seawiki	+	+	0	0
Jellyfish	0	0	0	0
Beach litter	+	+	0	0

Table 1: Impacts of different drivers (+, 0, −; positive, neutral, negative) on selected monitoring projects in Finland.

Selected monitoring projects

Hydrological monitoring – In service of infrastructure

Local volunteers were actively recruited into hydrological monitoring already in the beginning of the 20th century when the predecessor of the national hydrological service in Finland was initiated. The observers received small premiums for their services, but their key motivation was their own interest in hydrological phenomena (Kuusisto 2008a). For more than a hundred years these amateur station agents completed important observations on hydrological parameters such as water level, frost, snow thickness, water equivalent and ice cover.

Observers have usually been engaged with monitoring activities for a long period of time, with some monitoring sites having been managed by the same family since the 1910s. In the beginning, the number of observed parameters was large. Between 1913 and 1931, the volunteers also collected samples on water quality. There were 200 stations for the analysis of transparency and light attenuation coefficients at seven different wave lengths (Kuusisto 2008b), and 100 locations in which samples for organic and inorganic suspended sediment, dissolved inorganic and organic matters, alkalinity and dissolved oxygen demand were collected. The intensive programme continued until 1932, when the Great Depression forced the closure of many monitoring programmes. However, the hydrological network started after the Second World War.

The national hydrological monitoring is an example of the volunteers being very efficiently organised. From the beginning the system has taken into account the role of volunteers as a part of the entire system. Today, a remarkable part of the hydrological monitoring is automated, but the network of over 300 observers covers the entire country and is closely developed to support real-time forecasting systems (Ymparisto 2015). Trained observers take care of the pre-designed network of monitoring sites. This makes the system reliable and cost-efficient even in remote areas. The system is facing a great challenge as many of the elderly volunteers have stepped down and more marketing efforts are needed in order to recruit new amateurs to replace them, especially in rural areas.

Birds - Academic interest and hobby

In Finland regular bird monitoring started in the early 20th century. Though the main motivation for this was academic, safeguarding some of the diminishing bird populations was another goal as well. Up to the 1970s, voluntary birdwatchers were mainly skilled amateur naturalists living in urban areas. A central organisation for regional societies was established in the middle of the 1970s and the Bird Atlas project started activating birdwatchers in rural areas (Santaoja 2013). The central organisation joined the global network of bird organisations, BirdLife International, in 1992. Since then the number of birdwatchers has increased remarkably and today involves about 12,000 bird-watchers (Birdlife 2015).

Until the 1970s, the partnership between the birdwatchers and the Museum (Natural History Museum of Finland) was very tight. Since then the Museum has concentrated on programmed, traditional bird monitoring and left the collection and filing of other voluntary observations to the NGOs (regional ornithological societies and BirdLife Finland. They have maintained and collected the data since then and share it in the BirdLife database, which is open to all registered users. Both programmed monitoring and other data are collected from birdwatchers.

The majority of the official bird monitoring is now coordinated by the Museum. It is based on the work of voluntary birdwatchers who have committed to a regular monitoring, following the given instructions on when, where and how to monitor certain bird species or an area. The monitoring data is used in the European Bird Census Council's (EBCC) various projects, e.g. in the assessment of Pan-European Common Bird Indices. The monitoring data and the random observations of birdwatchers are made available as part of the Global Biodiversity Information Facility (GBIF), which aims to make the world's scientific biodiversity data freely and universally available via the Internet for the benefit of science, society and a sustainable future.

The reasons for increasing birdwatching are manifold. Today the literature and other methods for the identification of species are very developed. Besides strengthening regional societies, BirdLife has been active in publishing and communications and it has introduced a large set of new activities, such as the Big Garden Birdwatch in January, which is aimed for all citizens, and a competition called the Battle of the Bird Towers in May, which is a competition mainly for birdwatchers. The share of women has grown, and birdwatching is nowadays also a hobby for families. The current number of bird observations collected by BirdLife and its member societies is annually over 1 million in Finland. Bird monitoring is a global programme that has also managed to attract many observers from rural areas.

Game animals – Hobby and co-management

Official monitoring of the population abundance of game has been carried out in cooperation between the Ministry of Agriculture and Forestry, the Finnish Game and Fisheries Research Institute and hunter organisations since the 1970s. The objective of the data collection is to produce a scientific foundation for sustainable hunting. Researchers plan the census and organise it together with hunter organisations, volunteer hunters do the actual fieldwork (Rktl 2015a) and the public research institute conducts the analyses and reports to the national and local administrations.

Different game species groups have their own monitoring programmes. Forest and mixed-forest agricultural game species are monitored through wildlife triangle schemes. More than 30 forest game species are monitored. The population and breeding success of waterfowl are monitored with pair counts in May and brood counts in July. Specific census methods have been developed for moose, large carnivores, seals, beavers and wild forest reindeer.

It is important to note that in Finland about 300,000 citizens pay the annual hunting management fee, and the number has doubled between 1960 and 2000 (Saarsalmi et al. 2014). In 1960 a majority of the hunters owned their hunting land, but today 60 per cent hunt on rented land of their hunting club or on government land. A large proportion of the landless hunters live in towns. According to the enquiry (Rktl 2015b) the total amount of active voluntary work by hunters in Finland in 2008 comprised 290 man-years. Hunters were on standby for a total of 1,800 man-years to assist in moose, white-tailed deer and large game animal emergencies, like traffic accidents.

An estimated total of 40,000 hunters participated in voluntary work in 2008. The value of the voluntary work without overheads was estimated at 7.1 million euros. The Game Management Associations estimated that some 20,000 people performed voluntary work in game monitoring. Voluntary work in nationwide game monitoring schemes was estimated to be 89 man-years, of which observation of large carnivores made up 40 man-years. Hunters covered around 900,000 km with their cars to carry out these nationwide game monitoring schemes. Even though there have been more obligations for volunteers in Finland, these challenges have been overcome via good cooperation.

Algal Watch – Supporting other sources of information

Algal watch was initiated in 1998 to better inform the public about blue-green algal blooms in the Northern Baltic Sea. In 2000, the first group of trained volunteers, sea scouts, started monitoring coastal and archipelago areas in Southern Finland (Rapala et al. 2012). The work supplemented the data collected by commercial ferries in the Alg@line network (Finmari 2015) established in 1995 (Rantajärvi et al. 2003). The aim of the network was partly educational, but it also set in a practice for citizen monitoring on blue-green algal blooms, bladder wrack and water transparency. Later on, the system was extended and since 2011 citizens have been able to report their observations on blue-green algal blooms, bladder wrack density and Secchi depth by using a mobile phone application called Algae watch (Mmea 2015).

The application includes instructions and stamping, i.e. registration of position and timing of the observation is done by the GPS of the phone. It is also possible to take a photograph and send it simultaneously with the observation. During the performed pilot trials, no service misuse was detected (Kotovirta et al. 2014).

Users of the application were motivated by sending them a notification of the algae situation and reminders to contribute to observations in the future too. From the user data it can be seen that citizen activity decreased towards the end of the summer, although blooms were still present. This is most likely due to the timing of summer holidays in Finland, which usually end by the beginning of August (Kotovirta et al. 2014).

Lake&Seawiki – modern tools and social networking

Lake&Seawiki (Jarviwiki 2015) is a wiki service about Finnish lakes and coastal sea areas. The concept was developed in the Finnish Environment Institute (SYKE) to promote people's engagement in the protection and monitoring of their nearby waters and to allow non-professionals to upload observations on water temperature, ice situation, algal blooms etc. The service has been running since 2011 and anyone can contribute. The users of the service can take part in discussions and maintain their own observation sites. Service is well marketed and currently the first that is managed by the communication department of SYKE.

The service is running on open source software and has low operation costs. The moderation and upgrades of software require one person-month a year. Furthermore, an office hour helpdesk is needed in June-August. In 2014, almost 280,000 users visited the service, with 200 contributing during summer months and 30 during winter months. They produce more than 9,000 observations annually. The number of visitors has increased annually by 25% (Kettunen et al. 2014).

Still, a large part of people contributing to the service are citizens who have earlier been recording their observations in their private notebooks. The service has given them a platform and acts as an archive and a visualizer for their observations. Some time series clearly show the impact of climate change on ice breakup. Measurements received from experienced amateur observers are generally of good quality and thus complement nicely the other parts of the monitoring system. The service started receiving observations from desktop computers. Today, the share of mobile phone and tablet users is growing and is expected to speed up the growth of the service.

Jellyfish – Abundance unveiled for the 1st time

In 2010 the Finnish Environment Institute started to collect jellyfish observations from the public to determine jellyfish distribution in Finnish waters and the factors affecting the bloom formation in late summer. The proposal for the monitoring came from the energy industry, which needed data on jellyfish abundances and predictions on conditions in which jellyfish form blooms, since these can affect power plants by possibly clogging the cooling water intake pipes.

Using only public observations, the monitoring of jellyfish distribution and abundance has now been ongoing for 5 years. Citizens have reported their observations via a web form, which is planned to be developed into a mobile application in the near future. In the form citizens are asked to estimate the jellyfish abundance on a two-level scale (few, clearly less than <20 individuals m^{-2} or a bloom, >20 individuals m^{-2}), name the sea area where the observation was done, and if it was a coastal or an open sea observation. Wind and temperature estimates are also asked in the form.

This voluntary citizen science monitoring has revealed patterns in jellyfish abundance and distribution range in Finnish waters which would have otherwise been impossible to obtain. Further, the reported observations have also been used as a service when communicating with the power plants and other industry on the blooms close to their seawater intake areas. The challenge of this kind of monitoring is that it is highly dependent on the press releases informing citizens that observations are still (and continuously) needed. Without advertisements the number of observations is much smaller.

Beach litter – Global outsourced monitoring

Beach litter monitoring with the help of citizens started in Finland in 2012, when groups from Sweden, Finland, Estonia and Lithuania became partners in an EU-funded project. The aim was to implement the harmonised method for the first time around the Baltic Sea. The method used was based on a slightly modified UNEP protocol on beach litter survey (Cheshire, Adler & Barbière 2009). The length of the beach had to be adjusted to central Baltic conditions, as well as the timing of surveys. The survey was implemented by the Finnish partner in the project: a local NGO (Keep the Archipelago Clean).

For the survey to be successful, there has to be a dedicated group of people (at least two, but no more than 10) responsible for the survey during a certain time of the year on a certain beach. Each group has a contact person who collects the data and delivers it to the NGO in charge of the survey. In Finland the contact persons were people such as school teachers. Experience from the field has shown that the best commitment has come from schools, where the survey can be included as a part of environmental education.

The survey was a success in many ways, and the project was able to combine comparable data from macroscopic litter in different Baltic countries. The project received a lot of attention in national media, partly because Finland was categorised as the most littered country of the project. Some need of development/ improvement was also noted during the survey by the organiser. The type and location of the beach that will be surveyed has to be very carefully chosen. It is especially important when comparisons between countries are made. Each geographical area should have representative beaches from all beach categories: rural, urban or in between. It is especially important to identify what pressures are causing littering on the survey beach so that management is targeted correctly. Geographical expertise combined with local information of water currents, upwelling areas and other hydrographical aspects that may have an effect on the distribution of litter should also be included in the planning phase.

By using local citizens, the beach litter survey has proven to work so well that it is presently included in the Finnish monitoring plan for beach litter. The collaboration with the NGO is continuing, and new areas for monitoring are planned together with authorities.

Lessons learned

Society under change

There are some drivers in society that have had a strong influence on monitoring during the last few decades. One is urbanization, i.e. migration of people from rural areas to population centres. In Finland, this started as late as the 1960s, but the impact was stronger with the delay. Together with the rapid ageing of the Finnish population, it has gradually diminished the potential of getting new local volunteers for traditional hydrological and game animal monitoring. The lack of voluntary labour has worsened the situation with regards to tasks that require a constant standby or presence near the rural observation sites. Contrary to these continuous monitoring tasks, short-term monitoring efforts, such as wildlife triangle schemes, have revived again after years of downturn. This is explained by the intensive campaigns of hunter organisations. Short-term tasks are also better suited to the urban life-rhythm. The recent digitalisation of information systems has also made it possible to base the regulation of hunting on real-time data, which has encouraged voluntary monitoring work. Unlike the other traditional monitoring systems, birdwatching schemes seem to have benefitted from urbanisation. This has mainly followed from the unification and decentralisation of birdwatch organisations in the mid' 1970s (Santaoja 2013). The organisations have also been able to tackle the problem of ageing by developing new types of operational models based on ideas of competitions and other kinds of gamification, for example. Also, the educational and communicational material and tools supporting the birdwatch have greatly improved.

Geographical scale and stage in policy formation have changed

For the past 20 years, a strong driver changing both the ecological and environmental monitoring has been globalisation. It has changed the geographical scale of monitoring. After Finland joined the EU, the top-down regulations and the number of legal obligations have grown tremendously. Between 2000 and 2015, the existing ecological and environmental monitoring was redesigned to be compatible with EU regulations, which has also somewhat changed the citizen science in Finland. After 2008, however, the national economy has weakened and the environmental administration has been forced to reduce costs and reconsider the entire monitoring system. In the latest strategy, the Ministry of Environment (2011) has indicated new guidelines for environmental monitoring in 2020. According to the guidelines, the imperatives are to reduce costs while improving timeliness and usability. The tools suggested by the strategy are automation and digitalisation, remote sensing, increased use of applications of citizen science and increased co-operation between the public, private and voluntary sectors.

Operational models and the depth of engagement are different

Before, the traditional monitoring systems were top-down oriented. The volunteers were given a task to collect observations, briefed on the phenomenon and given forms to fill. Today, the operational models are more diverse and the depth of the engagement of volunteers varies (Haklay 2015). We see it in the birdwatch and game monitoring. Competitions and campaigns have remarkably increased the participation. However, the activity of the hobbyists is not constant. For example, during 2006-2010 57 per cent of the 1.2 million bird observations were made by 100 so-called superobservers. The first years of Lake&Seawiki have shown that people are most active in making observations and participating in wiki discussion while on vacation. In the beach litter watch we have seen that volunteers easily grow tired of observing if they do not have some additional motive.

All monitoring systems and the citizen science in them have their own life cycles. It seems probable that hydrological monitoring in its traditional form will be substituted by automation. However, it seems just as probable that Lake&Seawiki has also brought new means for extending the geographical coverage of snow and ice observations. Earlier, we designed our monitoring on a national basis. Our new systems are more generic and can also be taken into use internationally.

References

Birdlife. fi. 2015. *BirdLife Finland*. Available from: http://www.birdlife.fi/english/index.shtml [4 August 2015].

Cheshire, A., Adler, E., & Barbière, J. 2009. *UNEP/IOC guidelines on survey and monitoring of marine litter*, Nairobi: United Nations Environment Programme, Regional Seas Programme, University of Chicago Press, Chicago.

Finmari-infrastructure.fi. 2015. *Alg@line – FINMARI*. Available from: http://www.finmari-infrastructure.fi/ferrybox [3 August 2015].

Haklay, M. 2015. *Citizen Science and Policy: A European Perspective*, 1st edn. Available from: http://www.wilsoncenter.org/sites/default/files/Citizen_Science_Policy_European_Perspective_Haklay.pdf [4 August 2015].

Heikkila, E. 2003. Differential Urbanisation in Finland. *Tijd Voor Econ & Soc Geog Tijdschrift Voor Economische En Sociale Geografie, 94*(1): 49–63 (Web).

Jarviwiki.fi, Lake & Seawiki. Available from: http://www.jarviwiki.fi/wiki/Main_page?setlang=en [5 August 2015].

Kettunen, J., Silander, J., & Lindholm, M. 2014. Role of Citizens in the National Environmental Monitoring. In: Jørgensen, M., Brodersen, S., Dorland, J., & Copenhagen (Eds.) *6th Living Knowledge Conference Copenhagen*. Denmark, pp. 166–173.

Kotovirta, V., Toivanen, T., Järvinen, M., Lindholm, M., & Kallio, K. 2014. Participatory surface algal bloom monitoring in Finland in 2011–2013. *Environ Syst Res, 3*(1): 24.

Kuusisto, E. 2008a. 'Observers' in *The Water Cycle – Hydrological service in Finland 1908–2008*. In: Kuusisto, E. (Ed.) 1st edn., Finnish Environment Institute, Karisto Oy, Hameenlinna, pp. 106–110.

Kuusisto, E. 2008b. 'Water quality monitoring 1911–1931' in *The Water Cycle – Hydrological service in Finland 1908–2008*, Kuusisto, E. (Ed.) 1st edn., Finnish Environment Institute, Karisto Oy, Hämeenlinna, p. 40.

Luomus.fi. 2015. *Bird monitoring | LUOMUS*. Available from: http://www.luomus.fi/en/bird-monitoring [5 August 2015].

Ministry of the Environment. 2011. *Monitoring Strategy of the State of the Environment 2020*. Available from: https://helda.helsinki.fi/handle/10138/41382?show [4 August 2015].

Mmea.fi. 2015. *Environmental data – MMEA Testbed*. Available from: http://mmea.fi/cases/participatory-water-quality-measuring [3 August 2015].

Official Statistics of Finland (OSF). 2016. Population structure [e-publication]. ISSN=1797-5395. Helsinki: Statistics Finland [referred: 8.3.2016]. Access method: http://www.stat.fi/til/vaerak/index_en.html.

Rantajärvi, E., Ruokanen, L., Hällfors, S., Flinkman, J., Stipa, T., Suominen, T., Kaitala, S., & Maunula, P. 2003. 'Alg@line in 2003: 10 years of innovative plankton monitoring and research and operational information service in the Baltic Sea' in *Alg@line today*, Rantajärvi, E. (Ed.) 1st edn., Meri no. 48, Finnish Institute of Marine Research, pp. 9–16.

Rapala, J., Kilponen, J., Järvinen, M., & Lahti, K. 2012. Finland: Guidelines for Monitoring of Cyanobacteria and their Toxins. In: *Current Approaches to Cyanotoxin Risk Assessment, Risk management and Regulations in Different Countries*, Chorus, I. (Ed.), 1st edn., Federal Environment Agency (Umweltbundesamt), Germany, pp. 54–62.

Rktl.fi. 2015a. *FGFRI – Monitoring game abundance*. Available from: http://www.rktl.fi/english/game/monitoring_populations/ [5 August 2015].

Rktl.fi 2015b. *Hunters voluntary work*. Available from: http://www.rktl.fi/en/julkaisut/j/510.html [5 August 2015].

Saarsalmi, P., Koskela, T., Virtala, E., Murto, J., Pentala, O., Kauppinen, T., Karvonen, S., & Kaikkonen, R. 2014. *Terveyden ja hyvinvoinnin erot maalla ja kaupungissa vuonna 2013 – ATH-tutkimuksen tuloksia uuden kaupunki-maaseutu-luokituksen mukaan*, 1st edn., THL. Available from: http://www.julkari.fi/handle/10024/125351 [4 August 2015].

Santaoja, M. 2013. *For the love of Nature. Amateur Naturalists as Actors in Nature Conservation*, Acta Universitatis Tamperensis 1853, Tampere.

Shapin, S. 1994. *A social history of truth: civility and science in seventeenth-century England*, University of Chicago Press, Chicago.

Ymparisto.fi. 2015. *Environment > Hydrological situation and forecasts*. Available from: http://www.ymparisto.fi/en-US/Waters/Hydrological_situation_and_forecasts [3 August 2015].

CHAPTER 20

On the Contribution of Volunteered Geographic Information to Land Monitoring Efforts

Jamal Jokar Arsanjani* and Cidália C. Fonte†

*Department of Planning and Development, Aalborg University, A.C. Meyers Vænge 15, DK-2450 Copenhagen, Denmark, jja@plan.aau.dk, jamaljokar@gmail.com

†Department of Mathematics, University of Coimbra / INESC Coimbra, Coimbra, Portugal

Abstract

Land-related inventories are important sources of geoinformation for environmentalists, researchers, policy-makers, practitioners, and ecologists. Traditionally, a considerable amount of energy, time, and money have been dedicated to map global/regional/local land use datasets. While remote sensing images and techniques along with field surveying have been the main sources of data for determining land use features, field measurements of ground truth have always amplified the required time and money, as well as information credibility. Nowadays, volunteered geographic information (VGI) has shown its great contributions to different scientific disciplines. This was made possible thanks to Web 2.0 technologies and GPS-enabled devices, which have advanced citizens knowledge-based projects and made them user-friendly for volunteered citizens to collect and share their knowledge about geographical objects. OpenStreetMap as one of those leading VGI projects has shown its great potential for collecting and providing land use information. The collaboratively collected

How to cite this book chapter:
Arsanjani, J J and Fonte, C C. 2016. On the Contribution of Volunteered Geographic Information to Land Monitoring Efforts. In: Capineri, C, Haklay, M, Huang, H, Antoniou, V, Kettunen, J, Ostermann, F and Purves, R. (eds.) *European Handbook of Crowdsourced Geographic Information*, Pp. 269–284. London: Ubiquity Press. DOI: http://dx.doi.org/10.5334/bax.t. License: CC-BY 4.0.

land use features from diverse citizens could greatly back up the challenging element of land use mapping, which is in-field data gathering. Hence, in this literature we will look at the completeness, thematic accuracy and fitness for use of OpenStreetMap features for land mapping purposes over European countries. The empirical findings reveal that the degree of completeness varies widely ranging from 2% to 96% and overall and per-class thematic accuracies goes up to 80% and 96%, respectively compared to the European GMESUA datasets. Furthermore, more than 50% of land use features of eight European countries are mapped. This messages that the harnessing citizens' knowledge can play a great role in land mapping as an alternative and complementary data source.

Keywords

Land use mapping; Comparative assessment; Global Monitoring for Environment and Security Urban Atlas (GMESUA); OpenStreetMap

Introduction

Land cover (LC) and land use (LU) inventories contain geoinformation on the coverage and usage of our surrounding lands, respectively. LU and LC inventories are of high importance for many applications with regards to urban and regional planning, policy making, among others. These two concepts present two distinctive concepts, because LU maps explain human activities happening on the land, such as artificial surface construction, farming, and forestry that represent the usage of land (Ellis 2007; Wästfelt & Arnberg 2013), while LC maps present the physical cover on the ground (De Sherbinin 2002). Traditionally, applying image processing algorithms on remotely sensed data elaborated with ground-truth measurements and other complementary archive data have been the main source of collecting LU and LC features (Qi, Yeh, Li & Lin 2012; Saadat et al. 2011). Although remote sensing images and techniques often facilitate earth observation efforts, in-field surveying as well as personal interviews with local residents are required for the sake of results' validation, i.e. as ground-truth data coming from in-situ measurements play a critical role in delivering end products (Cihlar & Jansen 2001; De Leeuw et al. 2011). Therefore, we have to collect ancillary data as well in order to assign appropriate LU types to land parcels. As a result, LU mapping becomes even more complicated than LC mapping, and extensive data collection from local citizens, land managers, and evidence sources are vital for accurate LU mapping (Fritz et al., 2012).

From financial and temporal perspectives, a great deal of budget and time have been dedicated for producing LU and LC maps at global, regional, and local scales. Examples of global and regional scale with coarse resolution products include Global Land Cover (GLC)-2000 (Fritz et al., 2003), Moderate-resolution Imaging Spectroradiometer (MODIS (McIver & Friedl, 2002)), and GlobCover (Arino et al., 2012), CORINE 2000 (Büttner, Feranec, & Gabriel, 2002) and Global Monitoring for Environment and Security Urban Atlas (GMESUA (Seifert, 2009)) among others. In the case of GMESUA, high-resolution images including SPOT, RapidEye, and ALOS Images have been utilized to generate fine-scale maps of large metropolitan areas delivering GMESUA (Kong, Yin, Nakagoshi, & James, 2012). But, the accuracy of them has been the main concern as outlined by (Fritz et al., 2012; Herold, Mayaux, Woodcock, Baccini, & Schmullius, 2008). Thus, the necessity of having an alternative and complementary solution for mapping LU and LC features is evident. We believe that VGI could be of great importance, because the development of web technologies and large availability of GPS-enabled devices have resulted in the emergence of a large number of VGI platforms, which provide information about geographical objects from citizens (Fonte, Bastin, See, Foody, & Lupia, 2015). The majority of the VGI-like platforms offer very high-resolution satellite and aerial images (from 20 cm spatial resolution) through image libraries (e.g. Bing Maps) in their interfaces, which enable volunteers to visualize the whole globe with high detail so that they can map a large variety of features and attach respective attributes to them (Rouse, Bergeron, & Harris, 2007). In other words, a sort of visual analysis and interpretation of satellite images is applied. This convenient and straightforward way of visual interpretation of remote sensing images can be considered as an alternative solution for LU mapping and even achieving finer resolution LU maps than our current stored datasets at a global scale (Jokar Arsanjani, Mooney, Helbich, & Zipf, 2015). Undoubtedly, OSM has been a pioneer example of VGI and has shown its huge potential for being the Wikipedia of maps exactly as its motto. OSM is a unique platform for several reasons namely, it has attracted a huge amount of public attention and contributions (Ramm et al., 2011) by having exceeding 2.3 million users until today and continues to grow as outlined by Jokar Arsanjani, Helbich, et al. (2015). More importantly, OSM is highly democratic in receiving contributions through enabling any volunteer to add/edit/modify the existing features and sharing the whole data history freely and openly with the public in a structured way (Flanagin & Metzger, 2008; Koukoletsos, Haklay, & Ellul, 2012). Moreover, OSM collects geographic information in the form of GIS vector data such as points, polylines, and polygons and releases them based on different tags, which makes it quite user-friendly for end users (Jokar Arsanjani, Helbich, Bakillah, Hagenauer, & Zipf, 2013; Jokar Arsanjani, Mooney, Helbich, et al., 2015).

An extensive amount of analysis of road networks in OSM has been carried out (Ludwig, Voss, & Krause-Traudes, 2011; Mooney & Corcoran, 2012) and a few attempts in analyzing OSM for LU mapping has been conducted. We will assess the role of OSM in LU and LC mapping. Besides preparing a LU dataset from OSM contributions, we aim at a) measuring the completeness of OSM LU features, b) cross-comparing the thematic accuracy of the OSM LU features with the GMESUA data through a statistical assessment, c) assessing the fitness for use of OSM for LU and LC mapping.

Materials

OSM dataset

A snapshot of OSM features tagged as 'natural' and 'landuse' from November 2013 and February 2014 was collected. The features tagged with 'natural' describe a wide variety of physical features, which are categorized into different categories such as water bodies, forest, etc. as described in (Ramm, 2014). The term 'landuse' concerns the human use of land, which represents the purpose a land parcel is being used for.

Reference dataset

In this study, the pan-European GMESUA dataset serves as reference data, which comprises LU data for selected metropolitan areas exceeding 100,000 inhabitants. It is prepared for European needs and the contained information has been derived mainly from Earth Observation (EO) data supported by other reference data including commercial-off-the-shelf (COTS) navigation data and topographic maps. It has a minimum mapping unit (MMU) of 0.25–1 ha, and a minimum width of linear elements of 100 m with ± 5 m positional accuracy (European Union, 2011). It currently covers 305 urban regions within Europe. The minimum thematic accuracy for all classes is 80%. For more details see the Urban Atlas mapping guide (European Union, 2011). Table 1 represents the defined classes and their codes in GMESUA at different levels of details.

Study areas

In this study, the whole European continent was chosen as the study area for the regional scale analysis and ten random metropolitan areas were selected as case studies for the local scale analysis. These cities including their metropolitan areas are Berlin, Frankfurt am Main, Munich, and Hamburg, Bucharest, Rome, Stockholm, London, Budapest, and Vienna. Having multiple case studies from different countries would help to understand the heterogeneity of contributions in terms of quantity and quality.

		Classification Level (CL)		
Class type [Code]	CL0	CL1	CL2	CL3
	Land	Artificial surfaces [100]	Urban fabrics [110]	Continuous urban fabrics [111]
				Discontinuous urban fabrics [112]
				Isolated structures [113]
			Industrial, commercial, public, military, private and transport units [120]	Industrial, commercial, public, military and public units [121]
				Road and rail network and associated lands [122]
				Port areas [123]
				Airports [124]
			Mine, dump and construction sites [130]	Mineral extraction and dump sites [131]
				Construction sites [132]
				Land without current use [133]
			Artificial nonagricultural vegetated areas [140]	Green urban areas [141]
				Sports and leisure facilites [142]
		Agricultural + seminatural areas + wetlands [200]	-	-
		Forests [300]		
	Water	Water [500]	-	-

Table 1: Classification scheme applied in the preparation of GMESUA datasets as outlined in European Union (2011).

Methods

Quality of geodata should be considered internally and externally (Gervais, Bédard, Levesque, Bernier, & Devillers, 2009; Jokar Arsanjani, Barron, Bakillah, Helbich, & Arsanjani, 2013; van Oort, 2006). Internal quality reflects the specifications in the process of data production that address errors in the data. External quality measures the suitability of a dataset for a particular purpose and addresses its 'Fitness of Use' (FoU: (Devillers, Bédard, Jeansoulin, & Moulin, 2007; Guptill & Morrison, 1995)). The major standard organizations (e.g. ISO, ICA, FGDC, and CEN) have described their main criteria for data quality analysis and the following five criteria are common amongst them: (1) completeness, (2) positional accuracy, (3) thematic accuracy, (4) temporal accuracy, and (5) logical consistency (Guptill & Morrison, 1995). In this study, two major aspects of internal data quality namely completeness and thematic accuracy are considered and their external use is discussed.

Following Figure 1, first, OSM features tagged with 'landuse' and 'natural' are retrieved and merged together into a unique dataset. Second, overlaps and topological errors in the dataset are then resolved to assure the logical consistency of features. Third, the OSM features are re-classified and matched according to the GMESUA nomenclature. Fourth, the percentage of completeness for each country/city is determined to measure how complete a certain city is mapped. Finally, an error matrix between the OSM and GMESUA datasets is computed to measure the overall thematic accuracy of the OSM features along with a detailed per-class analysis.

Results and discussions

Completeness

Regional (European) scale

Figure 2 represents the measured completeness indices across European countries. This is calculated based on the total mapped area in each country relate to total area of the corresponding country. The values are diverse. While only 1.6% of land use features in Iceland are mapped, 96% of Bosnia and Herzegovina are mapped.

More than half of Belgium, Bosnia & Herzegovina, Germany, France, Luxemburg, the Netherlands, Romania, and Slovakia are mapped. Spatial distribution of the mapped features within Europe is displayed in Figure 3 by green cells. It should be noted that considering European countries with dissimilar population and physical patterns, these completeness values should not be used for judging the topology of citizen participations in OSM. For instance, Iceland with an area of 103,000 km^2 and nearly 300,000 inhabitants is the least

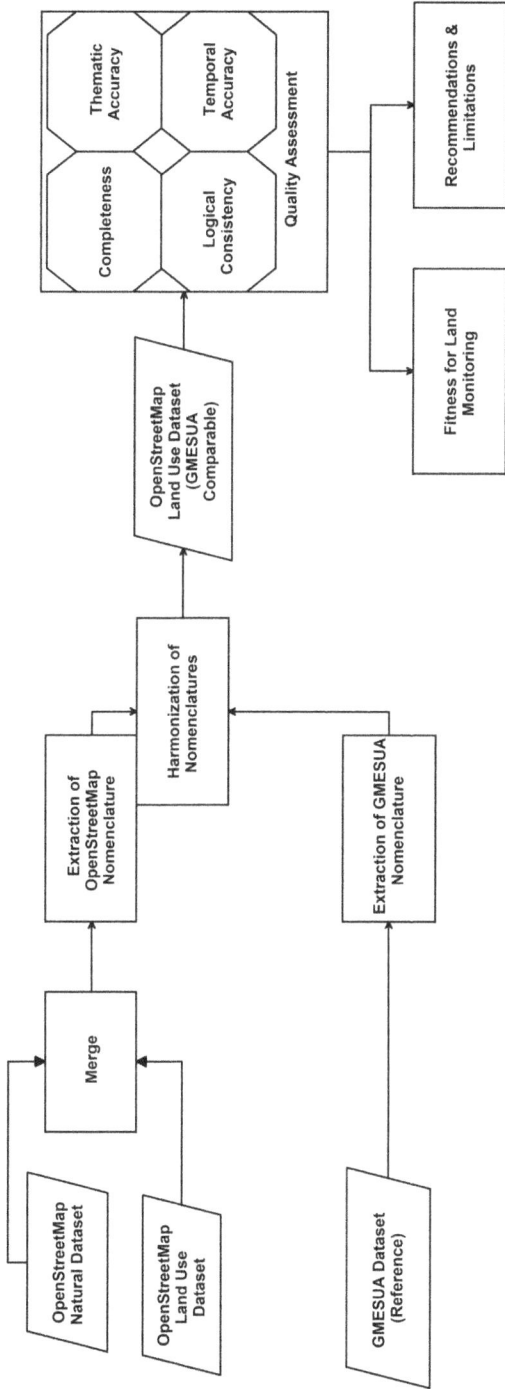

Figure 1: The flowchart of evaluating OSM land use features.

Completeness (%)

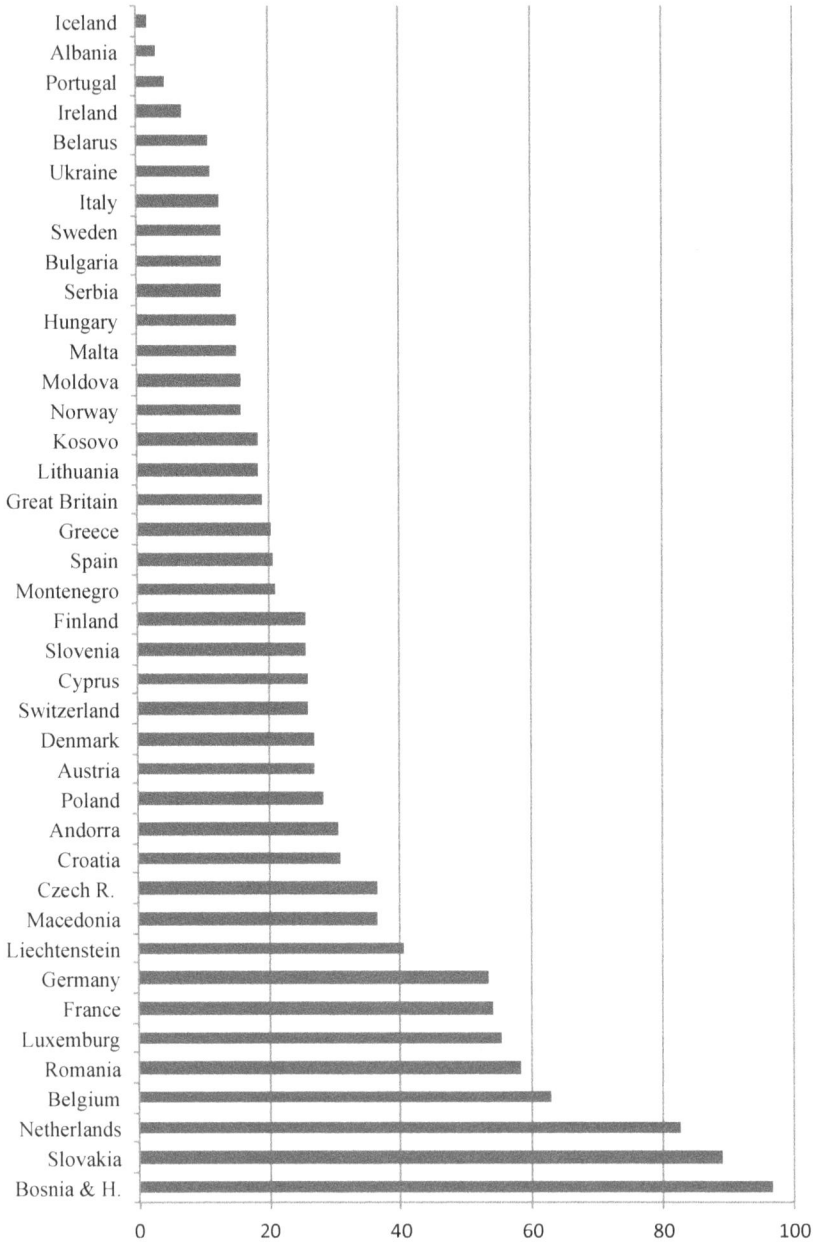

Figure 2: The calculated completeness index of OpenStreetMap land use features for European countries.

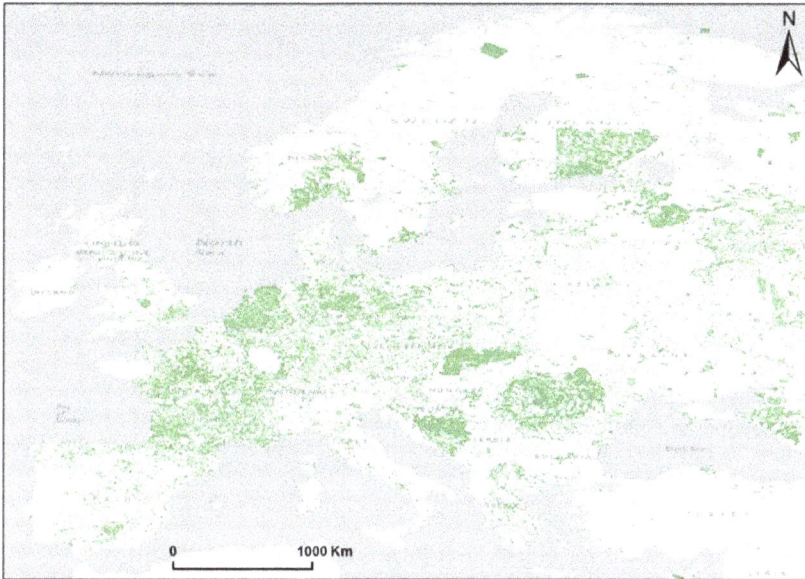

Figure 3: Spatial distribution of land use features from OpenStreetMap in Europe.

mapped country, which is not comparable with the Netherlands, holding an area of 41,500 km² and nearly 17 million inhabitants, corresponding to one the best mapped countries (82%). Likewise, while the completeness index for Sweden is reported as almost 13%, almost more than half of this country is covered by forests. This justifies the low completeness index value as minor residents live there or mappers do not prioritize mapping forests. This heterogeneity and inequality of public participation should be further investigated as outlined in (Jokar Arsanjani & Bakillah, 2015).

Local (metropolitan) scale

The degree of completeness at local level i.e. metropolitan area in several countries was checked and a wide range of values from 39% for Frankfurt to 100% for Bucharest was achieved. These values are shown in Figure 4.

Thematic accuracy

Apart from completeness, thematic accuracy is a key criterion to judge about the quality of the contributed LU features. This is meant to explore how properly the land parcels are tagged. Thematic accuracy is basically called 'accuracy

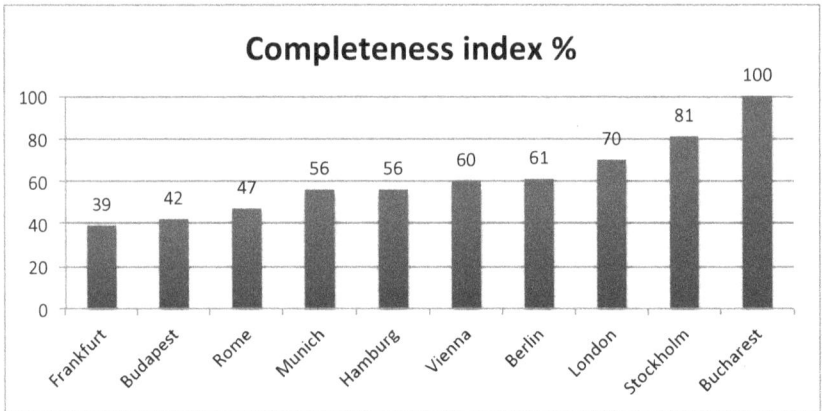

Figure 4: Completeness index of OpenStreetMap land use features for ten large metropolitan areas.

assessment' in the LU/LC classification studies, which reflects the difference between a target dataset against a reference dataset (Congalton, 1991; G. M. Foody et al., 2013; Giles M Foody, 2002). This is carried out through summarizing all data in a confusion matrix (i.e., error matrix) and calculating several indicators including 'overall/per class accuracies', 'Kappa index of agreement', 'user's accuracy' and 'producer's accuracy'" (Giles M Foody, 2002; Herold et al., 2008). In this study, a confusion matrix analysis is applied to reach these measures. A measure for the overall accuracy is calculated by dividing the number of identical pixels by the total number of pixels. However, it does not identify how well individual classes between the two datasets match. Hence, the user's accuracy and producer's accuracy should be calculated to measure the accuracy of each class (Herold et al. 2008). The user's accuracy indicates the probability that a pixel from the OSM LU map actually matches the GMESUA dataset, while the producer's accuracy refers to the probability that a specific LU type from the reference dataset is classified as such. These two measurements are not necessarily equal. For instance, if for a specific land type of 'farming', with accuracies achieved of 75% and 82% for user's accuracy and producer's accuracy respectively, it implies that as a user of the data, roughly 75% of all the pixels classified as 'farming' are the same in the reference dataset and, as a producer, only 82% of all 'farming' pixels are classified as such (Jokar Arsanjani et al., 2015).

In order to assess how well LU types in each city are mapped, Kappa index, overall accuracy, user's accuracy, and producer's accuracy, are calculated. Due to heterogeneous accuracies across cities, interpretation of the confusion matrix is discussed for each city separately in (Jokar Arsanjani, Mooney, Zipf, et al., 2015; Jokar Arsanjani & Vaz, 2015). Further to this, the geographical distribution of agreements and disagreements is visualized in (Jokar Arsanjani, Mooney, Zipf, et al., 2015). In general, land classes such as Isolated structures [113], Industrial, commercial, public, military and private units [121], Road and rail networks and associated land [122], Sport and leisure facilities [142], Agricultural+semi-natural+wetlands [200], Forests [300], and Water [500] show the highest level of agreement in the two datasets. In contrast, the remaining classes show disagreement, assuming that they are correctly reflected in the reference (GMESUA) dataset. This brings up the question whether OSM represents the right classification or the reference dataset. Finally, it can be concluded that the contributed OSM-LU features are heterogeneously distributed over inside/outside urban areas, which confirms the availability of LU features in both urban and rural areas.

Conclusions and recommendations

The recent emergence and rapid evolution of VGI platforms, such as OSM, has involved a massive number of citizens to collect and share geolocated information and attributes about geographical objects. This bottom-up process of collecting individuals' contributions has resulted in shaping big (geo)data, which has leveraged new applications such as indoor mapping (Goetz & Zipf, 2010), routing applications (Bakillah et al., 2014), tourism recommendations (Sun, Fan, Bakillah, & Zipf, 2013), and environmental monitoring (Fritz et al., 2012; Jokar Arsanjani & Vaz, 2015). Although the question on how to attract users and how to keep them active in the crowdsourcing activities is yet to be addressed, OSM has shown its continuing success in attracting more than 2.7 million users. Thus, a considerable potential in OSM exists and is yet to be further explored. Thus, in this study, we comparatively evaluated the completeness aspect of the contributed OSM-LU features across Europe as well as their thematic accuracy in ten large metropolitan areas to find out how reliable we could start exploiting them.

Results show that from a thematic accuracy perspective, the thematic quality of OSM features range from 'moderate' to 'substantial' rank of Kappa indices and overall accuracies. Per-class analysis of the LU types shows that, depending on the city, Isolated structures [113], Industrial, commercial, public, military and private units [121], Road and rail networks and associated land [122], Sport and leisure facilities [142], Agricultural+semi-natural+wetlands [200], Forests [300] and Water [500] reach the 'substantial' rank of accuracies, which means that these classes are highly useable. It should be noted that integrating ground-truth information with other reference data for accuracy assessment could be an alternative approach for producing hybrid LU datasets.

From a temporal accuracy perspective, archived images from within 2005-2010 have been used for LU mapping and this could have caused the above-mentioned disagreements, whereas the OSM-LU features have mainly been uploaded within since 2009, and therefore, some information from OSM might be even more close to reality than our reference data. Moreover, the MMU of the GMESUA datasets is 0.25–1 ha and, therefore, land parcels smaller than this MMU are ignored in the course of mapping, while in OSM even smaller parcels are mapped, i.e. a smaller MMU in OSM is possible. This means that in some parts while a polygon in GMESUA dataset is representing a specific LU type, the same area in OSM-LU dataset is covered by multiple small polygons showing multiple land types.

Concerning the volunteers' recognition of LU features, the citizens' perception of LU types should be further investigated to understand the way they visually interpret LU types from the online image libraries in OSM. As a final conclusion, the OSM-LU features message a promising data source for updating LU inventories. Certainly, the longer OSM exists, the more contributions will be received and consequently higher data quality can be achieved.

This study points out some other recommendations to the LU researchers, environmental scientists, policy makers, among others that will lead future research possibly in more suitable directions. Based on the presented completeness indices across Europe, as well as the accuracy values of the selected cities, the contributed OSM-LU features account for a potential alternative data source for mapping LU. Further studies on other areas must be conducted to explore the heterogeneity of completeness and thematic accuracy across space. Furthermore, applying data mining techniques and data fusion with national and regional datasets (e.g., GMESUA) for extracting the LU information of unmapped areas are of high importance. Additionally, the land types with the highest reliability can be separately incorporated into respective applications. This enables experts to: (a) possibly find ways to draw the attention of volunteer mappers to mapping LU features by highlighting their importance for more effective environmental monitoring, (b) possibly improve the OSM ontology of the LU dataset, (c) maximize the efficiency of OSM for LU mapping as users are not able to add further features in the urban areas, because the massive volume of mapped objects (e.g. POIs, roads, building, etc.) do not let users to have enough space for adding LU features.

Acknowledgments

The authors would like to acknowledge the support and contribution of COST Actions TD1202 "Mapping and Citizen Sensor" http://www.citizen-sensor-cost.eu and IC1203 "ENERGIC" http://vgibox.eu/.

References

Arino, O., Ramos Perez, J. J., Kalogirou, V., Bontemps, S., Defourny, P., & Van Bogaert, E. 2012. Global Land Cover Map for 2009 (GlobCover 2009). PANGAEA. DOI: http://dx.doi.org/10.1594/PANGAEA.787668

Bakillah, M., Lauer, J., Liang, S. H. L., Zipf, A., Jokar Arsanjani, J., Mobasheri, A., & Loos, L. 2014. Exploiting Big VGI to Improve Routing and Navigation Services. In *Big Data Techniques and Technologies in Geoinformatics*. CRC Press, pp. 177–192.

Büttner, G., Feranec, J., & Gabriel, J. 2002. *Corine land cover update 2000*. Retrieved from: http://www.fomento.es/nr/rdonlyres/de88aec2-d5fa-4e95-9694-939a326cd026/3128/021219techrep89.pdf.

Cihlar, J., & Jansen, L. J. M. 2001. From Land Cover to Land Use: A Methodology for Efficient Land Use Mapping over Large Areas. *The Professional Geographer*, *53*(2): 275–289. DOI: http://dx.doi.org/10.1111/0033-0124.00285

Congalton, R. G. 1991. A review of assessing the accuracy of classifications of remotely sensed data. *Remote Sensing of Environment*, *37*(1): 35–46. DOI: http://dx.doi.org/10.1016/0034-4257(91)90048-B

De Leeuw, J., Said, M., Ortegah, L., Nagda, S., Georgiadou, Y., & DeBlois, M. 2011. An Assessment of the Accuracy of Volunteered Road Map Production in Western Kenya. *Remote Sensing*, *3*(12): 247–256. DOI: http://dx.doi.org/10.3390/rs3020247

De Sherbinin, A. 2002. *A CIESIN thematic guide to land land-use and land land-cover change (LUCC)*. NY, USA. Retrieved from: http://sedac.ciesin.columbia.edu/binaries/web/sedac/thematic-guides/ciesin_lucc_tg.pdf.

Devillers, R., Bédard, Y., Jeansoulin, R., & Moulin, B. 2007. Towards spatial data quality information analysis tools for experts assessing the fitness for use of spatial data. *Int. J. Geogr. Inf. Sci.*, *21*(3): 261–282. DOI: http://dx.doi.org/10.1080/13658810600911879

Ellis, E. 2007. Land-use and land-cover change. In *Earth*. Retrieved from: http://www.eoearth.org/article/Land-use_and_land-cover_change.

European Union. 2011. *Mapping Guide for a European Urban Atlas*. Retrieved from: http://www.eea.europa.eu/data-and-maps/data/urban-atlas/mapping-guide/urban_atlas_2006_mapping_guide_v2_final.pdf/download

Flanagin, A. J., & Metzger, M. J. 2008. The credibility of volunteered geographic information. *GeoJournal*, *72*(3–4): 137–148. DOI: http://dx.doi.org/10.1007/s10708-008-9188-y

Fonte, C. C., Bastin, L., See, L., Foody, G., & Lupia, F. 2015. Usability of VGI for validation of land cover maps. *International Journal of Geographical Information Science*, *29*(7): 1269–1291. DOI: http://dx.doi.org/10.1080/13658816.2015.1018266

Foody, G. M. 2002. *Status of land cover classification accuracy assessment, 80*: 185–201.

Foody, G. M., See, L., Fritz, S., Van der Velde, M., Perger, C., Schill, C., & Boyd, D. S. 2013. Assessing the Accuracy of Volunteered Geographic Information arising from Multiple Contributors to an Internet Based Collaborative Project. *Transactions in GIS*, (828332): n/a–n/a. DOI: http://dx.doi.org/10.1111/tgis.12033

Fritz, S., Bartholomé, E., Belward, A., Hartley, A., Stibig, H. J., Eva, H., …, others. 2003. *Harmonisation, mosaicing and production of the Global Land Cover 2000 database (Beta Version)*. Office for Official Publications of the European Communities Luxembourg.

Fritz, S., Mccallum, I., Schill, C., Perger, C., See, L., Schepaschenko, D., …, Velde, M. Van Der. 2012. Geo-Wiki: An online platform for improving global land cover. *Environmental Modelling & Software*, *31*(0): 110–123. DOI: http://dx.doi.org/10.1016/j.envsoft.2011.11.015

Gervais, M., Bédard, Y., Levesque, M., Bernier, E., & Devillers, R. 2009. Data quality issues and geographic knowledge discovery. *Geographic Data Mining and Knowledge Discovery*: 99–115.

Goetz, M., & Zipf, A. 2010. Extending OpenStreetMap to Indoor Environments : Bringing Volunteered Geographic Information to the Next Level, (Hansen 2004).

Guptill, S. C., & Morrison, J. L. 1995. *Elements of spatial data quality*. Oxford [etc.]: Elsevier Science.

Herold, M., Mayaux, P., Woodcock, C. E., Baccini, A., & Schmullius, C. 2008. Some challenges in global land cover mapping : An assessment of agreement and accuracy in existing 1 km datasets, *112*: 2538–2556. DOI: http://dx.doi.org/10.1016/j.rse.2007.11.013

Jokar Arsanjani, J., & Bakillah, M. 2015. Understanding the potential relationship between the socio-economic variables and contributions to OpenStreetMap. *International Journal of Digital Earth*, *0*(0): 1–16. DOI: http://dx.doi.org/10.1080/17538947.2014.951081

Jokar Arsanjani, J., Barron, C., Bakillah, M., Helbich, M., & Arsanjani, J. J. 2013. Assessing the Quality of OpenStreetMap Contributors together with their Contributions. In *16th AGILE International Conference on Geographic Information Science*. Leuven, Belgium, pp. 14–17. Retrieved from: http://www.agile-online.org/Conference_Paper/CDs/agile_2013/Short_Papers/SP_S4.2_Arsanjani.pdf.

Jokar Arsanjani, J., Helbich, M., Bakillah, M., Hagenauer, J., & Zipf, A. 2013. Toward mapping land-use patterns from volunteered geographic information. *International Journal of Geographical Information Science*, *27*(12): 2264–2278. DOI: http://dx.doi.org/10.1080/13658816.2013.800871

Jokar Arsanjani, J., Helbich, M., Bakillah, M., & Loos, L. 2015. The emergence and evolution of OpenStreetMap: a cellular automata approach. *International Journal of Digital Earth*, *8*(1): 76–90. DOI: http://dx.doi.org/10.1080/17538947.2013.847125

Jokar Arsanjani, J., Mooney, P., Helbich, M., & Zipf, A. 2015. An exploration of future patterns of the contributions to OpenStreetMap and development of a Contribution Index. *Transactions in GIS*, *19*(6): 869–914. DOI: http://dx.doi.org/10.1111/tgis.12139

Jokar Arsanjani, J., Mooney, P., Zipf, A., & Schauss, A. 2015. Quality assessment of the contributed land use information from OpenStreetMap versus authoritative datasets. In: Jokar Arsanjani, J., Zipf, A., Mooney, P., & Hel-bich, M., (Eds.) *OpenStreetMap in GIScience: experiences, research, applica-tions.* Springer.

Jokar Arsanjani, J., & Vaz, E. 2015. An assessment of a collaborative mapping approach for exploring land use patterns for several European metropo-lises. *International Journal of Applied Earth Observation and Geoinforma-tion*, 35(0): 329–337. DOI: http://dx.doi.org/10.1016/j.jag.2014.09.009

Kong, F., Yin, H., Nakagoshi, N., & James, P. 2012. Simulating urban growth processes incorporating a potential model with spatial metrics. *Ecological Indicators, 20*: 82–91. DOI: http://dx.doi.org/10.1016/j.ecolind.2012.02.003

Koukoletsos, T., Haklay, M., & Ellul, C. 2012. Assessing Data Completeness of VGI through an Automated Matching Procedure for Linear Data. *Trans-actions in GIS*, (Goodchild 2007), no–no. DOI: http://dx.doi.org/10.1111/j.1467-9671.2012.01304.x

Ludwig, I., Voss, A., & Krause-Traudes, M. 2011. A Comparison of the Street Networks of Navteq and OSM in Germany. In: Geertman, S., Reinhardt, W., & Toppen, F. (Eds.) *Advancing Geoinformation Science for a Changing World SE – 4.* Springer Berlin Heidelberg, pp. 65–84. DOI: http://dx.doi.org/10.1007/978-3-642-19789-5_4

McIver, D., & Friedl, M. 2002. Using prior probabilities in decision-tree clas-sification of remotely sensed data. *Remote Sensing of Environment, 81*(2–3): 253–261. DOI: http://dx.doi.org/10.1016/S0034-4257(02)00003-2

Mooney, P., & Corcoran, P. 2012. OpenStreetMap, *16*(4): 561–579. DOI: http://dx.doi.org/10.1111/j.1467-9671.2012.01306.x

Qi, Z., Yeh, A. G.-O., Li, X., & Lin, Z. 2012. A novel algorithm for land use and land cover classification using RADARSAT-2 polarimetric SAR data. *Remote Sensing of Environment, 118*(0): 21–39. Retrieved from: http://www.sciencedirect.com/science/article/pii/S0034425711003877.

Ramm, F. 2014. *OpenStreetMap Data in Layered GIS Format.* Retrieved from: http://www.geofabrik.de/data/geofabrik-osm-gis-standard-0.6.pdf.

Ramm, F., Names, I., Files, S. S., Catalogue, F., Features, P., Features, N., …, Cars, C. 2011. OpenStreetMap Data in Layered GIS Format, pp. 1–21.

Rouse, L. J., Bergeron, S. J., & Harris, T. M. 2007. Participating in the Geospa-tial Web: Collaborative Mapping, Social Networks and Participatory GIS. In: Scharl, A., & Tochtermann, K., (Eds.), *The Geospatial Web*. Springer London, pp. 153–158. Retrieved from: http://dx.doi.org/10.1007/978-1-84628-827-2_14

Saadat, H., Adamowski, J., Bonnell, R., Sharifi, F., Namdar, M., & Ale-Ebra-him, S. 2011. Land use and land cover classification over a large area in Iran based on single date analysis of satellite imagery. *ISPRS Journal of Pho-togrammetry and Remote Sensing*, 66(5): 608–619. Retrieved from: http://www.sciencedirect.com/science/article/pii/S0924271611000517.

Seifert, F. 2009. Improving Urban Monitoring toward a European Urban Atlas. In: *Global Mapping of Human Settlement.* CRC Press. DOI: http://dx.doi.org/10.1201/9781420083408-c11

Sun, Y., Fan, H., Bakillah, M., & Zipf, A. 2013. Road-based travel recommendation using geo-tagged images. *Computers, Environment and Urban Systems*. DOI: http://dx.doi.org/10.1016/j.compenvurbsys.2013.07.006

van Oort, P. 2006. *Spatial data quality: From Description to Application*. Wageningen University.

Vaz, E., Walczynska, A., & Nijkamp, P. 2013. Regional challenges in tourist wetland systems: an integrated approach to the Ria Formosa in the Algarve, Portugal. *Regional Environmental Change*, *13*(1): 33–42. DOI: http://dx.doi.org/10.1007/s10113-012-0310-9

Wästfelt, A., & Arnberg, W. 2013. Local spatial context measurements used to explore the relationship between land cover and land use functions. *International Journal of Applied Earth Observation and Geoinformation*, *23*(0): 234–244. DOI: http://dx.doi.org/10.1016/j.jag.2012.09.006

CHAPTER 21

Discussing the Potential of Crowdsourced Geographic Information for Urban Areas Monitoring Using the Panoramio Initiative: A Case Study in Rome, Italy

Flavio Lupia* and Jacinto Estima[†]

*Council for Agricultural Research and Economics – CREA, Roma, Italy,
flavio.lupia@crea.gov.it
[†]NOVA Information Management School (IMS),
Universidade Nova de Lisboa (UNL), Lisbon, Portugal,
D2011086@novaims.unl.pt

Abstract

During the last decade, Crowdsourced Geographic Information or Volunteered Geographic Information has attracted the attention of the research community to explore this vast amount of data and extract useful information for various applications. Among these, geotagged photos shared publicly online have been explored as potential source for Land Use/Cover (LULC) mapping creation and validation. In this work, we performed an evaluation of the adequacy of the geotagged photos available from the Panoramio initiative for monitoring LULC in urban areas, with a study conducted in the urban area of Rome, Italy. We investigated the temporal distribution of the photos for the time range 2007–2013 with different resolution (year, season and month) as well as the spatial distribution in the study area. Then, we evaluated the representativeness of the

---

How to cite this book chapter:
Lupia, F and Estima, J. 2016. Discussing The Potential Of Crowdsourced Geographic Information For Urban Areas Monitoring Using The Panoramio Initiative: A Case Study In Rome, Italy. In: Capineri, C, Haklay, M, Huang, H, Antoniou, V, Kettunen, J, Ostermann, F and Purves, R. (eds.) *European Handbook of Crowdsourced Geographic Information*, Pp. 285–294. London: Ubiquity Press. DOI: http://dx.doi.org/10.5334/bax.u. License: CC-BY 4.0.

Panoramio dataset for each LULC class by using the Urban Atlas database as a reference and computing the number and density of photos per square km for each class. Finally we discussed the main limitations of the dataset for LULC monitoring in urban areas and we proposed alternative approaches useful to overcome some of the identified limitations.

Keywords

Geotagged photographs, Panoramio, Volunteered geographic information.

Introduction

Land Use/Cover Change (LULCC) is one of the most relevant phenomena caused by humans and linked with global environmental and climate change. LULCC trends are characterized by loss of natural areas and by the expansion of urbanization due to population increase. Within urban areas, landscape and soil functions are threatened by the expansion of artificial surfaces and therefore mapping and monitoring LULCC is crucial to support proper planning decisions. Nevertheless, the creation and update of geographic information occur with low frequency being realized institutionally by mapping agencies and being particularly expensive (Goodchild 2008).

The geographic information created and shared by the crowd has been increasing significantly over the last decade and, today, might be a potential alternative, to some extent, to the official map-making. This Crowdsourced Geographic Information phenomenon, also called Neogeography (Tuner 2006), Volunteered Geographic Information (Goodchild 2007), and more recently Ambient Geographic Information (Stefanidis, Crooks & Radzikowski 2011), have attracted the attention of the research community to explore this vast amount of data and extract useful information for various applications (e.g. Arsanjani et al. 2013). In particular, geotagged photos have been explored for Land Use/Cover (LULC) applications (Estima & Painho 2013; Estima & Painho 2014; Lupia, Estima & Painho 2015) at different geographic scales showing their potential for this kind of application, despite some issues related with this type of data such as the positional accuracy or the content of photos.

The main contribution of this paper is to analyse the potential use of geotagged photos in monitoring LULC in urban areas by discussing also the main limitations and suggesting improvements and possible solutions to overcome them. We focused on the Panoramio initiative with a case study in the urban area of Rome, Italy (Figure 1). We explored the temporal and spatial characteristics of the Panoramio dataset by assessing the representativeness of the photos in each LULC class and using the last version of the Urban Atlas database (EEA 2012) as a reference.

Methodology

Datasets and study area

We performed our analysis in the inner area of Rome (Italy) covering 343 km²
and delimited by the *Grande Raccordo Anulare (GRA)*, the highway encircling
the urban area. The city has experienced, during the last fifty years, relevant
land use changes with phenomena such as soil sealing and urban sprawl that
have modelled the actual spatial structure.

The Urban Atlas (UA) 2006 dataset was used as a reference for LULC data.
The UA was produced in 2009 through the Global Monitoring for Environment
and Security (GMES) program. The LULC classes are based on the Corine Land
Cover with a detailed characterization of the artificial classes by providing the
degree of sealing in percentage for some subclasses (Sealing Layer). The geometric
resolution is 1:10,000 with a minimum mapping unit of 0.25 ha (EEA 2012).

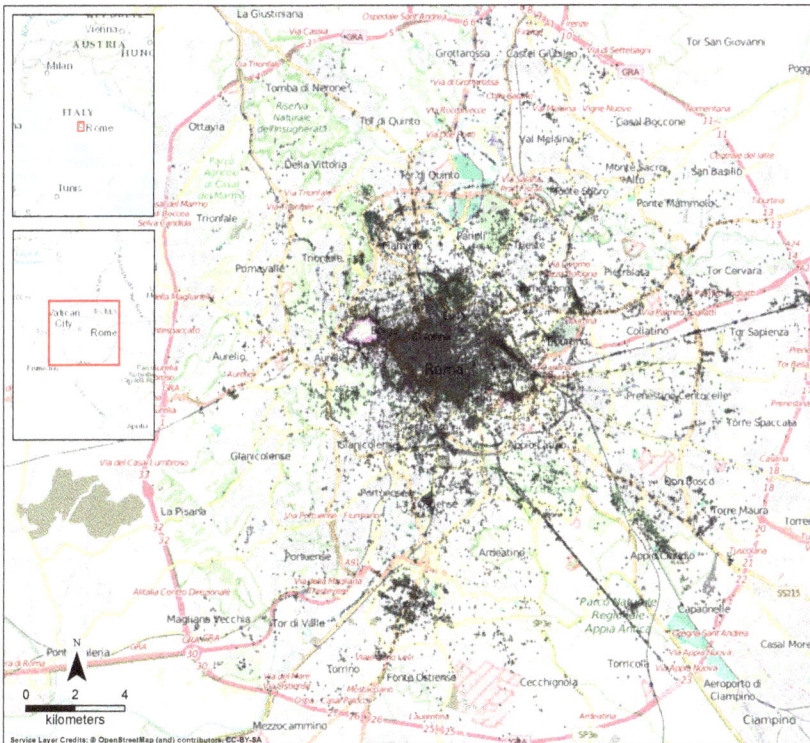

Figure 1: The study area: the urban area of Rome delimited by the round
shaped highway Grande Raccordo Anulare (in light red). Points depict the
spatial distribution of the Panoramio geotagged photos extracted for the time
range 2007–2013.

Urban Atlas class	Area (km²)	%
Continuous Urban Fabric	36.42	10.61%
Discontinuous Dense Urban Fabric	41.74	12.16%
Discontinuous Medium Density Urban Fabric	15.24	4.44%
Discontinuous Low Density Urban Fabric	8.55	2.49%
Discontinuous Very Low Density Urban Fabric	1.84	0.54%
Isolated Structures	0.88	0.26%
Industrial, commercial, public, military and private units	57.47	16.74%
Fast transit roads and associated land	1.94	0.57%
Other roads and associated land	32.00	9.32%
Railways and associated land	4.41	1.28%
Airports	1.00	0.29%
Mineral extraction and dump sites	0.64	0.19%
Construction sites	5.60	1.63%
Land without current use	3.40	0.99%
Green urban areas	28.53	8.31%
Sports and leisure facilities	13.43	3.91%
Total Artificial surfaces	**253.09**	**73.72%**
Agricultural + Semi-natural areas + Wetlands	79.79	23.24%
Forests	7.51	2.19%
Water bodies	2.91	0.85%
Grand total	**343.30**	**100.00%**

Table 1: Area and percentage over the total of the Urban Atlas classes in the study area.

The study area (see Table 1) is covered by Artificial surfaces for almost two-thirds (73.72%), followed by Agricultural + Semi-natural areas + Wetlands (23.24%), Forests (2.19%) and Water bodies (0.85%).

The dataset containing all the publicly available geotagged photos from the Panoramio initiative was created by downloading the metadata for all the available photos within the study area. This task was performed by using a script to contact the Panoramio servers through their public Application Programming Interface (API) and collect the available metadata for each photo (e.g. latitude, longitude, photo ID, user ID, upload date, etc.). The resulting dataset was composed by a total of 26,908 georeferenced photographs for the time interval 16 October 2005 – 11 August 2014. The dataset was then converted to a GIS format, a point shapefile in this case, using the latitude and longitude attributes of each photo.

As the year of 2014 was not complete and there was a very small number of photos during the first years of the initiative (183 total photos for 2005 and 2006) and this could bias the results in terms of temporal analysis, we decided to remove them from the final collection of photos. Therefore, a subset containing 24,367 photos for the period 2007-2013 was extracted and used for the subsequent analysis. The subset excluded also the 1,035 pictures inside the Vatican City State for which LULC data from UA were not available.

Data analysis

To assess the potential of the Panoramio dataset for monitoring LULC in the urban area of Rome, the following method was used:

1) Analysis of the temporal distribution of the photos. We used the "upload date" tag of the Panoramio dataset to understand the temporal distribution within the study area by using three resolution: month, season and year. Monthly and seasonal temporal distributions were evaluated by using the average number of photos for the time range 2007–2013.
2) Analysis of the spatial distribution of the photos within the study area. We observed the spatial distribution of photos within the study area to verify uniformity or clustering both through visual inspection and by computing number and density of photos (number of photos per km^2) for some spatial units.
3) Analysis of the spatial distribution of the photos within each LULC class. We computed the number and density of photos for each UA class to assess the degree of coverage for every LULC class inside the study area.

Results and discussion

Temporal distribution

Results by year show an increase of the number of photos, after the start of the initiative, with maximum values in 2011 (4,144) and 2012 (4,379) and a yearly average of 3,481 for the period 2007-2013 (Figure 2-a). The distribution of the average number of photos by month has the highest values in February, October and November and a minimum in September (Figure 2-c). By observing the average distribution per season the majority of photos are uploaded in winter, on the contrary the lowest values are in summer (Figure 2-b). A possible explanation to this temporal trend could be that a large part of photos are taken by tourists from other countries during their summer vacation, while the uploading phase is postponed to the winter time because they don't have high speed internet connection to share the photos immediately.

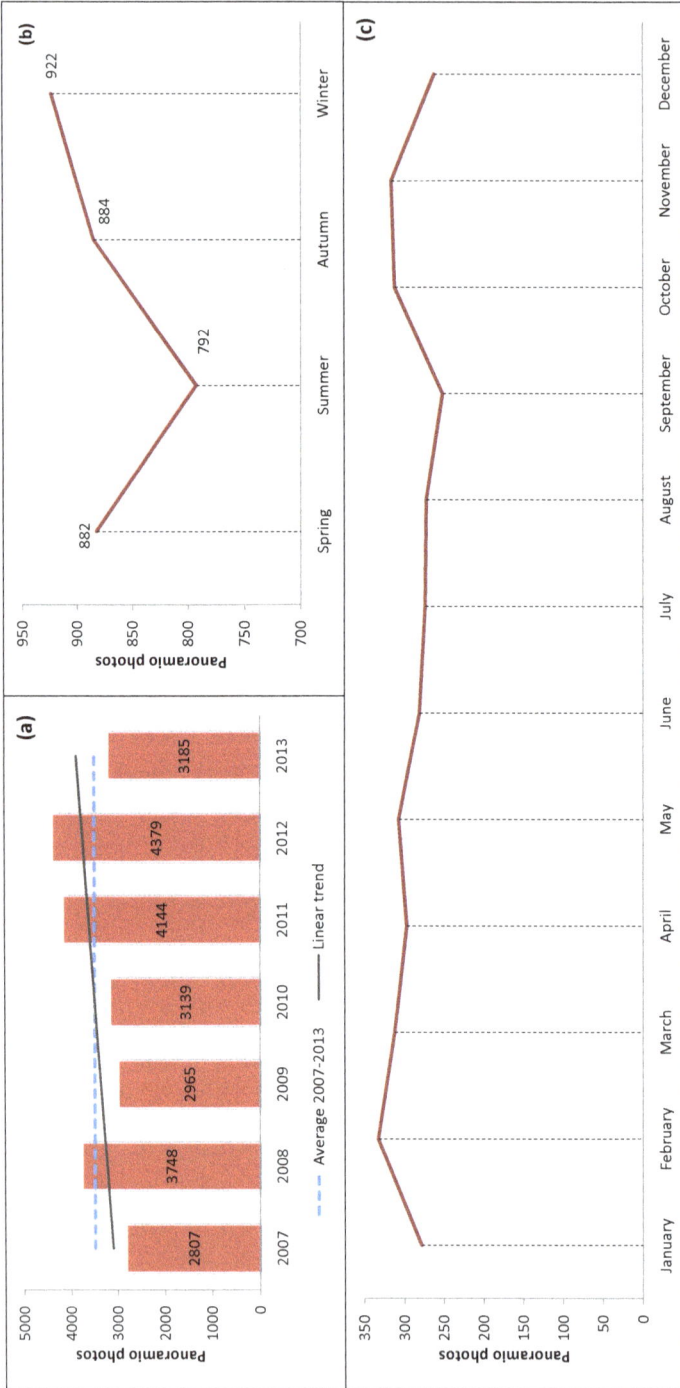

Figure 2: Number of photos per year with lines of the average and the linear increasing trend for the period 2007–2013 (a); seasonal (b) and monthly (c) average of photos for the period 2007–2013.

Spatial distribution within the study area

A visual analysis of the spatial distribution of the photos show a strong concentration in the urban centre where the main tourist attractions are located, while moving outward the concentration decrease strongly (Figure 1). Over the whole study area the average density is 71 photos/km². However, this value changes abruptly across the study area where photos create clusters of different size and shape. Photos can be concentrated along linear features, for example, the cluster along the South-East direction (Figure 3) is centered on the famous ancient road *Via Appia Antica* (633 photos/km² inside a 50 meters buffer around the centreline of the road). Another example is the Vatican area. Although Vatican is not considered for this study we calculated the density of photos to understand the impact of tourist attractions to the availability of data; as expected, this small area (0.53 km²) has an extremely high density (1,957 photos/km² ca.).

Spatial distribution over UA classes

In terms of number of photos the majority is concentrated inside Artificial surfaces (22,713 photos, representing 93.27%), followed by Agricultural +

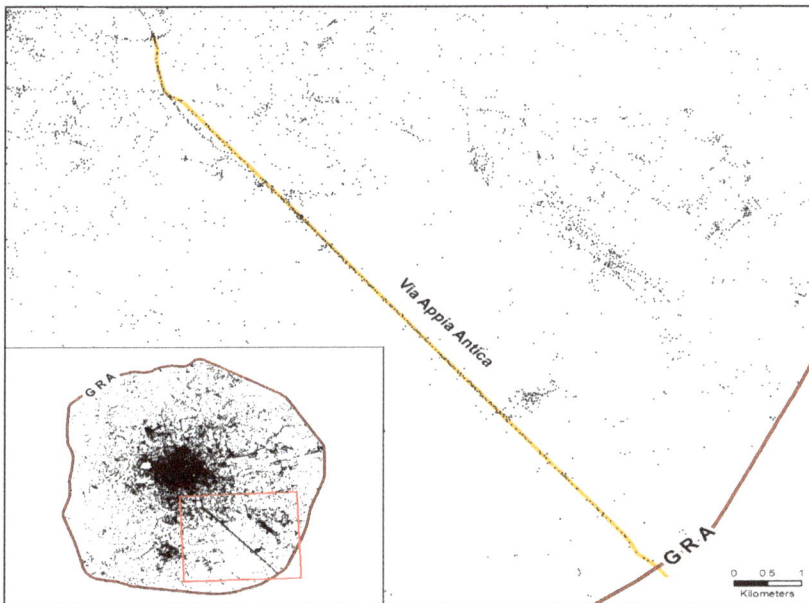

Figure 3: Concentration of the Panoramio photos along the famous ancient road Via Appia Antica. The number and the density of photos were computed for the area delimited with a 50 m wide buffer along the centreline of the road.

Semi-natural areas + Wetlands (1,007 photos, representing 4.14%), Water bodies (598 photos, representing 2.46%) and Forests (35 photos, representing 0.14%). Two-thirds of the photos belonging to the Artificial surfaces are distributed in the following subclasses: Industrial, commercial, public, military and private units (6,862 photos, representing 28.18%), Other roads and associated land (5,401 photos, representing 22.18%) and Continuous Urban Fabric (4,324 photos, representing 17.76%), see Figure 4-b.

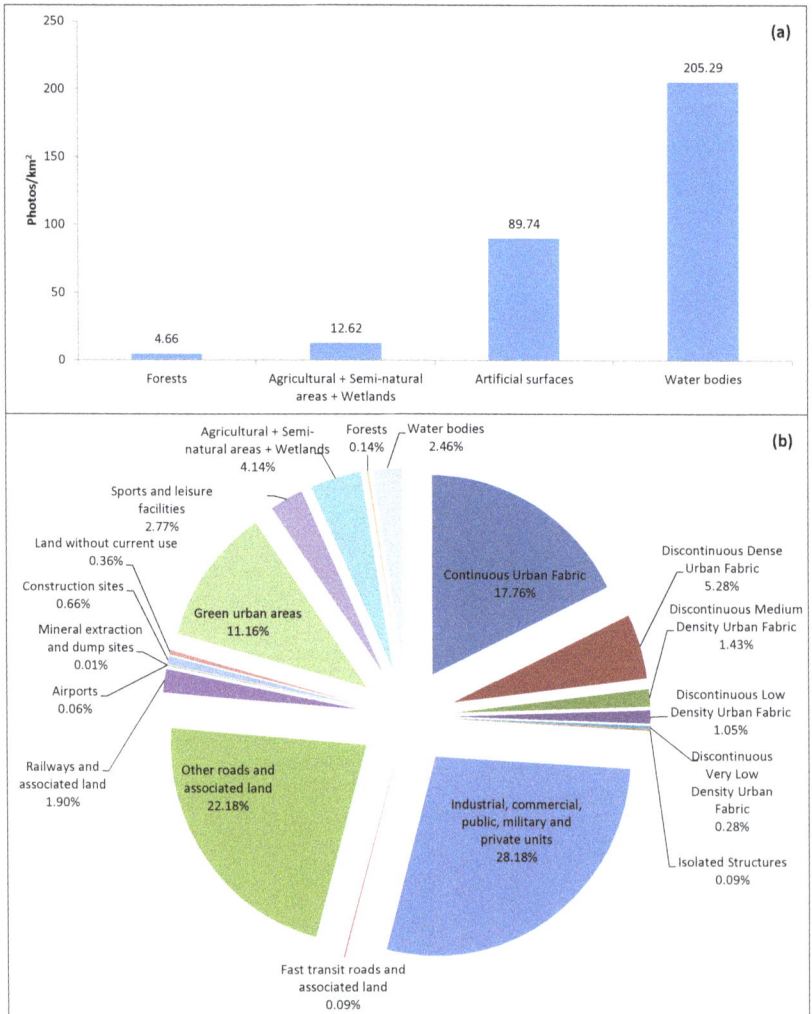

Figure 4: Density (number of photos per km^2) (a); number of photos over the total (in percentage) for each Urban Atlas class (b).

In terms of density (number of photos per km²), Water bodies have the highest density (205.29), followed by the Artificial surfaces (89.74), Agricultural + Semi-natural areas + Wetlands (12.62) and Forests (4.66), see Figuere 4-a. Within the Artificial surfaces the following subclasses have the highest values of density: Other roads and associated land (168.8), Industrial, commercial, public, military and private units (119.4), Continuous Urban Fabric (118.7), Railways and associated land (104.8) and Green urban areas (95.3). The density and the number of the photos within the UA classes confirm a strong unevenness with a predominance of potential information in the Artificial surfaces and, surprisingly, in the Water bodies, which correspond to the Tiber River. The latter result can be explained with the large number of photos (598) spread over a very small surface of the study area (2.91 km², 0.85%). Tiber River is an important landmark with several relevant tourist attractions along its banks, but also a place monitored for environmental aspects. In fact, 76 out of 598 (12.71%) photos were published by a public authority during field observations during the period 2007-2013 and 190 out of 736 (25.82%) during the period 16/10/2005 – 11/08/2014.

Conclusions

In this paper, we analysed the potential of geotagged photos from the Panoramio initiative as a source of information for LULC monitoring in urban areas, with a case study in the city of Rome.

Similarly to what has been reported in Estima and Painho (2013, 2014), the most positive aspects of this dataset are the amount of available photos and their temporal distribution. On the opposite side, this dataset showed some limitations for urban land use monitoring analysis. Some LULC classes have a better coverage of photos compared to others, generally, artificial areas and areas where important landmarks and tourist attractions are located. The potential use of photos may be not homogeneous among different urban areas, with famous touristic places having usually more photos than urban areas that do not have any famous landmarks. Uneven temporal distribution might be also found in some places and LULC classes as special events attracting a high number of people occur in particular dates. The metadata downloaded from Panoramio include the date when photos have been uploaded that in most cases is not the same date when they were actually taken. This issue bias the temporal characteristic of photos and affect their reliability if one needs to consider the temporal aspect. Finally, the actual content of photos that in some cases do not show a subject related to LULC.

There are few solutions to address some of these issues. Downloading additional metadata from the initiative, such as the Exif information, not available currently through the Panoramio public API, would solve the date mismatch once it integrates the date when the photo was taken and add, in some cases, even more information (e.g. the zoom level). Also the integration of photos from other available and similar initiatives such as Flickr, Instagram, among others, could increase the reliability of this type of data in some aspects.

Acknowledgment

The work was carried out thanks to the partial contribution of the Cost Actions TD 1202 and IC 1203.

References

Arsanjani, J., Helbich, M., Bakillah, M., Hagenauer, J., & Zipf, A. 2013. Toward mapping land-use patterns from volunteered geographic information. *International Journal of Geographical Information Science, 27*(12): 2264–2278. DOI: http://dx.doi.org/10.1080/13658816.2013.800871

European Environment Agency. 2012. Mapping Guide for a European Urban Atlas. Available at: http://www.eea.europa.eu/data-and-maps/data/urban-atlas/mapping-guide [Last accessed 1 August 2015].

Estima, J., & Painho, M. 2013. Flickr Geotagged and Publicly Available Photos: Preliminary Study of Its Adequacy for Helping Quality Control of Corine Land Cover. In: *Computational Science and Its Applications – ICCSA 2013.* Springer Berlin Heidelberg, pp. 205–220. DOI: http://dx.doi.org/10.1007/978-3-642-39649-6_15

Estima, J., & Painho, M. 2014. Photo Based Volunteered Geographic Information Initiatives: A Comparative Study of their Suitability for Helping Quality Control of Corine Land Cover. *International Journal of Agricultural and Environmental Information Systems, 5*(3): 73–89. DOI: http://dx.doi.org/10.4018/ijaeis.2014070105

Goodchild, M. F. 2007. Citizens as sensors: the world of volunteered geography. *GeoJournal, 69*(4): 211–221. DOI: http://dx.doi.org/10.1007/s10708-007-9111-y

Goodchild, M. F. 2008. Commentary: whither VGI? *GeoJournal, 72*(3–4): 239–244. DOI: http://dx.doi.org/10.1007/s10708-008-9190-4.

Lupia, F., Estima, J., & Painho, M. 2015. Analisi esplorativa del potenziale delle fotografie georeferenziate condivise pubblicamente per il monitoraggio dei cambiamenti in aree urbane. In: *"Recuperiamo terreno – Politiche, azioni e misure per un uso sostenibile del suolo".* Milano 6 maggio 2015.

Stefanidis, A., Crooks, A., & Radzikowski, J. 2011. Harvesting ambient geospatial information from social media feeds. *GeoJournal, 78*(2): 319–338. DOI: http://dx.doi.org/10.1007/s10708-011-9438-2

Turner, A. J. 2006. *Introduction to Neogeography.* "O'Reilly Media, Inc.".

CHAPTER 22

AtrapaelTigre.com: enlisting citizen-scientists in the war on tiger mosquitoes

Aitana Oltra*, John R.B. Palmer[†] and Frederic Bartumeus[*,†,‡]

*CEAB-CSIC, Cala Sant Francesc 14, 17300 Blanes, Spain, aoltra@ceab.csic.es
[†]CREAF, Cerdanyola del Vallès, 08193 Barcelona, Spain
[‡]ICREA, Pg. Lluís Companys 23, 08010 Barcelona, Spain

Abstract

This chapter describes *AtrapaelTigre.com*, a citizen science project focusing on the Asian tiger mosquito in Spain. Commonly known for its aggressive biting during the day, the tiger mosquito represents a global environmental problem. It is an invasive species and a vector for dengue, chikungunya and other diseases, making it a serious public health risk. It is also an everyday nuisance and a threat to tourism and related industries. The management of invasive species, and particularly disease vectors, requires integrated programs that combine public communication and education with research, surveillance and control. *AtrapaelTigre.com* aims at achieving this by engaging citizen scientists to raise awareness and collect data on tiger mosquito adults and their breeding sites with a smartphone app (*Tigatrapp*) and a multi-proxy data validation system that combines expert, crowd, and app-user input. Lessons learned during the first year of implementation in Spain, in 2014, have guided our current strategies with respect to both tiger mosquitoes and the formal integration of citizen

How to cite this book chapter:
Oltra, A, Palmer, J R B and Bartumeus, F. 2016. *AtrapaelTigre.com*: enlisting citizen-scientists in the war on tiger mosquitoes. In: Capineri, C, Haklay, M, Huang, H, Antoniou, V, Kettunen, J, Ostermann, F and Purves, R. (eds.) *European Handbook of Crowdsourced Geographic Information*, Pp. 295–308. London: Ubiquity Press. DOI: http://dx.doi.org/10.5334/bax.v. License: CC-BY 4.0.

science into the research, surveillance and control of invasive species and disease vectors generally. We address the challenges of implementing such frameworks and discuss their fitness for use in public health systems. The goal of *AtrapaelTigre.com* is not only to enhance participation and raise awareness, but also to promote novel research and a more informed and cost-effective management of the tiger mosquito across Spain.

Keywords

citizen science, volunteered geographic information, invasive species, disease vectors, Asian tiger mosquito

Introduction

AtrapaelTigre.com ('Catch-the-Tiger') is a citizen science project with a *Volunteered Geographic Information* (VGI) (Goodchild 2007) component that enlists ordinary people in the research, surveillance and control of Asian tiger mosquitoes (*Aedes albopictus*) (Skuse 1894) in Spain. The tiger mosquito is an invasive species from Southeast Asia that has spread worldwide and become common in developed landscapes (Hawley 1988). It is well known for biting aggressively during the day, and importantly, it is a vector of such diseases as dengue and chikungunya (Paupy et al. 2009). The species is relatively easy to recognize from its behavior and appearance, and it was first detected in Spain in 2004 (Aranda et al. 2006). The tiger mosquito is now established along the Spanish Mediterranean coast (Alarcón-Elbal et al. 2014), where it threatens public health and degrades the quality of life, while also harming the tourism sector (Roiz et al. 2007), which peaks in the summer, just when the species is most active.

The efficacy of public management programs is limited because tiger mosquitoes breed not only in public spaces, but also in small water containers in private areas, such as the plates people place under their flower pots on balconies and patios. Removing the water from these containers prevents the development of larvae and may significantly reduce the presence of adults at a given place. This can be especially effective, since the lifetime dispersal range of the species is only around 600 m (Hawley 1988). Therefore, awareness-raising and education campaigns are key tools for control programs. This, combined with the ease with which tiger mosquitoes can be identified and the extent to which they are often well known to the communities in which they are prevalent, makes the species a good target for citizen science. A number of projects have recently begun using public participation to monitor mosquitoes elsewhere (e.g. Mückenatlas in Germany, iMoustique in France) (Kampen et al. 2015). Although invasive species are common citizen science targets (e.g. Dickinson et al. 2010), it is much less common to specifically target disease vectors like the tiger mosquito that affect humans.

AtrapaelTigre.com has two specific objectives: (1) to explore new methodologies for acquiring data for tiger mosquito research, surveillance and control through public participation, and (2) to raise public awareness and promote household control actions. The project was initiated in 2013 as a pilot in a small region of Spain with a limited target group of participants. The know-how and network acquired served to extend the approach in 2014 to the whole territory of Spain. Here we present the existing system (data collection, validation and visualization) and the lessons learned in 2014, during the first year of implementation in Spain. We also placed the project and future challenges in a broader conceptual framework useful for other citizen science projects targeting invasive species and disease vectors.

Data collection and validation approach

Data collection is done using the smartphone app *Tigatrapp*, available on Google Play and iTunes. *Tigatrapp* is free and open source software,[1] and it may be redistributed or modified under the GNU General Public License (version 3). Participation is anonymous, but participants must consent first to the privacy policy and terms of use. With *Tigatrapp*, ordinary people can collect and send geolocalized reports of tiger mosquitoes and their breeding sites (Figure 1). To do this, they need to learn how to identify the tiger mosquito (including basic taxonomy and life cycle), and this information is provided by the app and the project website, as well as in workshops and talks organized throughout the mosquito season. *Tigatrapp* reports include (Figure 1): i) location, obtained by the app directly from the device's GPS receiver or its network connections or from the user selecting the location on a map, ii) key taxonomic traits of the reported mosquito or characteristics of the reported breeding site based on a small survey (i.e. user level validation), iii) photographs (compulsory for breeding sites but optional for adult mosquitoes, which are often hard to photograph), and iv) optional complementary notes.

The app also collects 5 randomly-timed, anonymous samples of user locations on a daily basis (although users have the option to switch this feature off). This background location system (Figure 1) is used to estimate sampling effort and territorial exposure, making it possible to adjust the report analysis for the fact that users are not randomly distributed across the landscape. In other words, the background location system helps in identifying the extent to which reports from a given area are being driven by the density of users, the density of mosquitoes, or both. In order to protect privacy, all background location information is masked on users' devices by placing locations on a predetermined

---

[1] https://github.com/MoveLab.

DATA COLLECTION APP TIGATRAPP	
REPORTS	
MOSQUITOES	**BREEDING SITES**
Location Taxonomic survey Photographs (optional) Notes (optional)	Location Sites survey Photographs Notes (optional)
OTHER DATA	
BACKGROUND LOCATION	**MISSIONS**
Approximate location 5 times/day (optional)	Variable information sent/collected to/from a specific region (optional)

Figure 1: Data collected with *Tigatrapp*. Background location and missions are not yet implemented in the iOS version of the app due to financial constraints.

grid of 0.05 degrees latitude and longitude. Only the grid cell identifier is transmitted from the device to the server, making it impossible to determine the actual location within the cell. In addition, these locations are identified only by a code that is randomly assigned on the user's device, without any additional information or any way to link the location to a given user's reports.

Finally, the app incorporates the concept of missions (Figure 1), which are extra voluntary activities, notes or surveys that are sent as an incoming notification to *Tigatrapp*. These can be set to appear on the devices of all users or only those in a given area. Missions allow the implementation of extra activities on the go, and direct communication with participants, according to specific needs.

The two types of reports (mosquito adults and breeding sites) and the sampling effort (covered area) can be visualized on a webmap embedded in the project website. The coverage map is also a useful tool for increasing user engagement, as people are able to see the extent to which the app is being used across Spain and, indeed, throughout the world. Raw data is also available to some scientists (even outside the project) and to some public administrations responsible for tiger mosquito control and additional data will be made available to the public through public-access repositories and other means in the future.

To validate reports, a multi-proxy system that combines expert, crowd and app-user validation has been implemented (Figure 2). Expert and crowd validation are based on the analysis of report photographs and are done using two different on-line platforms: a) a custom-built platform for experts and b) Crowdcrafting.org (*Tigafotos* project) for the public. App-user validation is based on the user's responses to the survey contained in each submitted report. For expert validation, the experts analyze the photographs to classify each report into one of five categories (Figure 2) based on the assessed probability of its being accurate (i.e. the probability that the user actually observed a tiger mosquito or a tiger mosquito breeding site). Expert validation is done before publishing report photographs on the webmap and thus, also serves to filter out pictures that are sensitive (e.g. bites on people's bodies), include personal information (e.g. whole body or face pictures), or are unrelated to the project (although there have been few of these). Experts can also decide to publish the app-user's note (if available) or to add their own note in the report pop-up on the webmap. Expert validation results are used as the main filter and

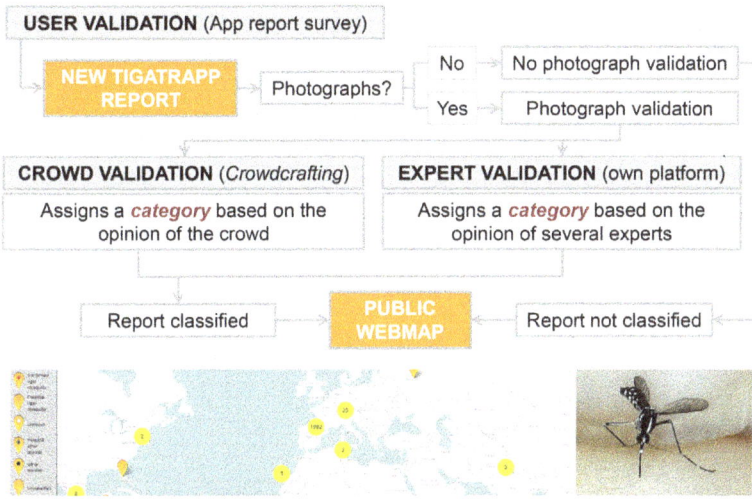

Figure 2: Diagram of the multi-proxy validation system (top), and categories used for the validation of mosquito reports (bottom). In the middle, a map screenshot and a citizen scientist's picture of a tiger mosquito.

classification method in the public webmap. However, crowd (if available) and app user validation results are also displayed in the report pop-ups. Reports without pictures are displayed as 'unclassified'.

Lessons from the first year of implementation (2014)

During the 2014 data collection period (late spring to late winter 2014), almost 7,000 people downloaded *Tigatrapp* and registered as users (Figure 3). In total, ~2,900 reports (including mission answers) were sent from ~1,300 unique user identifiers. Both app download and data collection dynamics were strongly influenced by media appearances. Most reports (~60%) were of mosquito sightings, followed by mission answers (~30%) and breeding sites (~10%). It is not clear why breeding site reports have been so much less frequent than adult mosquito reports. It could be that people are less motivated to report breeding sites than to report adult mosquitoes (which may have just bitten them), or that the complexity of the breeding site concept (e.g. indirect cause-effect due to mosquito life cycles) or the requirement that breeding site reports include photographs makes their reporting more challenging. Whatever the reason, the low number of breeding site reports in 2014 led us to focus the expert validation for that year on only the adult mosquito reports. In contrast, our strategy for 2015 has been to improve breeding site reporting by working with public mosquito management agencies to increase outreach and ensure that breeding site reports lead to tangible results in urban public spaces (see next Section).

Participation in crowd validation ('*Tigafotos*' project in Crowdcrafting.org) was low (~300 validated photographs out of ~1,200) and heterogeneous in time. Being hosted in a separate platform, it is difficult to compare participation trends to *Tigatrapp* participation. On one hand, hosting a crowd based photograph validation system on an international crowdsourcing platform has clear benefits in terms of participation, visibility and ease of implementation. On the other hand, this approach has the drawback of relying on participants who are disconnected from the project and its objectives. Crowd engagement in the *Tigafotos* project might be improved by embedding it also in the *AtrapaelTigre.com* website (Crowdcrafting.org allows for that), improving the project's design, making it multi-lingual, and adding other gamification and engagement elements.

Although photographs are the basis for report validation (expert and crowd), photograph attachments are optional for mosquito reports. This allows users to report mosquitoes without having to catch them. However, it may be that this makes it too easy for users to avoid making the effort of taking a picture, even when they are able to. In 2014, only around 30% of the adult mosquito reports actually included a photograph (Figure 3). Of those, expert validation approximately assigned ~40% to the "Unknown" category, ~50% to 'Confirmed Asian

tiger mosquito' or 'Possible tiger mosquito', and ~10% to 'Possible other species" or "Other species' (Figure 2). Improving users' skill in taking photographs
of mosquitoes might well result in more valuable data by moving reports out
of the 'Unknown' category. We are now using *Tigatrapp* missions and embedded information, social media (Twitter and Facebook), and the project blog
to systematically train and encourage users to take more and better pictures.
These strategies seem to be improving the fitness for use of data considerably,
since the number of adult reports with pictures increased to 60% (30% in 2014)
and the number of reports that could be classified as "Confirmed Asian tiger
mosquito" or "Possible tiger mosquito" in 2015 was ~1,700, almost the same
as the total number of mosquito reports (with or without pictures) received
in 2014, i.e. ~1,740 (Figure 3). We are also developing quantitative methods to
make reports without photographs more useful for scientific and management
purposes. For example, by combining responses to the taxonomic survey in
each report with knowledge about the user based on the quality and quantity
of previous reports, we may better assess the probability that a given report
corresponds to an actual tiger mosquito. Other methods, like taxonomic validation of georeferenced mosquitoes sent by post (Kampen et al. 2015) may be
explored in the future, and the ultimate goal will be to compare the results of
several independent validation methods, along with semi-automated and intelligent algorithms based on prior knowledge.

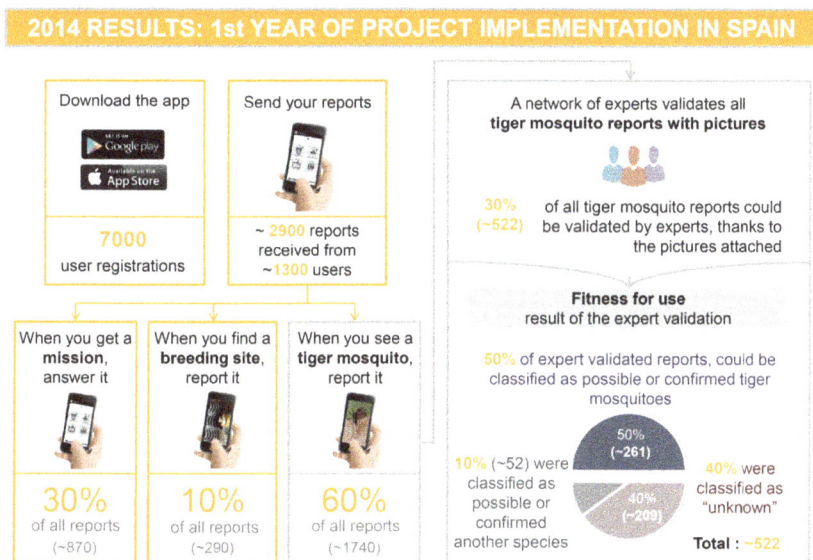

Figure 3: Summary infographic of the results obtained during 2014, the first
year of project implementation in Spain (numbers are approximate).

Integrating citizen science and VGI in the management of invasive species and disease vectors

Our experience with *AtrapaelTigre.com* has made it clear that using citizen science to target invasive species or disease vectors requires at least three basic domains of expertise: i) communication and education, ii) surveillance and control and iii) research (Figure 4). These domains and their interrelations acquire a whole new dimension as a consequence of the citizen scientists' involvement and the need to implement control measures that complement current environmental management and public health system polices. Here we discuss how to improve the project in each of these domains, and the main challenges.

Communication and education

The project has 3 communication and education objectives: 1) spread the word to gain new participants and wider geographic coverage, 2) keep the interest of participants and the media, and 3) inform participants so that they not only provide useful data but also take control actions at places out of the scope of public administrations (e.g. their houses). For instance, by spotting and recognizing breeding sites, citizens become aware of the importance of water removal in their backyards. To accomplish these objectives, the project actively disseminates information through a blog, Facebook and Twitter accounts, press releases, and talks and workshops for different audiences. The project also collaborates with public administrations and private stakeholders and encourages these entities to include information about *AtrapaelTigre.com* in their own outreach campaigns (e.g. flyers, websites, media appearances). In 2015, we have coordinated communication actions with the Barcelona Public Health Agency (ASPB) and public entities in Valencia and the Canary Islands, amongst others. We are not explicitly assessing trends in public awareness or the population behavior change related to project actions. However, several indicators demonstrate a good performance in terms of communication: media appearances increased each year (see project website for a full list) and sessions in the project website have more than quadrupled between 2014 and 2015, based on estimates from Google Analytics website data between July and December.

From a communication perspective, we have developed a two-fold strategy. We offer a regular stream of new, interesting, and in-depth scientific outreach material related to the targeted species and to other mosquito disease vectors aspects. At the same time, we use project results to demonstrate and explain how citizen scientist participation can help to improve the surveillance and control of the species in the short term. This is a challenge, but of high importance, since most people are likely to participate, not out of scientific interest, but because they are affected by the presence of the species and have a personal interest in its local eradication.

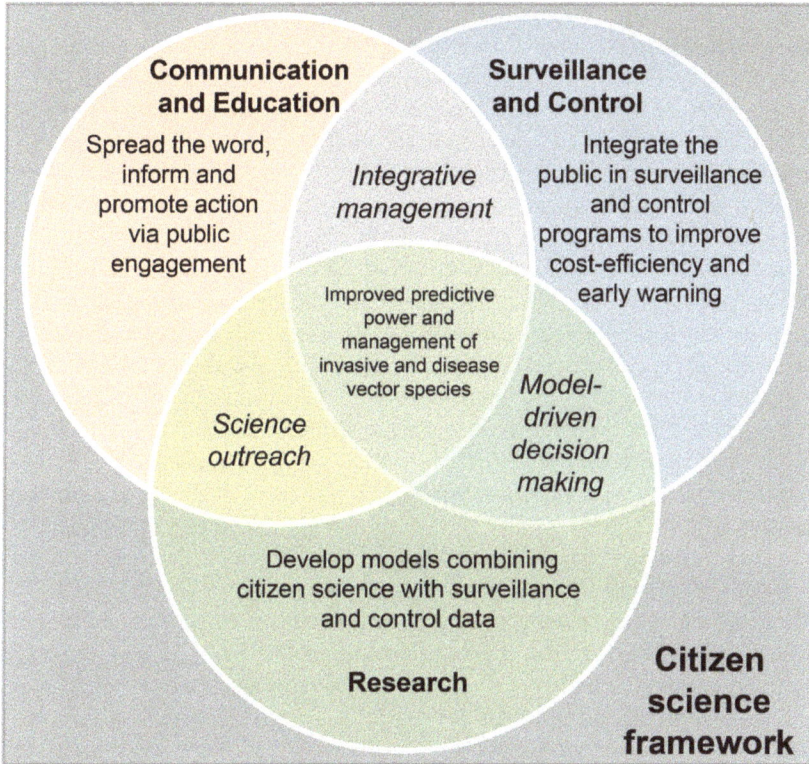

Figure 4: Identified expertise domains for improved predictive power and management of invasive species and disease vectors under a citizen science framework.

Surveillance and control

Tiger mosquito surveillance and control programs in Spain currently involve a traditional integrative management approach that incorporates communication and education. The citizen science framework (Figure 4) goes beyond this by exploiting new technologies (apps, webmaps, and social media) that enable massive and systematic calls-to-action while making the resulting data immediately available to management services and the general public. It has been demonstrated elsewhere that public participation (even more through the use of new technologies) can advance the detection of an invasive species even 2 years before traditional monitoring programs (Scyphers et al. 2014).

The usefulness of participative frameworks as early warning systems (the primary role of surveillance) is confirmed in the case of *AtrapaelTigre.com*. By enlisting a large team of citizen scientists while also engaging with the network of management agencies and other tiger mosquito stakeholders in Spain, we have been able to gather critical information and pass it to the actors responsible for surveillance in the implicated regions. These actors are then able to decide whether to further investigate and activate relevant environmental and public health protocols. For example, the first-ever report of tiger mosquitoes in Andalusia came from a citizen scientist via *Tigatrapp* in 2014. After detecting this as a credible alarm, we contacted specialists there, who corroborated the presence of the species in the field (Delacour-Estrella et al. 2014). Similarly, citizen scientists using *Tigatrapp* were also the first to detect tiger mosquitoes in the Catalan pre-Pyrenees, a discovery confirmed by rural agents and passed through social networks to raise awareness (pers. comm.). Citizen science systems like *AtrapaelTigre.com* should not be seen as substitutes for active surveillance (e.g. targeted sampling methods) by specialists. The detection of the species in the Basque Country in 2014, for instance, came about only through active surveillance (Delacour-Estrella et al. 2015), demonstrating the extent to which the two approaches are complimentary. Indeed, efficiency of combining passive (e.g. data gathered by the general public) and active surveillance for mosquitoes in Europe is increasingly apparent (Kampen et al. 2015).

Despite their importance, early warnings are only part of the story. There are many regions in Spain where the tiger mosquito has already become established and well known. The challenge for *AtrapaelTigre.com* in these regions is to build up a participatory system for management and control, reducing the public health risk and improving life quality. To this end, we have contacted actors and stakeholders responsible for control programs in these areas, and developed tools (e.g. interactive web-interfaces) to make citizen science data more accessible and useful for them. This step is costly but it has made the project much more powerful, as the data from citizen scientists can be immediately used to improve management and control in affected areas where this joint collaboration is established. For instance, in 2015, the ASPB incorporated part of its team directly into the *AtrapaelTigre.com* expert validation system, and it is using citizen science data from the project to improve tiger mosquito control in the city of Barcelona. A similar strategy was followed by stakeholders in the city of Valencia in the same year, and all signs are that this type of involvement is highly beneficial. The numbers of breeding sites' reports have doubled in 2015 (although the total number is still much smaller than for mosquito reports). In Barcelona, 20% of ~280 adult and breeding sites' reports received in 2015 in and around the city, were considered useful by the ASPB for management purposes and were incorporated into their already long-lasting *Public space surveillance and control program*. In Valencia, with a more recent history of surveillance programs starting in 2014, 40% of the detected positive breeding sites in the city were thanks to citizen's reports (breeding sites and adults). Indeed, without discounting the importance of engagement with the citizen scientists themselves, our renewed efforts to communicate with stakeholders are proving crucial for the long-term maintenance of the whole participatory system.

Research

Often citizen science is challenged by sampling bias (based, for example, on the distribution of users across territory, or the involvement of restricted social layers) and large variations in the quality of the data (Dickinson et al. 2010). However, similar problems are also present in field data collected by professional scientists (e.g. biodiversity estimates, density estimates of a population). Mosquito surveillance in many European countries often relies on ovitrap networks (networks of traps in which females are prone to lay their eggs). Such trap networks can detect the presence of tiger mosquitoes but do not provide good estimates of abundance, can generate false negatives, can be biased by placement and exposure time, and are generally limited to recently colonized or highly populated areas, leaving large gaps in territorial coverage. An important challenge for *AtrapaelTigre.com* (and, by extension, to other citizen science projects) is to demonstrate how the combination of citizen scientist data and ovitrap surveillance (Kampen et al. 2015) can improve predictions of the distribution (current and potential), risk factors, and spreading dynamics of the targeted species.

The key is to let the strengths of each approach compensate for the other's weaknesses and to use reliable results from each as a means of cross-calibration. For example, there is now a large amount of ovitrap data available and *Tigatrapp* data covering the same areas of Spain, as well as *Tigatrapp* data for areas and times for which ovitrap data is lacking. We are using the ovitrap data in the overlapping areas to calibrate models built from the *Tigatrapp* data, and we are then using these calibrated models to make estimates about the areas and times for which the ovitrap data is absent. We expect such novel modelling approaches to produce more robust conclusions that contribute to cost-effective management strategies.

Conclusions

Once the infrastructure and basic implementation of a citizen science project has been put in place, attention turns to sustaining long term participation and obtaining sound scientific conclusions from the volunteered data. For projects like *AtrapaelTigre.com*, that focus on invasive species or disease vectors, participants are often motivated more by management and control goals than scientific interest, making it important to work closely with environmental and public health agencies and related stakeholders. At the same time, the project must be built on a solid scientific foundation and this requires novel approaches to data validation and analysis.

AtrapaelTigre.com uses three independent methods for validating citizen scientist reports, as well as a background location feature that makes it possible to estimate sampling effort and correct bias. Moreover, *Tigatrapp*'s passive citizen scientist data is combined with active ovitrap surveillance, with each data type complementing the other and allowing for cross-calibration.

Our 2014 and 2015 results suggest that these methods are promising, but still must be improved. We hope to increase more the quality and quantity of citizen scientists' photographs and to develop new quantitative methods for assessing the reliability of reports without attached photographs. We also hope to improve the multi-proxy validation system, as each of the three validation methods has its own set of drawbacks: the crowd and app-user validation methods are capable of handling massive data but are prone to error, while expert validation appears more accurate but is more costly and better suited to limited quantities of data. The goal is to find ways for the expert validation of a manageable part of the data to inform and improve the crowd and app-user validation for the rest. Finally, another important step will be to formally frame all of these methods in a semi-automated, intelligent alert system that incorporates prior knowledge and directs interesting findings back to stakeholders and the general public.

Final note: from *AtrapaelTigre.com* to Mosquito Alert

On February 2016, at the time of editing this manuscript, the project incorporated a new target species, the yellow fever mosquito (*Aedes aegypti*) and changed name accordingly: from *AtrapaelTigre.com*, specific for the tiger mosquito, to Mosquito Alert (available at www.mosquitoalert.com). The yellow fever mosquito has a similar appearance and behavior to the Asian tiger mosquito and is currently considered the primary vector of yellow fever, chikungunya, dengue and Zika viruses (ECDC 2016), having raised a lot of international concern in the recent American Zika outbreak (Kindhauser et al. 2016). The species was historically present in Spain, and could be reintroduced again through the island of Madeira, where it is known to be the cause of a dengue outbreak in 2012 (ECDC 2016). In this sense, the incorporation of this new species for the project in Spain is relevant in public health terms and follows primarily an early warning system strategy.

Acknowledgments

AtrapaelTigre.com has been funded by the Spanish Foundation for Science and Technology-Ministry of Economy and Competitiveness (FCT-12-3730, FCT-13-7019), Bloom, Lokímica and Plan Estatal I+D+I. (CGL2013-43139-R), in collaboration with "la Caixa" Banking Foundation. The research leading to these results has received also funding from RecerCaixa. We want to thank all of the participants, students, volunteers, and professionals who have made the project possible. We would like to thank the Barcelona Public Health

Agency, the Valencia City Council and the company lokímica for assessing the performance of *AtrapaelTigre.com* in their surveillance and control programs. The images shown in the *Tigatrapp* screenshot in Figure 1 include Mosquito, designed by Monika Ciapala, Water, designed by Alessandro Suraci, Pot, designed by factor[e] design initiative, Location, designed by Alex Auda Samora, Map designed by Jonathan Higley, and Images designed by Simon Henrotte, all obtained from the Noun Project under the Creative Commons – Attribution (CC BY 3.0) license. Figure 3 was done with the help of the project communication team: A. Ramon, M. Torres and J. L. Ordóñez.

References

Alarcón-Elbal, P. M., Delacour-Estrella, S., Ruiz-Arrondo, I., Collantes, F., Delgado, J. A. Morales-Bueno, J., Sánchez-López, P. F., Amela, C., Sierra-Moros, J. M., Molina, R., & Lucientes, J. 2014. Updated distribution of *Aedes albopictus* (Diptera: *Culicidae*) in Spain: new findings in the mainland Spanish Levante, 2013. *Memórias do Instituto Oswaldo Cruz, 109*(6): 782–786. DOI: http://dx.doi.org/10.1590/0074-0276140214

Aranda, C., Eritja, R., & Roiz, D. 2006. First record and establishment of the mosquito *Aedes albopictus* in Spain. *Medical and Veterinary Entomology, 20*(1): 150–152. DOI: http://dx.doi.org/10.1111/j.1365-2915.2006.00605.x

Dickinson, J. L., Zuckerberg, B., & Bonter, D. N. 2010. Citizen science as an ecological research tool: Challenges and benefits. *Annual Review of Ecology, Evolution, and Systematics, 41*: 149–72. DOI: http://dx.doi.org/10.1146/annurev-ecolsys-102209-144636

Kampen, H., Medlock, J. M., Vaux, A. G. C., Koenraadt, C. J. M., van Vliet, A. J. H., Bartumeus, F., Oltra, A., Sousa, C. A., Chouin, S., & Werner, D. 2015. Approaches to passive mosquito surveillance in the EU. *Parasites & Vectors, 8*: 9. DOI: http://dx.doi.org/10.1186/s13071-014-0604-5

Delacour-Estrella, S., Collantes, F., Ruiz-Arrondo, I., Alarcón-Elbal, P. M., Delgado, J. A., Eritja, R., Bartumeus, F., Oltra, A., Palmer, J. R. B., & Lucientes, J. 2014. Primera cita de mosquito tigre, *Aedes albopictus* (Diptera, *Culicidae*), para Andalucía y primera corroboración de los datos de la aplicación Tigatrapp. *Anales de Biología, 36*: 93–96. DOI: http://dx.doi.org/10.6018/analesbio.36.16

Delacour-Estrella, S., Barandika, J. F., García-Pérez, A. L., Collantes, F., Ruiz-Arrondo, I., Alarcón-Elbal, P. M., Bengoa, M., Delgado, J. A., Juste, R. A., Molina, R., & Lucientes, J. 2015. Detección temprana de mosquito tigre, *Aedes albopictus* (Skuse, 1894), en el País Vasco (España). *Anales de Biología, 37*: 25–30. DOI: http://dx.doi.org/10.6018/analesbio.37

European Center for Disease Prevention and Control *Aedes aegypti*. Available at: http://ecdc.europa.eu/en/healthtopics/vectors/mosquitoes/Pages/aedes-aegypti.aspx [Last accessed 16 March 2016].

Goodchild, M. F. 2007. Citizens as Sensors: The World of Volunteered Geography. *GeoJournal*, 69(4): 211–221. DOI: http://dx.doi.org/10.1007/s10708-007-9111-y

Hawley, W. A. 1988. The biology of *Aedes albopictus*. *Journal of the American Mosquito Control Association*, (Supplement 1): 1–39

Kindhauser, M. K., Allen, T., Frank, V., Santhana, R., & Dye, C. 2016. Zika: the origin and spread of a mosquito-borne virus. *Bulletin of the World Health Organization*. DOI: http://dx.doi.org/10.2471/BLT.16.171082

Paupy, C., Delatte, H., Bagny, L., Corbel, V., & Fontenille, D. 2009. *Aedes albopictus*, an arbovirus vector: from the darkness to the light. *Microbes and Infection*, *11*(14–16): 1177–1185. DOI: http://dx.doi.org/10.1016/j.micinf.2009.05.005

Roiz, D., Eritja, R., Melero-Alcibar, R., Molina, R., Maruès, E., Ruiz, S., Escosa, R., Aranda, C., & Lucientes, J. 2007. Distribucion de *Aedes* (Stegomyia) *albopictus* (Skuse, 1894) (Diptera, Cuclicidae) en España. *Boletín Sociedad Entomológica Aragonesa, 40*: 523–526

Scyphers, S. B., Powers, S. P., Adkins, J. L., Drymon, J. M., Martin, C. W., Schobernd, Z. H., Schofield, P. J., Shipp, R. L., & Switzer, T. S. 2014. The role of citizens in detecting and responding to a rapid marine invasion. *Conservation Letters, 8*(4): 242–250. DOI: http://dx.doi.org/10.1111/conl.12127

CHAPTER 23

Crowdsourcing geographic information for disaster management and improving urban resilience: an overview of recent developments and lessons learned

João Porto de Albuquerque*,†, Melanie Eckle†,
Benjamin Herfort† and Alexander Zipf†

*Centre for Interdisciplinary Methodologies, University of Warwick, UK,
j.porto@warwick.ac.uk, joao.porto@geog.uni-heidelberg.de
†GIScience Chair, Heidelberg University, Germany

Abstract

In the past few years, crowdsourced geographic information (also called volunteered geographic information) has emerged as a promising information source for improving urban resilience by managing risks and coping with the consequences of disasters triggered by natural hazards. This chapter presents a typology of sources and usages of crowdsourced geographic information for disaster management, as well as summarises recent research results and present lessons learned for future research and practice in this field.

Keywords

Crowdsourced Geographic Information, Volunteered Geographic Information, Disaster Management, Social Media, OpenStreetMap

How to cite this book chapter:
de Albuquerque, J P, Eckle, M, Herfort, B and Zipf, A. 2016. Crowdsourcing geographic information for disaster management and improving urban resilience: an overview of recent developments and lessons learned. In: Capineri, C, Haklay, M, Huang, H, Antoniou, V, Kettunen, J, Ostermann, F and Purves, R. (eds.) *European Handbook of Crowdsourced Geographic Information*, Pp. 309–321. London: Ubiquity Press. DOI: http://dx.doi.org/10.5334/bax.w. License: CC-BY 4.0.

Introduction

The potential of Crowdsourced Geographic Information (CGI) as a new information source for disaster risk management has been paradigmatically shown during the earthquake that hit Haiti in 2010. Due to a lack of official data, information gathered from social media, via SMS and from OpenStreetMap became crucial for disaster response. In the past few years, CGI found their way into different disaster situations and scenarios (Horita et al. 2013). The use of geographic information for disaster risk management has attracted great interest both in research and practice, mainly because of the possibility to tap into the 'collective intelligence' or the 'wisdom of the crowds' to improve urban resilience, i.e. to improve the capacity of urban areas to better managing disaster risks and coping with the effects of extreme events.

In general, CGI in the context of disaster risk management can be categorised according to the information source into the following types:

1) *Social media*: Information produced by people about the event in usual social media platforms *(e.g. Twitter, Flickr, Instagram, Facebook)*, such as from eyewitness that exchange and disseminate information about a disaster event.
2) *Crowd sensing*: Information collected from dedicated applications and platforms (e.g. Ushahidi) that are aimed specifically at producing information for disaster risk management.
3) *Collaborative mapping*: Information about geographic features of disaster-affected or disaster-prone areas, which is produced by volunteers using mapping platforms (e.g. OpenStreetMap, Wikimapia), e.g. as derived from satellite imagery.

Although there is a growing body of research related to each of these CGI types in different phases and tasks of disaster management, existing research studies usually focus on a particular type of CGI and are not able to relate to relevant developments associated with other CGI types. The goal of this chapter is to present to a holistic view of this field by means of a typology that is able to distinguish the main features and potentials of each CGI type for disaster risk management. This typology is valuable not only to summarise recent research results, but also to identify more integrated directions for future research on CGI towards improving disaster management and urban resilience. The next sections are thus dedicated to exploring these issues for each of the aforementioned CGI types in turn, followed by a conclusion.

Social Media

The first type of geo-information produced by the 'crowd' in the context of disasters is related to the use of existing social media platforms to exchange information. Social media has been defined as 'a group of Internet-based applications that build on the ideological and technological foundations of Web 2.0, and that allow the creation and exchange of User Generated Content' (Kaplan & Haenlein 2010). As such, these platforms allow users to easily share self-produced content within a network of contacts and/or for the general public in a variety of forms: texts via blogs (from 'web log') or short messages in 'microblogging' (e.g. Twitter), web pages and forums, photos, videos, etc. Popular social media platforms include Twitter, Facebook, Flickr, YouTube, Instagram etc.

As people are increasingly familiar with and ordinarily use social media in their day-to-day life, they naturally tend to uptake these platforms in the occurrence of a disaster for communicating their experience and/or urgent needs. Indeed, in different catastrophic events of the past few years – from the wildfires in Southern California, USA in 2007, over the Earthquake in Haiti in 2010, up to the recent super typhoon in the Philippines 2013 – social media has enabled the affected population to produce information about extreme events and their catastrophic impacts (Sakaki, Okazaki & Matsuo 2010; Crooks et al. 2013; De Longueville et al. 2010).

In the field of disaster risk management, a large part of the existing research focused on the analysis of short messages of the Twitter platform, the so-called Tweets (Steiger et al. 2015). For instance, Sakaki et al. (2010) and Crooks et al. (2013) investigated the use of Twitter for detecting and estimating the trajectory of earthquakes in real time. De Longueville et al. (2010) proposed the use of VGI as a sensor for detecting forest fire hot spots, based on previous work that analysed the application of Twitter as a source of spatiotemporal information for wildfire events in France. In contrast, Fuchs et al. (2013) showed that event detection based on peaks of Twitter activity did not work for the 2013 floods in Germany and presented an analysis of spatiotemporal clusters. Bakillah et al. (2014) applied graph clustering to support the detection of geolocated communities in Twitter after the typhoon Haiyan in the Philippines. Furthermore, a number of studies are concerned about developing tools for visualising social media data in order to enable make-sensing and location- based knowledge discovery (MacEachren et al. 2011; Terpstra & de Vries 2012; Croitoru et al. 2013; Spinsanti & Ostermann 2013).

Another group of studies seek to identify useful information from social media that could be valuable for improving situation awareness (Yin et al. 2012). For instance, Vieweg et al. (2010) and Starbird et al. (2010) analysed Twitter messages during the flooding of the Red River Valley in the United States and Canada in 2009, seeking to discern activity patterns and extract useful information.

Most of the existing work in the area has sought to make sense of social media data as a stand-alone source by analysing aggregated patterns, e.g. by defining thresholds for the size of spatiotemporal clusters of messages that would serve as signals for crisis events of earthquakes (Sakaki et al. 2010, Crooks et al. 2013), wildfires (De Longueville et al. 2010, Slavkovikj et al. 2014) or disease surveillance (Gomide et al. 2011, Bernardo et al. 2013). However, with such an approach the actual content of social media messages is largely ignored, and with this, much of their potential to improve the current knowledge about the unfolding situation is lost. Furthermore, although event detection is useful for sudden-onset crises for which there do not exist any other related data, in many concrete cases, there are additional information sources available. As pointed out by Lazer et al. (2014), one should not see 'big data' as a substitute for all existing data, but rather take the challenge of doing innovative analytics by using data from all traditional and new sources.

This is in line with a nascent research stream that uses VGI in combination with other geodata sources in the field of disaster management (Albuquerque et al. 2015; Schnebele, Cervone & Waters 2014; Triglav-Čekada & Radovan 2013; Spinsanti & Ostermann 2013). For instance, Albuquerque et al. (2015) leveraged authoritative sensor data of water gauges to show that Tweets close to flooded areas are more probable to contain useful information for disaster management (see Figure 1).

Building upon these initial results, an important direction for future research endeavours is the development of improved analytical methods that are able leverage several different data sources in order to provide event detection, visualisation and information extraction from crowdsourced geo-information of social media that are better matched to the needs of decision makers in the field of disaster management.

Crowd Sensing

A second type of activity related to the use of new collaborative technologies for disasters is the emergence of the so-called 'crowd sensing' (Ma et al. 2014). This activity involves citizens on the Web that can act as sensors and share their observations. Differently from crowdsourced information derived from social media covered in the previous section, here the term 'crowd sensing' is used to describe approaches that rely upon dedicated software platforms for gathering specific and structured data, as well as for exploiting the interpretive and analytic skills and local knowledge of citizens.

These approaches are also related to the concept of citizen science, which is described by Haklay (2013) as 'scientific activities in which non-professional scientists voluntarily participate in data collection, analysis and dissemination of a scientific project'. As such, people using platforms for 'citizens as sensors' or 'citizen scientists' get engaged for accomplishing a set of tasks in a coordinated and purposeful manner. These tasks mostly involve some kind of data collection for different types of scientific investigations, the most famous examples being bird watching and other types of environmental observations.

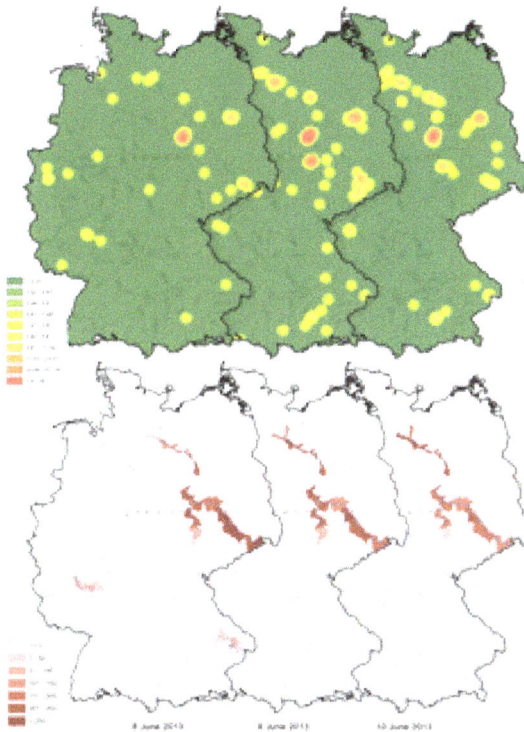

Figure 1: Distribution of Twitter messages that were sent during the 2013 Elbe Floods in Germany (top) in contrast with flooded catchments as indicated by river gauges (bottom) (adapted from de Albuquerque et al. 2015).

In the context of disasters, several 'crowd sensing' platforms were created including dedicated mobile applications for disaster management and earth observation (Ferster & Coops 2013). Using volunteers to perform a specific task, such as environmental monitoring, collectively make a Citizen Observatory (CO), where data can be collected, collated and published (Degrossi, et al. 2014; Liu et al. 2015). Thus, the term Citizen Observatory can be understood as a software platform used by citizens to produce volunteered information about a specific topic through different devices (e.g. web, mobile app and SMS), and allow their visualisation.

An important software platform for implementing Citizen Observatories is called Ushahidi[1] (which means 'testimony' in Swahili). This platform was first developed in the context of election monitoring in Kenya and later developed

---

[1] Available at: http://www.ushahidi.com [Accessed February 12th 2014].

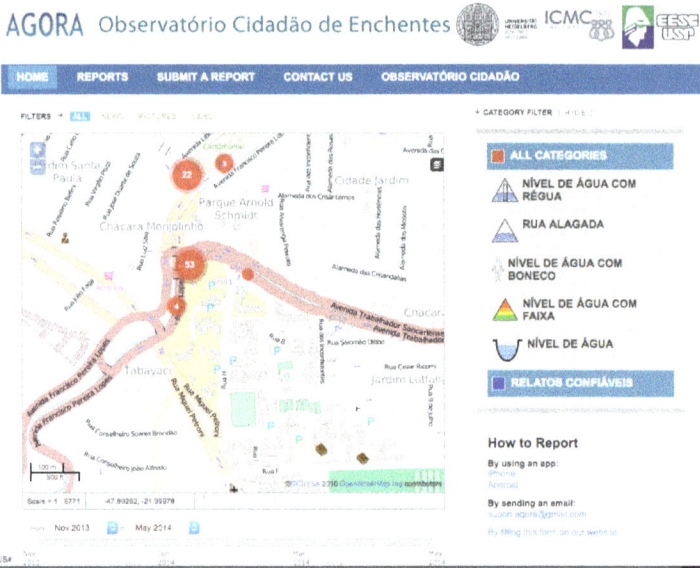

Figure 2: Flood Citizen Observatory prototype (adapted from Degrossi et al. 2014).

as an open-source toolbox that can be deployed in several situations to collect data from on-the-ground volunteers via web site and mobile application, but also for remote volunteers to collaborative categorise information collated from many sources, including social media (discussed in the previous section). One application example that is built upon the Ushahidi platform is the prototype Flood Citizen Observatory implemented in Brazil (Figure 2) for allowing citizens to report about the local conditions of river levels, flooded areas, as well as consequences of flooding (Degrossi et al. 2014; Horita et al. 2015).

While crowd sensing and citizen observatories can be potentially used to provide useful information about the impacts caused by extreme events and their victims, one important issue is to be addressed is how to motivate people to contribute with valuable information. Another important point to be addressed is how to validate and integrate information from volunteers with other sources of data for effectively improving decision-making related to disaster risk management.

Collaborative Mapping

The third type of crowdsourced geo-information comprises a specific type of information and collaboration platform: the collaborative edition of geographic features to fulfil internet-based interactive maps. The well-known platforms Wikimapia[2] and OpenStreetMap (OSM)[3] fall into this category, as well as the 'crowdsourcing' component of the popular GoogleMaps platform, so-called GoogleMapMaker[4] (which, unlike the previous ones, does not have an open data policy and thus does not provide users with full access to the collected data).

A distinctive feature of this type of activity is the collaborative collection of a very specific type of data – namely, georeferenced data about features like streets and roads, buildings etc. – and the structuring of this information in form of a map. In doing so, the volunteer community seeks to produce a map that is as complete and detailed as possible, leveraging the local knowledge of a wide base of users to collaboratively fill the gaps. Recent research works have shown that, at least for the regions with the most active communities in OSM, the results achieved a quality level that is comparable to official and commercial maps (Neis, Zielstra & Zipf 2011; Haklay 2010).

The maps produced by volunteers in this way are clearly of great relevance in the context of disasters. High quality and precise maps are an important resource for a number of tasks in disaster management, being used from emergency planning up to the coordination of relief efforts. In several disaster events of the past few years, the volunteer community has been very actively engaged in producing collaborative maps to assist disaster management, especially the community of OpenStreetMap. By digitising the infrastructure, and especially important, also the level of damage (where it can be detected), they create a situation map that can be used by the emergency responders directly in the field.

After the devastating earthquake that struck Haiti in January 2010, for instance, there was a very significant response of the international OSM community (Neis et al. 2010; Zook et al. 2010). In the aftermath of that severe quake, good-quality maps were not available to guide the relief efforts, and the standard map services in the web (e.g. GoogleMaps) lacked adequate coverage and had to be updated to reflect the current status of the many blocked roads and streets. A few hours after the quake, the volunteers of the OpenStreetMap (OSM) community around the world started mapping remotely the affected regions based on satellite imagery, seeking to trace the outlines of streets, buildings and places of interest. Later, high-resolution and very up-to-date post-disaster satellite images were made freely available and the OSM community could then resort to those in order to also record the damage of buildings and blockages in streets and roads. Such imagery is a very crucial source of information for the mappers to be used, because it allows also volunteers to contribute from all over the world, not only people that are directly at the affected areas. For large-scale disaster events with many international contributors this can

---

[2] Available at: http://wikimapia.org [Accessed February 12th 2014].
[3] Available at: http://openstreetmap.org [Accessed February 12th 2014].
[4] Available at: https://www.google.com/mapmaker [Accessed February 12th 2014].

Figure 3: Elements of the critical infrastructure from OpenStreetMap in the Phillipines, which was affected by Typhoon Haiyan 2013 (adapted from Reimer et al. 2014).

generate highly-detailed maps extremely quickly. The information produced was then available for guiding relief efforts, not only allowing better visual orientation through the interactive maps, but also for importing the data into GPS devices for local orientation, as well as using the database behind OSM for providing more sophisticated services. For instance, an emergency routing service was developed to allow quick identification of the best routes for relief efforts based on the up-to-date situation mapped by the OSM community (Neis, Singler & Zipf 2010).

Ever since 2010, numerous OSM contributors provided their support in mapping events in the aftermath of a disaster, producing the so-called Crisis Maps. As a result, within the OSM community a initiative called Humanitarian OpenStreetMap Team (H.O.T.)[5] was launched to organise the many crisis mapping actions of the OSM community and is also in contact with other relevant humanitarian organisations. This engagement attracted serious interest in academic circles as well as on the side of humanitarian-aid organisations (Harvard Humanitarian Initiative 2011; United Nations Office for the Coordination of Humanitarian Affairs 2013). One significant example of the use of such information could be attested in a more recent major catastrophic event: the typhoon Haiyan that hit the Philippines in 2013 (Reimer et al. 2014). In this case, the international OSM community was also very active and collaborated in a coordinated way with humanitarian organisations such as the American

---

[5]   Available at: http://hot.openstreetmap.org/ [Accessed February 12th 2014].

Red Cross and UN-OCHA (Office for Coordination Affairs of the United Nations), for example, for extracting information about elements at risk of the so-called Critical Infrastructure (see Figure 3), i.e. critical elements that must particular attention in a disaster management such as schools, hospitals, fuel stations etc. (Reimer et al. 2014; Schelhorn et al. 2014; Herfort et al. 2015).

However, while one main advantage of OSM is that their contributors mainly focus on their well-known local surroundings (Goodchild 2007; Neis & Zipf 2012), Crisis Maps originate largely from mappers who work remotely. Therefore, due to the fact that OSM is a crowdsourced map and that a main part of the data in Crisis Maps originates exclusively from contributors that work remotely, humanitarian-aid agencies and first responders have doubts about the quality of the OSM data and therefore sometimes refrain from utilising it (Harvard Humanitarian Initiative 2011). Furthermore, activity areas of remote mappers generally lack official cross-reference data, making it difficult to apply usual quality assessment methods, which are based on comparisons with reference data. In this manner, an important direction for future research is to develop methods for assessing and improving the quality of the geo-information produced by remote volunteers, especially considering the particular needs and requirements of the field of disaster risk management.

Conclusion

Crowdsourced geographic information (CGI) holds a big potential not only for coping with the effects of disaster events, but also for implementing preventive measures for improving the resilience of urban areas against natural hazards and extreme events. We presented and discussed three main types of crowd-sourced geo-information that can be explored for this purpose: social media, crowd sensing and collaborative maps.

CGI of these different types can be incorporated into disaster risk management in many different ways. As shown in the previous sections, the most important usage of CGI in this context is improve situation awareness in the monitoring of unfolding events, i.e. to complement conventional information sources with first-hand geographic information from the crowd shared in social media, citizen sensing platforms and/or collaborative maps. In this manner, it is possible to get more fine-grained and up-to-date spatial information about what is happening on the ground. Clearly, this information is of great value for creating maps to support emergency agencies in disaster relief, both in field missions and in emergency operation centres. Although CGI is becoming more and more used for this purpose, significant challenges remain in filtering and prioritising useful and valuable information amidst the large stream of non-relevant data. Since most existing studies are still focused on a single source of CGI, the integration and fusion of the different types of CGI with other authoritative data sources and processes of emergency agencies is a still underexplored topic that should be addressed in future research efforts in this area.

Furthermore, the use of CGI in mitigation and preparation phases should be emphasised in future studies. This could be done for instance by leveraging initial examples of using CGI from collaborative maps to support activities in disaster risk management, such as in the identification of critical infrastructures to support emergency planning (Herfort et al. 2015; Schelhorn et al. 2014), for instance for performing evacuation simulations (Bakillah et al. 2012, Goetz & Zipf 2012) and estimating the vulnerability of urban areas based on synthetic information about the potentially affected population (Bakillah et al. 2014).

References

de Albuquerque, J. P., Herfort, B., Brenning, A., & Zipf. A. 2015. A Geographic Approach for Combining Social Media and Authoritative Data towards Identifying Useful Information for Disaster Management. *International Journal of Geographical Information Science*: 1–23. Available at: http://www.tandfonline.com/doi/abs/10.1080/13658816.2014.996567.DOI:http://dx.doi.org/10.1080/13658816.2014.996567

Bakillah, M., Andrés Domínguez, J., & Zipf, A. 2012. Multi-agents Evacuation Simulation Data Model with Social Considerations for Disaster Management Context. The 8th International Conference for Geo-Information for Disaster Management. GI4DM. Enschede. The Netherlands.

Bakillah, M., Liang, S., Mobasheri, A., Arsanjani, J. J., & Zipf, A. 2014. Fine-Resolution Population Mapping Using OpenStreetMap Points-of-Interest. *International Journal of Geographical Information Science,* (April 24): 1–24. Available at: http://www.tandfonline.com/doi/abs/10.1080/13658816.2014.909045. DOI: http://dx.doi.org/10.1080/13658816.2014.909045

Crooks, A., Croitoru, A., Stefanidis, A., & Radzikowski, J. 2013. "#Earthquake: Twitter as a Distributed Sensor System." *Transactions in GIS, 17*(1): 124–147. Available at: http://doi.wiley.com/10.1111/j.1467-9671.2012.01359.x. DOI: http://dx.doi.org/10.1111/j.1467-9671.2012.01359.x

De Longueville, B., Annoni, A., Schade, S., Ostlaender, N., & Whitmore, C. 2010. Digital Earth's Nervous System for Crisis Events: Real-Time Sensor Web Enablement of Volunteered Geographic Information. *International Journal of Digital Earth, 3*(3) (September): 242–259. Available at: http://www.tandfonline.com/doi/abs/10.1080/17538947.2010.484869. DOI: http://dx.doi.org/10.1080/17538947.2010.484869

Degrossi, L. C., Albuquerque, J. P., Fava, M. C., & Mendiondo, E. M. 2014. Flood Citizen Observatory : A Crowdsourcing-Based Approach for Flood Risk Management in Brazil. In: *26th International Conference on Software Engineering and Knowledge Engineering (SEKE 2014).*

Ferster, C. J., & Coops, N. C. 2013. A Review of Earth Observation Using Mobile Personal Communication Devices. *Computers & Geosciences, 51*: 339–349. Available at: http://www.sciencedirect.com/science/article/pii/S0098300412003184.

Goetz, M., & Zipf, A. 2012. Using Crowdsourced Indoor Geodata for Agent-Based Indoor Evacuation Simulations. *ISPRS International Journal of Geo-Information, 1*(2): 186–208. MDPI. DOI: http://dx.doi.org/10.3390/ijgi1020186

Goodchild, M. F. 2007. Citizens as Sensors: The World of Volunteered Geography. *GeoJournal, 69*(4): 211–221.

Haklay, M. 2013. Citizen Science and Volunteered Geographic Information: Overview and Typology of Participation. In: Sui, D., Elwood, S., & Goodchild, M. (Eds.) *Crowdsourcing Geographic Knowledge*. Dordrecht: Springer Netherlands, pp. 105–122. Available at: http://www.springerlink.com/index/10.1007/978-94-007-4587-2. DOI: http://dx.doi.org/10.1007/978-94-007-4587-2.

Haklay, M. 2010. How Good Is Volunteered Geographical Information? A Comparative Study of OpenStreetMap and Ordnance Survey Datasets. *Environment and Planning B: Planning and Design, 37*(4): 682–703. Available at: http://discovery.ucl.ac.uk/150445/.

Harvard Humanitarian Initiative. 2011. Disaster Relief 2.0. Washington D.C., Berkshire.

Herfort, B., Eckle, M., Albuquerque, J. P., & Zipf, A. 2015. Towards Assessing the Quality of Volunteered Geographic Information from OpenStreetMap for Identifying Critical Infrastructures. In: *Proceedings of the ISCRAM 2015 Conference – Kristiansand, May 24–27*, edited by Palen, Büscher, Comes, and Hughes. Kristiansand, Norway, pp. 1–8.

Horita, F. E. A., Degrossi, L. C., Assis, L. F. F. G., Zipf, A., & Albuquerque, J. P. 2013. The Use of Volunteered Geographic Information and Crowdsourcing in Disaster Management: A Systematic Literature Review. In: *Proceedings of the Nineteenth Americas Conference on Information Systems, Chicago Illinois, August 15–17, 2013*. Atlanta, GA, USA: AIS, pp. 1–10.

Horita, F. E. A., Albuquerque, J. P., Degrossi, L. C., Mendiondo, E. M., & Ueyama, J. 2015. Development of a Spatial Decision Support System for Flood Risk Management in Brazil That Combines Volunteered Geographic Information with Wireless Sensor Networks. *Computers & Geosciences, 80*: 84–94. Available at: http://www.sciencedirect.com/science/article/pii/S0098300415000746. DOI: http://dx.doi.org/10.1016/j.cageo.2015.04.001

Kaplan, A. M., & Haenlein, M. 2010. Users of the World, Unite! The Challenges and Opportunities of Social Media. *Business Horizons, 53*(1): 59–68. Available at: http://ideas.repec.org/a/eee/bushor/v53y2010i1p59-68.html.

Liu, H-Y., Kobernus, M., Broday, D., & Bartonova, A. 2015. A Conceptual Approach to a Citizens' Observatory - Supporting Community-Based Environmental Governance. *Environmental Health: A Global Access Science*

Source, 14(1): 107. Available at: http://www.ehjournal.net/content/13/1/107. DOI: http://dx.doi.org/10.1186/1476-069X-13-107

Ma, H., Zhao, D., & Yuan, P. (2014). Opportunities in mobile crowd sensing. IEEE Communications Magazine, 52(8), 29–35. http://doi.org/10.1109/MCOM.2014.6871666

Neis, P., Singler, P., & Zipf, A. 2010. Collaborative Mapping and Emergency Routing for Disaster Logistics – Case Studies from the Haiti Earthquake and the UN Portal for Afrika. In: *Geospatial Crossroads @ GI_Forum 2010. Proceedings of the Geoinformatics Forum Salzburg.* Salzburg, Austria, pp. 239–248.

Neis, P., Zielstra, D., & Zipf, A. 2011. The Street Network Evolution of Crowdsourced Maps: OpenStreetMap in Germany 2007–2011. *Future Internet,* 4(1): 1–21. DOI: http://dx.doi.org/10.3390/fi4010001. Available at: http://www.mdpi.com/1999-5903/4/1/1/

Neis, P., & Zipf, A. 2012. Analyzing the Contributor Activity of a Volunteered Geographic Information Project — The Case of OpenStreetMap. *ISPRS International Journal of Geo-Information, 1*(2): 146–165. Available at: http://www.mdpi.com/2220-9964/1/2/146/htm. DOI: http://dx.doi.org/10.3390/ijgi1020146

Reimer, A., Neis, P., Rylov, M., Schellhorn, A., Sagl, G., Resch, B., Albuquerque, J.P., & Zipf, A. 2014. Erfahrungsbericht Crisis Mapping Zum Taifun Hayan. In: *Gemeinsame Jahrestagung, Geoinformatik 2014.* Hamburg.

Sakaki, T., Okazaki, M., & Matsuo, Y. 2010. Earthquake Shakes Twitter Users. In: *Proceedings of the 19th International Conference on World Wide Web – WWW '10*, 851. New York, New York, USA: ACM Press. Available at: http://dl.acm.org/citation.cfm?id=1772690.1772777. DOI: http://dx.doi.org/10.1145/1772690.1772777

Schelhorn, S. J., Herfort, B., Leiner, R., Zipf, A., & Albuquerque, J. P. 2014. Identifying Elements at Risk from OpenStreetMap: The Case of Flooding. In: Hiltz, S. R., Pfaff, M. S., Plotnick, L., & Shih, P. C. (Eds.) *Proceedings of the 11th International ISCRAM Conference.* University Park, Pennsylvania, USA,: ISCRAM, pp. 508–512.

Schnebele, E., Cervone, G., & Waters, N. 2014. Road Assessment after Flood Events Using Non-Authoritative Data. *Natural Hazards and Earth System Science, 14*(4): 1007–1015. Available at: http://www.nat-hazards-earth-syst-sci.net/14/1007/2014/nhess-14-1007-2014.html. DOI: http://dx.doi.org/10.5194/nhess-14-1007-2014

Spinsanti, L., & Ostermann, F. 2013. Automated Geographic Context Analysis for Volunteered Information. *Applied Geography, 43*: 36-44. DOI: http://dx.doi.org/10.1016/j.apgeog.2013.05.005

Steiger, E., Albuquerque, J. P., & Zipf, A. 2015. Twitter as a Location Based Social Network – An Advanced Systematic Literature Review on Spatiotemporal Analyses of Twitter Data. *Transactions in GIS.* DOI: http://dx.doi.org/10.1111/tgis.12132

Triglav-Čekada, M., and D. Radovan. 2013. "Using Volunteered Geographical Information to Map the November 2012 Floods in Slovenia." *Natural*

Hazards and Earth System Science, 13(11) (November 5): 2753–2762. Available at: http://www.nat-hazards-earth-syst-sci.net/13/2753/2013/nhess-13-2753-2013.html. DOI: http://dx.doi.org/10.5194/nhess-13-2753-2013

UN OCHA. 2012. *Humanitarianism in the Network Age.*

Zook, M., Graham, M., Shelton, T., & Gorman, S. 2010. "Volunteered Geographic Information and Crowdsourcing Disaster Relief: A Case Study of the Haitian Earthquake." *World Medical & Health Policy*, 2(2) (July 21): 7. Available at: http://www.psocommons.org/wmhp/vol2/iss2/art2. DOI: http://dx.doi.org/10.2202/1948-4682.1069

PART V

VGI in mobility

CHAPTER 24

Crowdsourcing for individual needs – the case of routing and navigation for mobility-impaired persons

Alexander Zipf*, Amin Mobasheri, Adam Rousell and Stefan Hahmann

GIScience research group, Institute of Geography, Heidelberg University, Heidelberg, Germany
*zipf@uni-heidelberg.de

Abstract

Routing and navigation web services are becoming widely used, and make use of both commercial and VGI datasets. It is now becoming widely acknowledged that a 'one fits all' method of generating and presenting routes is not applicable. In particular, the accessibility of places for the mobility impaired has become a key focus with several services addressing topics such as how accessible locations of interest are and how to best generate routes for people who need to consider additional factors. Though datasources such as OpenStreetMap (OSM) are well suited for such topics, several issues including the quality of the underlying data remain. Through the use of quality assessment tools it is possible to identify areas with inadequate data completeness with regards to the information needed for the mobility impaired and thus encourage the enrichment of these areas through specialised tagging applications. Such data can then be used in routing and navigation services which focus on ensuring that routes being generated and presented fit the personal requirements of the traveller.

How to cite this book chapter:
Zipf, A, Mobasheri, A, Rousell, A and Hahmann, S. 2016. Crowdsourcing for individual needs - the case of routing and navigation for mobility-impaired persons. In: Capineri, C, Haklay, M, Huang, H, Antoniou, V, Kettunen, J, Ostermann, F and Purves, R. (eds.) *European Handbook of Crowdsourced Geographic Information*, Pp. 325–337. London: Ubiquity Press. DOI: http://dx.doi.org/10.5334/bax.x. License: CC-BY 4.0.

Introduction

Routing and navigation services for vehicles and people are based on geospatial data. Due to the availability of GPS sensors within handheld devices, systems have been able to localise themselves precisely on the road. These services increasingly make use of public transport data and integrate near real-time traffic data. Also, through the development of new algorithms that employ hierarchical methods, routing has become faster, especially for calculating long distance routes. Nowadays, routing services are able to provide routes based on several criteria such as distance, road type, and traffic to an acceptable degree of accuracy. In addition, further research prototypes have identified a number of extra criteria. With the arrival of crowdsourcing and VGI (in particular the OpenStreetMap project), a new generation of route planning services using such services has emerged. As an example, OpenRouteService.org (Neis & Zipf, 2008) used OSM as data source for deriving optimal routes between two locations for different modes of transport including car, several types of bikes, and pedestrians. The potential of crowdsourced geographic information for routing and navigation can be highlighted by the fact that the OSM based OpenRouteService[1] was able to provide pedestrian and bicycle routing across several countries even before Google offered these features. This is because of the different way crowdsourced information like OSM is being collected. Volunteers do this on the ground (commonly on foot or bicycle) leading to a higher representation of this particular kind of data than offered by commercial providers before. This particular richness of VGI also offers new possibilities for even more specific information needs and specialised applications.

Besides the support of these mainstream route planning needs, routing services that serve people with special requirements (such as wheelchair routing) are being designed and developed. However, in order to provide the appropriate data required for such services that meet the specific information requirements of the respective users, new sources of information are required. For instance, in addition to road features, a wheelchair routing service would need to consider sidewalk data, such as curbs, surface type and incline information, to name a few. Müller et al. (2010) therefore suggested several extensions to the OSM tagging schema including new tags, as well as identifying a selection of relevant already existing ones. This supported OSM mappers in the collection of such information. These tags have been added to the OSM schema and are described in the OSM wiki[2]. Such detailed information is however, still not generally available or complete in the OpenStreetMap dataset. Therefore, special crowdsourcing tools and services are necessary in order to better support the collection of geospatial data relevant for routing and navigation of people with limited mobility.

---

[1] http://www.openrouteservice.org
[2] http://wiki.openstreetmap.org/wiki/Wheelchair_routing

To address the incompleteness of data it is necessary to develop and pilot-test methods and tools for collectively gathering and sharing spatial information for improving accessibility. The power of online maps and mobile devices can help foster an awareness of barriers for individuals with limited mobility and encourage the removal of such barriers. As this is highly relevant for an increasing proportion of our society, the European Commission has decided to fund projects in this domain. One example is the CAP4Access[3] project. The agenda of research and development in this field includes the design and implementation of tools and methods for (a) quality assessment, i.e. checking the completeness of OSM data with regard to required information; b) tagging, i.e. describing and discussing locations and routes within the built environment according to their accessibility; (c) route planning and navigation; (d) raising awareness and preparing effective measures at local level for eliminating barriers. Target groups include people requiring enhanced accessibility, grassroots initiatives supporting people with disabilities, policy-makers, planners and service providers with responsibility for the built environment, and the general public.

Required and existing services

Quality assessment and enrichment of crowd-sourced data

Ensuring the quality of crowdsourced data is of particular importance in order to ensure that the results of routing and navigation services offered are accurate and ultimately useful. Generally, routing and navigation services for people with limited mobility benefit from information regarding new obstacles provided by involved communities. However, crowdsourced data has significant differences from traditional geospatial data which are often created by specifically dedicated organizations and experts, and are generated according to standardised structures and languages. Especially in the case of routing tools, completeness of data and spatial accuracy is of great concern to ensure proper routing. The development of a data quality assessment component is therefore a crucial objective.

There are a number of different geo-data quality elements that one might need to check before using any kind of datasets in their project. These elements include positional accuracy, attribute accuracy, completeness, logical consistency and temporal accuracy (van Oort, 2006). A discussion on all data quality elements are out of the scope of this chapter. However, as an example we provide information on one of the important data quality elements for routing and navigation services – completeness. Completeness is defined as errors of omission (measure of the absence of data), and errors of commission

---

[3] http://myaccessible.eu/

(measure of the presence of extra data) (van Oort, 2006). The completeness of a dataset can be suitable for a specific task but not for another. So, when completeness has to be measured, the concept of fitness for use comes in mind.

In order to check the completeness of a dataset three tests could be performed:

- *missing object/line*: here we find missing information at object level, whether a line or polygon is missing in the dataset. This should be done by using a reference dataset if possible.
- *missing attribute*: for those objects (point/line/polygon) that are available we need to know which attributes are missing (based on a list of attributes that are important and used by the routing and navigation system). Missing attributes are counted as inconsistencies and reported. This check can be performed by automated means of intrinsic data check, which means that only the dataset itself is investigated (Barron et al. 2013).
- *missing value*: From those existing attributes, some may be incomplete in terms of missing values. Here we check those attributes/fields that lack value

There are several tools that could be used for checking the completeness of an OpenStreetMap dataset. For example, OSMatrix[4] is a tool for visualising mapping progress/quality on various metrics (Roick et al. 2011, 2012). By using OSMatrix, one can calculate the number of certain object features in OSM (e.g. sidewalks) at various timestamps. For example, Figure 1 shows a snapshot of visual and statistical information regarding the total number of sidewalk information (tags in OpenStreetMap related to sidewalk information) for a selected region in Heidelberg, in Jan 2016. The sample area is divided into hexagonal cells with the size of 1 km. Each cell shows a value representing the aggregation of the total number of tags in OSM related to sidewalk attributes in that hexagon. OSMatrix could also be used in order to derive statistical and visual information regarding the completeness of OSM data at the attribute level (e.g. sidewalk width, incline, etc.).

As another example of tools, OSM Quality Assurance Editor[5] can be used to understand the completeness of certain object features in OSM data using an object-based approach. Figure 2 shows a region in Heidelberg (Germany) where sidewalk information is missing (the road features contain no information regarding the presence or absence of sidewalks attached to the road). This information is given per object, meaning that one could select an object and view its properties as opposed to the provision of an aggregated region-wide value. This is a large benefit in comparison to OSMatrix is in that tool the statistical information is aggregated and provided for each cell, while information regarding route objects inside the cells cannot be realised.

---

[4] http://alborz.geog.uni-heidelberg.de/osmatrix/
[5] http://editor.osmsurround.org/

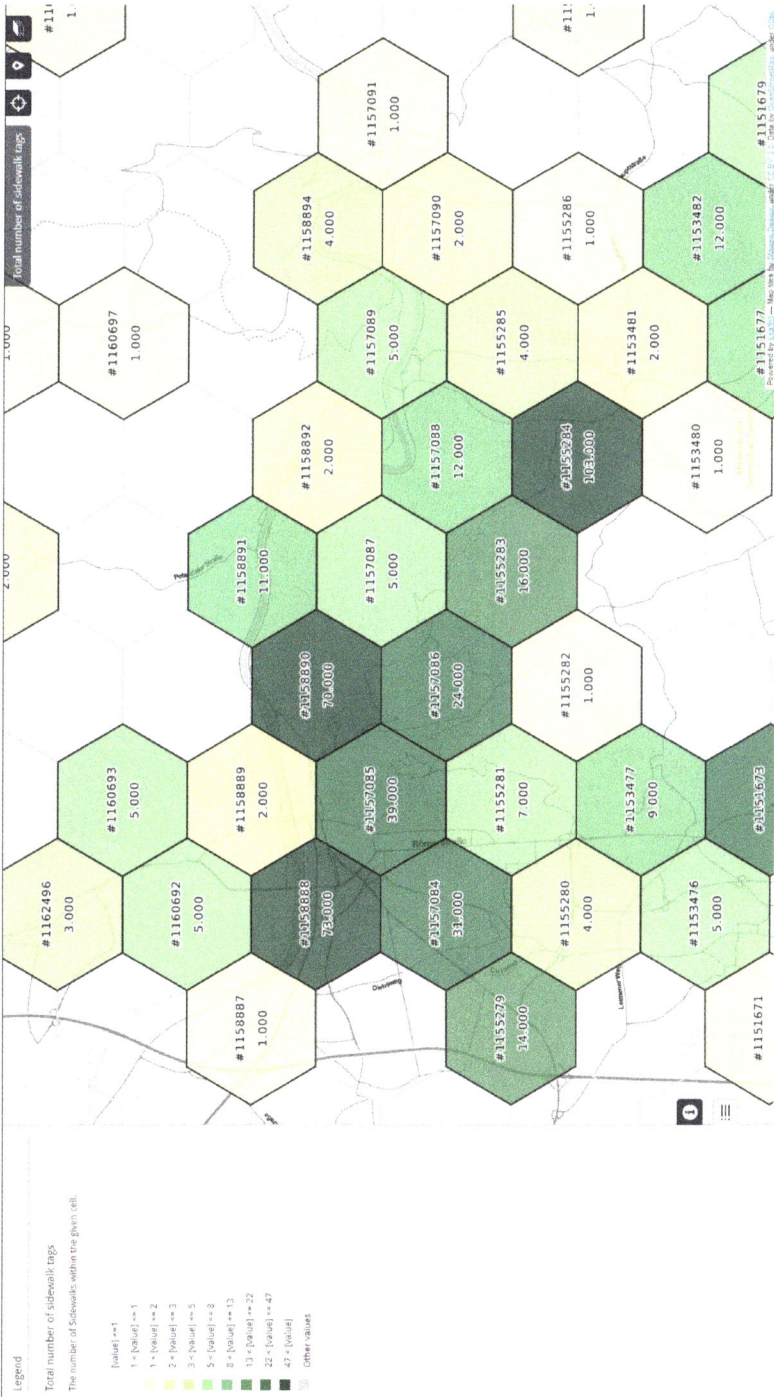

Figure 1: Mapping the completeness statistics of total number of sidewalk information using OSMatrix.

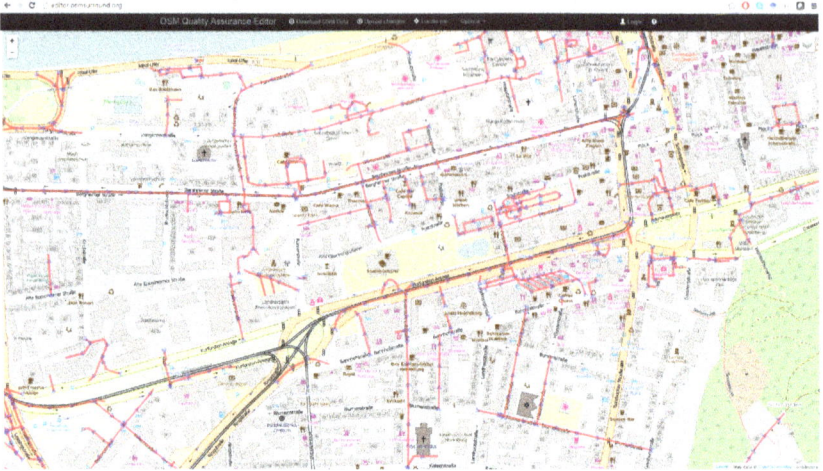

Figure 2: An example of selected road features that have no information regarding existence of sidewalk.

In order to provide an effective routing and navigation service for people with limited mobility, it is crucial to have data regarding sidewalk features and their attributes such as surface texture, width, incline, etc. The quality assessment tools can aid in the understanding of the level of incompleteness of such information within OpenStreetMap data, as well as identifying the places where such information is missing. In order to better inform the user about the potentially non-perfect data for wheelchair routing, Neis (2014) developed an initial prototype of a reliability index that attempts to measure some aspects regarding the information quality of the selected route.

With regards to data availability, Figure 2 shows that most road and streets in the selected region in Heidelberg are missing sidewalk information. Therefore, in order to collect and enrich OSM with sidewalk information a tagging system is suggested. Tagging systems are developed to support the collection of user-generated data. In the case of the restricted mobility topic, this data focusses on the accessibility of places, points of interest and roads. The collective tagging approach has already been applied within the accessibility theme through the Wheelmap[6] platform. Wheelmap is a map for wheelchair-accessible places. Locations are rated and portrayed according to a traffic light system based on their accessibility status (e.g. accessibility of restaurants for people on wheelchairs). One of the disadvantages of Wheelmap with regards to collecting accessibility information for improving routing and navigation of people with

_____

[6] http://wheelmap.org/

restricted mobility is that it is not capable of collecting accessibility information for linear objects such as sidewalks.

Since sidewalk information is crucial to routing services that need to provide the most accessible route from one location to another, other tagging systems should be used that provide the capability of enriching sidewalk information such as (for example) availability of a sidewalk, sidewalk width, incline and surface texture. For this purpose, any OSM editor that could be used for editing line features (e.g. Vespucci[7] OSM editor, JOSM, etc.) can be employed.

Routing

A routing service tailored to special needs is a core service to improve the mobility of people with various types of disabilities (Neis and Zielstra 2014). A routing system mostly consists of two core components: a graph network representing the underlying street datasets, and a routing engine that uses this graph to generate feasible routes.

There are a number of available routing services described on the OSM Wiki pages[8] that make use of OSM data. However, of the 13 route services documented there only three currently provide the functionality to be extended for to address specialised requirements, such as wheelchair routing. These three are OpenRouteService, Routino[9] and OpenTripPlanner[10]. OpenRouteService. org (ORS; Neis & Zipf 2008) is built according to open standards from the Open Geospatial Consortium (OGC) meaning that it can easily be integrated in other applications or regional web sites. In order to use the crowdsourced data from OSM for wheelchair routing, applications such as ORS have been (and will need to further be) extended so that they can be used by persons with limited mobility with various profiles and parameters. Figure 3 shows the result of a route plan where a difference between the pedestrian profile and the wheelchair profile occurs due to a pedestrian bridge that is only accessible via steps which are not feasible for wheelchair users. A first prototype for wheelchair routing based on OpenRouteService was developed earlier by Müller et al. (2010)[11].

The main challenge to serve special route planning requirements is the need for very specific data (e.g. sidewalk data), which is not available for large areas, e.g. country-wide. This especially includes data about sidewalks (e.g. width, surface, smoothness, incline and existence of sidewalks) and also the position and height of sloped/dropped curbs. Such data is not usually published in authoritative products due to their scope, even though the data is often available

---

[7] http://wiki.openstreetmap.org/wiki/Vespucci
[8] http://wiki.openstreetmap.org/wiki/Routing/online_routers
[9] http://www.routino.org/
[10] http://www.opentripplanner.org/
[11] http://rollstuhlrouting.de

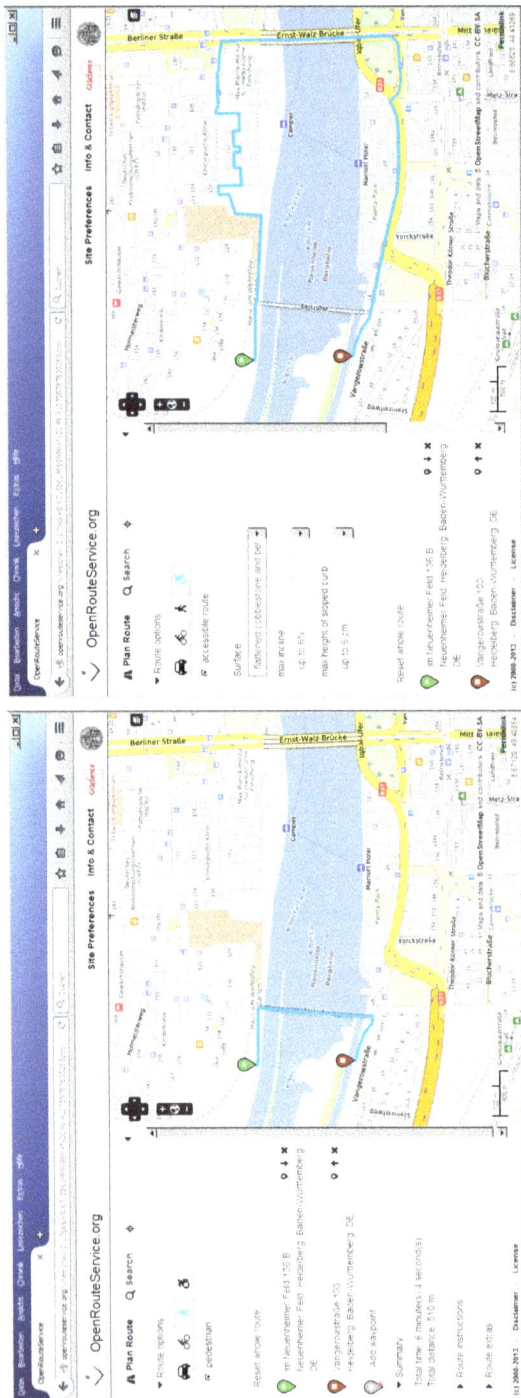

a) Route plan for pedestrian profile

b) Route plan for wheelchair profile with the same start and end points

Figure 3: Showing a routing situation where the wheelchair profile leads to a different route.

within the organisation. Even if these datasets were made available however, the heterogeneity between data structures would make integration problematic. Considering this situation, VGI becomes an important data source as local knowledge from the users can be harnessed. Care needs to be taken however to provide feedback (in the form of changes to existing services) to the volunteers to ensure that their motivation is maintained. In terms of the sidewalks themselves, the OSM community has decided that within their dataset, apart from defined exceptions, sidewalks should not be represented by separate features, but instead through attributes held by their associated roads.

Another challenge relating to routing (in particular for pedestrians) is that existing routing algorithms can only process data that is provided in a line format. Within OSM, many traversable surfaces (such as city squares and parks) are represented as polygon areas which most routing algorithms simply do not know how to handle. Several methods could be implemented to generate linear representations of these spaces such as using the polygon outline (figure 4a), or generating internal linear structures through grids (figure 4c), skeletons (figure 4d) or lines-of-sight.

Navigation

Although routing systems provide a valuable service with regards to getting around unknown environments, the determination of a route is only one step in an overall wayfinding process. As well as being able to calculate a route through space that is suitable for an individual's preferences, this information needs to be conveyed in a format that allows them to successfully traverse the intended route. Many navigation systems exist which attempt to provide direction instructions to travellers, often in the form of distance and street names. It is well documented that when describing directions (particularly for pedestrians) landmarks form an essential component of the descriptive (Duckham et al. 2010, Winter et al. 2005). Proper selection of landmarks in navigation of people with restricted mobility is challenging since it requires special considerations compared to normal pedestrians. For instance, the fact that people on wheelchair have less visibility than those who are standing is an issue that must be considered.

A number of methods have been identified that extract landmarks that can be used for navigational instructions, some of which make use of VGI datasets (such as Dräger and Kroller 2012). Key considerations when producing such methods include the scalability, performance in terms of determination speed, and the suitability of the actual landmarks identified.

With regards to navigation services for mobility impaired users, it is important to take into account the individual requirements of each user and then feed these through to the underlying route generation system. The system should also allow users to report any obstacles that they encounter which were not included in the route generation due to the incompleteness of the underlying

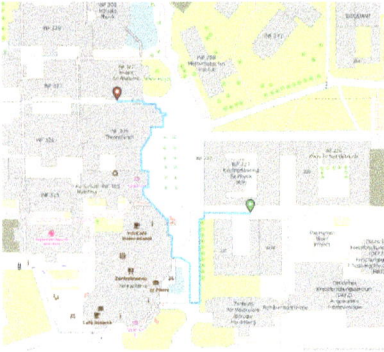

a) Planned route using the outline of an open space area

b) Desired route going through the open space area

c) Grid (orange) based approach as a subgraph of the polygon (light blue)

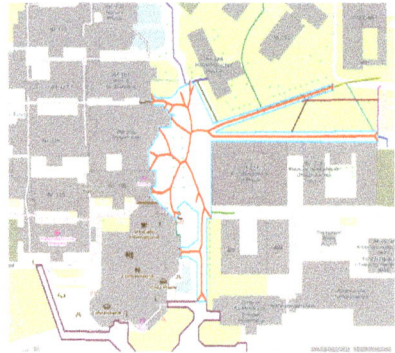

d) Skeleton (orange) based approach as a subgraph of the polygon (light blue)

Figure 4: Different routing approaches through polygons (grid, skeleton, line of sight).

dataset. This in turn not only empowers the user in terms of them making a difference to the system, but it also increases the overall completeness of the source dataset.

One of the main challenges posed currently is how VGI and social media can be used to identify suitable landmarks that can be used in navigational instructions. Not only does this require the fusion of freely available data from various sources (i.e. OSM, FourSquare and Twitter), it also generates technical problems such as providing a fast lookup service for possible candidates from a population. Overall, what we want is a system that can quickly provide an instruction for any route similar to *"just after xxx, turn left"* where *xxx* is the most salient object in relation to the turning point.

Discussion and Conclusion

Routing and navigation services are becoming more personalized and near real-time, by recommending routes that take into account various preferences and up-to-date information. This has long been argued for in the field of adaptive and personalized Location Based Services (LBS) (Malaka and Zipf 2000, Zipf 2002). In contrast to former times, such developments can now be realised much easier due to the wealth of rich data that is made available through VGI and the Social Web.

In this chapter, the potentials of using crowdsourced geo-information for routing and navigation with a focus on people with restricted mobility were presented. Although the completeness of OpenStreetMap data with regard to specific information for those user groups such as relevant barriers or details on sidewalks still is low, there are developments under way to improve the situation both in an automated way as well as by harnessing the power of the crowd, so we can expect more and more areas where enough information is available. We explored the OSMatrix Web-Service and presented how such a portal visualizing intrinsic quality indicators can be used in order to help understand the data and attribute availability for certain features (e.g. sidewalks). Furthermore, we see the need for further research in the area of (semi-) automated data integration from different user generated data sources in order to further enrich the routing graph with relevant objects and attributes.

As an example related to both pedestrian and wheelchair navigation, we discussed the problems on deriving routing graphs dealing with with open spaces (e.g. city squares and parks). For this possible ideas for generating routing graph were explained. These are further investigated and evaluated with respect to efficiency and quality in our current research. It was discussed that while routing services provide valuable information, they are only the first step in the overall wayfinding process. A navigation service is an additional component that receives routing information as input, processes, translates and communicates the information in a format that allows the individuals to traverse the relevant route. For this, the derivation and selection of landmarks from OSM and other crowdsourcing platforms is an interesting research topic. Regarding navigation services for mobility impaired users, it was concluded that individual requirements of each user need to be collected and considered in the process of route generation. Extraction of relevant information from both the users' context as well as the environment from VGI and crowdsourced data could lead to a richer real-time routing and navigation service that could benefit from the availability of temporally detailed road network data with conditions such as speed or other detailed semantic attributes.

Acknowledgements

The research leading to these results has received funding from the European Community's Seventh Framework Programme (FP7/2007-2013) under grant agreement n° 612096 (CAP4Access). The work has been partly supported by funding from the Klaus-Tschira-Foundation (KTS), Heidelberg.

References

Barron, C., Neis, P., & Zipf, A. 2013. A Comprehensive Framework for Intrinsic OpenStreetMap Quality Analysis. Transactions in GIS. DOI: http://dx.doi.org/10.1111/tgis.12073

Dräger, M., & Koller, A. 2012. Generation of landmark-based navigation instructions from open-source data. In: *Proceedings of the 13th Conference of the European Chapter of the Association for Computational Linguistics*. Avignon, France: Association for Computational Linguistics, pp. 757–766.

Duckham, M., Winter, S., & Robinson, M. 2010. Including landmarks in routing instructions. *Journal of Location Based Services*, 4(1): 28–52.

Grossner, K., & Glennon, A. 2007. Volunteered geographic information: Level III of a digital earth system. In: *Proceedings of the Position paper presented at the Workshop on Volunteered Geographic Information, Santa Barbara, CA, USA*, pp. 13–14.

de Longueville, B., Ostlander, N., & Keskitalo, C. 2009. Addressing vagueness in volunteered geographic information (VGI) – A case study. *International Journal of Spatial Data Infrastructures Research. Special Issue GSDI-11*.

Malaka, R., & Zipf, A. 2000. DEEP MAP – Challenging IT research in the framework of a tourist information system. In: Fesenmaier, D., Klein, S., & Buhalis, D. (Eds.) *Information and Communication Technologies in Tourism 2000*. Proceedings of ENTER 2000, Barcelona. Spain. Springer Computer Science, Wien, New York, pp. 15–27.

Müller, A., Neis, P., & Zipf, A. 2010. Ein Routenplaner für Rollstuhlfahrer auf der Basis von OpenStreetMap-Daten. Konzeption, Realisierung und Perspektiven. AGIT 2010. In: *Proceedings of Symposium für Angewandte Geoinformatik*. Salzburg. Austria.

Neis, P. 2014. Measuring the Reliability of Wheelchair User Route Planning based on Volunteered Geographic Information. *Transactions in GIS*.

Neis, P., & Zipf, A. 2008. OpenRouteService.org is three times "Open": Combining OpenSource, OpenLS and OpenStreetMaps. *GIS Research UK (GIS-RUK 08)*. Manchester, UK.

Neis, P., & Zielstra, D. 2014. Generation of a tailored routing network for disabled people based on collaboratively collected geodata. *Applied Geography*, 47: 70–77.

Roick, O., Hagenauer, J., & Zipf, A. 2011. OSMatrix – Grid based analysis and visualization of OpenStreetMap. SOTM-EU 2011. State of the Map EU. Scientific Track. Wien.

Roick, O., Loos, L., & Zipf, A. 2012. Visualizing spatio-temporal quality metrics of Volunteered Geographic Information – A case study for OpenStreetMap. Geoinformatik 2012. Mobilität und Umwelt. Braunschweig. Germany.

van Oort, P. V. 2006. Spatial Data Quality: from Description to Application, *PhD dissertation*, Wageningen Universiteit, Wageningen, The Netherlands.

Winter, S., Raubal, M., & Nothegger, C. 2005. Focalizing Measures of Salience for Wayfinding. In: Meng, L., Reichenbacher, T., & Zipf, A., (Eds.) *Map-based Mobile Services*. Springer Berlin Heidelberg, pp. 125–139.

Zipf, A. 2002. User-Adaptive Maps for Location-Based Services (LBS) for Tourism In: Woeber, K., Frew, A., & Hitz, M. (Eds.) *Proc. of the 9th Int. Conf. for Information and Communication Technologies in Tourism, ENTER* 2002. Innsbruck, Austria. Springer Computer Science. Heidelberg, Berlin.

CHAPTER 25

Smart Timetable Service Based on Crowdsensed Data

Károly Farkas

Budapest University of Technology and Economics, Hungary, farkask@hit.bme.hu

Abstract

Rapid technological development and the introduction of smart services make it possible for modern cities to offer an enhanced perception of city life for their inhabitants. For instance, a smart timetable service of the city's public transportation lines updated in real-time can decrease unnecessary waiting times at stops and increase the efficiency of travel planning. However, the implementation of such a service in a traditional way requires the deployment and maintenance of some costly sensing and tracking infrastructure. Fortunately, for this purpose mobile crowdsensing can be a viable and almost free of charge alternative. In this case, the crowd of passengers and their mobile devices are used to gather data.

In this chapter, we place emphasis on the introduction of a crowdsensing based smart timetable service, which has been developed as a prototype smart city application. The front-end interface of this service is called TrafficInfo. It is a simple and easy-to-use Android application which visualizes public transport information of the given city on Google Maps in real-time. The live updates of transport schedule information rely on the automatic stop event detection of public transport vehicles. TrafficInfo is built upon an Extensible Messaging and Presence Protocol (XMPP) based communication framework which was designed to facilitate the development of crowd assisted smart city applications.

How to cite this book chapter:
Farkas, K. 2016. Smart Timetable Service Based on Crowdsensed Data. In: Capineri, C, Haklay, M, Huang, H, Antoniou, V, Kettunen, J, Ostermann, F and Purves, R. (eds.) *European Handbook of Crowdsourced Geographic Information*, Pp. 339–351. London: Ubiquity Press. DOI: http://dx.doi.org/10.5334/bax.y. License: CC-BY 4.0.

The chapter introduces this generic framework shortly, then describes the prototype smart timetable service.

Keywords

Smart cities, Crowdsensing, Public transportation, XMPP, GTFS

Introduction

More and more modern cities offer smart services, which are services using modern infrastructure and/or providing value added functions to ease the everyday life of inhabitants. Unfortunately, the traditional way of introducing a new service usually implies a huge investment to deploy and maintain the necessary background infrastructure. One of the most popular city services is public transportation. Maintaining and continuously improving such a service are imperative in modern cities. However, the implementation of even a simple feature that extends the basic service functions can be expensive. For instance, let us consider the replacement of static timetables with a live public transport information service updated in real-time. It requires the deployment of a vehicle-tracking infrastructure consisting of among others GPS sensors, communication infrastructure, back-end systems and front-end user interfaces, which can be a cost intensive investment.

An alternative approach to collect real-time tracking data is exploiting the power of the crowd via participatory sensing or often called mobile crowdsensing, which does not call for such an investment. In this scenario (see Figure 1), the passengers' mobile devices and their built-in sensors, or the passengers themselves via reporting incidents, are used to generate the monitoring data for vehicle tracking. Moreover, they send instant route information to the service provider in real-time. The service provider then aggregates, cleans, analyzes the data gathered, and derives and disseminates the real-time updates. The sensing task is carried out by the built-in and ubiquitous sensors of the smartphones either in participatory or opportunistic way depending on whether the user is involved or not in data collection. Every traveler can contribute to this data-harvesting task. Thus, passengers waiting for a ride at the stop can report the line number with a timestamp of every arriving public transport vehicle during the waiting period. On the other hand, onboard passengers can be used to gather and report actual position information of the moving vehicle and detect halt events at the stops.

In this chapter, we focus on the introduction of a crowdsensing based smart timetable service, which has been developed as a prototype smart city application. The front-end interface of this service, called TrafficInfo, is a simple and easy-to-use Android application. It visualizes live public transport information of the given city on Google Maps. TrafficInfo is built upon an Extensible

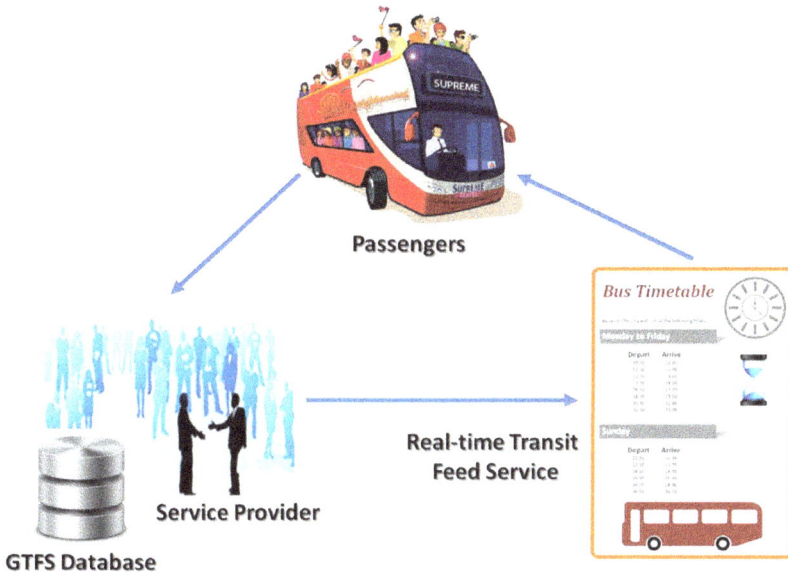

Figure 1: Live public transport information service based on mobile crowdsensing.

Messaging and Presence Protocol (XMPP) (Saint-Andre 2011) based communication framework (Szabo & Farkas 2013). This framework was designed to facilitate the development of crowd assisted smart city applications (and will be introduced shortly in the upcoming section). Following the publish/subscribe (pub/sub) communication model the passengers subscribe in TrafficInfo to traffic information channels according to their interest. These channels are dedicated to different public transport lines or stops. Hence, the passengers are informed about the live public transport situation. For instance, they can see the actual vehicle positions, deviation from the static timetable, crowdedness information, travel conditions, etc.

To motivate user participation in data collection an initial service is offered to the passengers, which is a static public transportation timetable. It is built on the General Transit Feed Specification (GTFS) (Google Inc. 2006) based transit schedule data and provided by public transport operators. GTFS is the best practice for providing such information, and is available in 350 cities attracting more than 6.5 million users. According to the GTFS developer page, currently GTFS data is available for 879 transit agencies worldwide. TrafficInfo basically presents this static timetable information to the users which is then updated in real-time, if appropriate crowdsensed data is available. To this end, the application collects the following information: position data; the timestamped halt events, detected automatically, of the public transport vehicles at the stops; simple annotation data entered by the user, such as reports on crowdedness and travel conditions. After analyzing the data gathered live updates are generated and TrafficInfo refreshes the static information with these updates.

The rest of this chapter is structured as follows. A quick overview of crowd assisted transit-tracking systems is provided in the next section. Then, our generic framework to facilitate the development of crowdsensing based services is introduced shortly. Next, we describe the prototype smart timetable service. Finally, we give a short summary.

Crowd Assisted Transit-tracking Systems and Approaches

This section gives an overview of crowd assisted transit-tracking solutions.

Moovit (Moovit Developers 2014) is meant to be a live transit app on the market providing real-time information about public transportation. Moovit has been successful only in those cities where it has already a mass of users, just like in Paris, and not successful in cities where its user base is low, e.g. in Budapest. In order to create a sufficiently large user base Moovit provides, besides live data, schedule based public transportation information as an initial service, too. The source of this information is the company who operates the public transportation network. Moovit partially relies on GTFS.

Several other mobile crowdsensing based transit-tracking ideas have been published recently. For instance, Zhou, Zheng and Li (2012) propose a bus arrival time prediction system based on bus passengers' participatory sensing. The proposed system uses movement statuses, audio recordings and mobile cell tower signals to identify the vehicle and its actual position. Thiagarajan et al. (2010) propose a method for transit tracking using the collected data of the accelerometer and the GPS sensor on the users' smartphone. Bedogni, Di Felice and Bononi (2012) use smartphone sensors data and machine learning techniques to detect motion type, e.g. traveling by train or by car. EasyTracker (Biagioni et al. 2011) provides a low cost solution for automatic real-time transit tracking and mapping based on GPS sensor data gathered from mobile phones, which are placed in transit vehicles. It offers arrival time prediction, as well.

These approaches focus on the data to offer enriched services to the users. The focus of our work, in turn, is on how to introduce such enriched services incrementally. Namely, how one can create an architecture and service model, which allows incremental introduction of live updates from participatory users over static services that are available in competing approaches. Hence, our work, in essence, complements the above ones.

Generic Framework for Crowdsensing Based Smart City Applications

In this section, our generic framework (Szabo & Farkas 2013) to aid the development of crowdsensing based smart city applications is described shortly. This framework is based on the XMPP publish/subscribe architecture. TrafficInfo is implemented on top of this framework.

Communication Model

XMPP (Saint-Andre 2011) is an open technology for real-time communication using Extensible Markup Language (XML) message format. XMPP allows sending of small information pieces from one entity to another in quasi real-time. It has several extensions, like multi-party messaging or the notification service. The latter realizes a publish/subscribe (pub/sub) communication model, where publications sent to a node are automatically multicast to the subscribers of that node. This pub/sub communication scheme fits well with most of the mobile crowdsensing based applications. In these applications, the users' mobile devices are used to collect data about the environment (publish) and the users consume the services updated on the basis of the collected data (subscribe).

Hence, we use XMPP and its publish/subscribe communication model in our generic framework to implement interactions. In this model, we defined three roles, like *Producer*, *Consumer* and *Service Provider* (see Figure 2). These entities interact with each other via the core service, which consists of event based pub/sub nodes.

Producer: The Producer acts as the original information source in the model producing raw data streams and plays a central role in data collection. He is the user who contributes his mobile's sensor data.

Consumer: The Consumer is the beneficiary of the provided service(s). He enjoys the value of the collected, cleaned, analyzed, extended and disseminated information. The user is called as *Prosumer*, when he acts in the service as both Consumer and Producer at the same time.

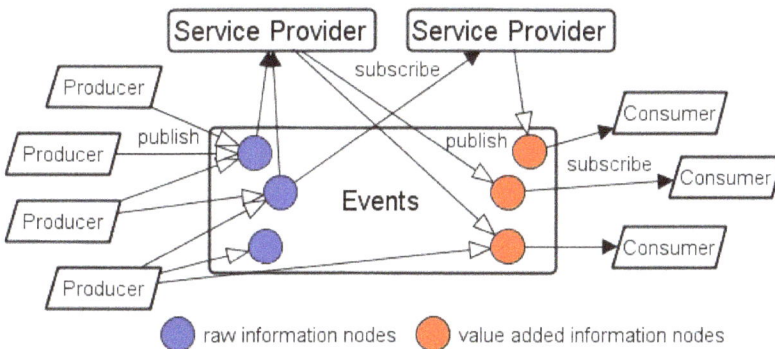

Figure 2: Crowdsensing model based on publish/subscribe communication.

Service Provider: The Service Provider introduces added value to the raw data collected by the crowd. Thus, he intercepts and extends the information flow between Producers and Consumers. A Service Provider can play several roles at the same time, as he collects (Consumer role), stores and analyzes Producers' data to offer (Service Provider role) value added service. Moreover, multiple Service Providers can act concurrently and offer different value added services to different Consumers.

In the model, depicted in Figure 2, Producers are the source of original data by sensing and monitoring their environment. They publish (marked by arrows with empty arrowhead) the collected information to event nodes (raw information nodes are marked by blue dots). On the other hand, Service Providers intercept the collected data by subscribing (marked by arrows with black arrowhead) to raw event nodes and receiving information in an asynchronous manner. They extend the crowdsensed data with their own information or extract cleaned-up information from the raw data to introduce added value to Consumers. Moreover, they publish their service to different content nodes. Consumers who are interested in the reception of the added value/service just subscribe to the appropriate content node(s) and collect the published information also in an asynchronous manner.

Framework Architecture

This model can be directly mapped to the XMPP publish/subscribe model as follows (see Figure 3):

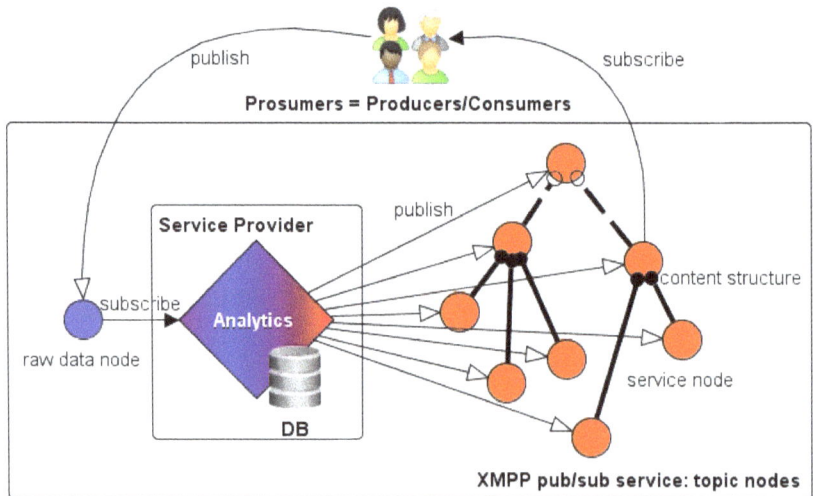

Figure 3: Mobile crowdsensing – the publish/subscribe value chain using XMPP.

Service Providers establish raw pub/sub data nodes, which gather Producers' data, for the services they offer.

- Consumers can freely publish their collected data to the corresponding nodes with appropriate node access rights, too. However, only the owner or other affiliated Consumers can retrieve this information.
- Producers can publish the collected data or their annotations to the raw data nodes at the XMPP server only if they have appropriate access rights.
- Service Providers collect the published data and introduce such a service structure for their added value via the pub/sub subscription service, which makes appropriate content filtering possible for their Consumers.
- Prosumers publish their sensor readings or annotations into and retrieve events from XMPP pub/sub nodes.
- Service Providers subscribed to raw pub/sub nodes collect, store, clean up and analyze data and extract/derive new information introducing added value. This new information is published into pub/sub nodes on the other side following a suitable structure.

The pub/sub service node structure can benefit from the aggregation feature of XMPP via using collection nodes, where a collection node will see all the information received by its child nodes. Note, however, that the aggregation mechanism of an XMPP collection node is not appropriate to filter events. Hence, the Service Provider role has to be applied to implement scalable content aggregation. Figure 3 shows XMPP aggregations as dark circles at the container node while empty circles with dashed lines represent only logical containment where intelligent aggregation is implemented through the service logic.

Smart Timetable Service

In this section, the architecture of the prototype smart timetable service is delineated first, then TrafficInfo, its front-end Android interface together with the developed automatic stop event detector is described.

Service Architecture

The prototype smart timetable service architecture has two main building blocks, such as the generic crowdsensing framework described in the previous section and the front-end application called TrafficInfo (see Figure 4). The framework can be divided into two parts, a standard XMPP server and a GTFS Emulator with an Analytics module.

XMPP Server

The XMPP server maps the public transport lines, stored in GTFS (Google Inc. 2006) format, to a hierarchical pub/sub channel structure. Thus, the GTFS

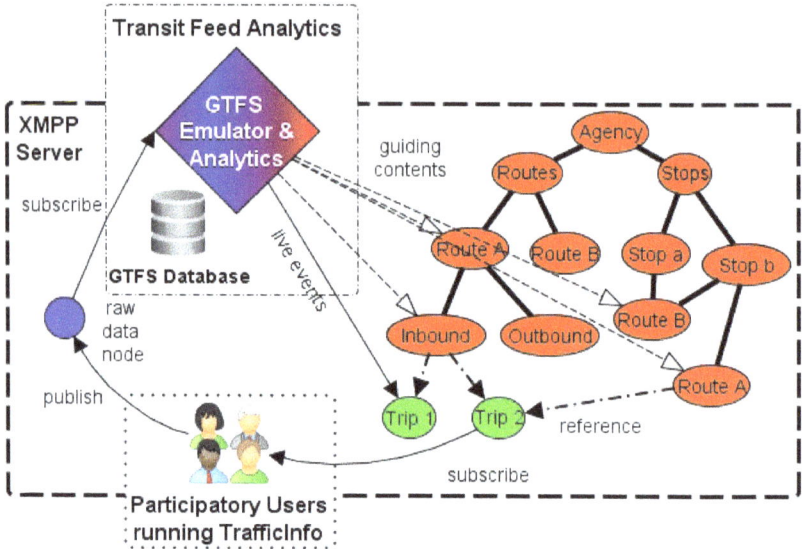

Figure 4: Smart timetable service architecture.

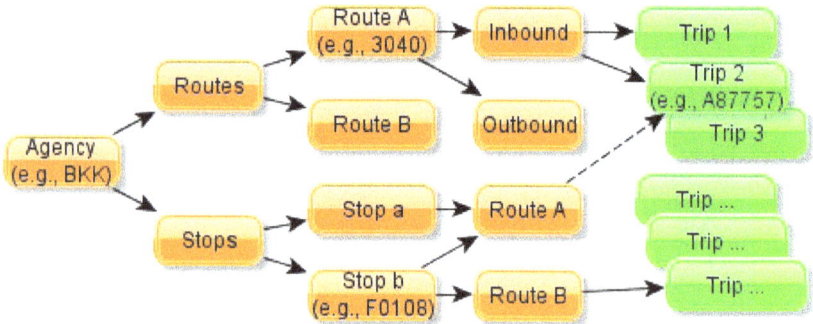

Figure 5: Publish/subscribe model for GTFS feeds.

database is turned into an XMPP pub/sub node hierarchy. This node structure facilitates searching and selecting transit feeds according to user interest. The pub/sub node model for content filtering in a transport information feed is depicted in Figure 5.

The root of the pub/sub tree is the *Agency* node referring to the public transport operator. Transit information and real-time event updates are handled in the *Trip* nodes at the leaf level. The inner nodes in the node hierarchy contain only persistent data and references relevant to the trips. The users can access the transit data via two ways, based on *Routes* or *Stops*. When the user wants to see a given trip (vehicle) related traffic information the route based filtering is applied. On the other hand, when the forthcoming arrivals at a given stop (location) are of interest, the stop based filtering is the appropriate access way.

For instance, the leaf node with trip ID 'BKK-Routes-3040-Inbound-A87757' (cf. the bracketed labels in the nodes of Figure 5) handles transit feed and its real-time updates. It is related to *Trip 2* in the *inbound* direction and belongs to *Route A* of *Agency BKK* (operator at Budapest, Hungary). On the other hand, node 'BKK-Routes-3040' stores persistent transit information with regard to *Route A* (e.g. route name, short name, stops, head-signs). References to all the currently active *inbound* trips are found in node 'BKK-Routes-3040-Inbound'. Similarly, node 'BKK-Stops-F0108' stores persistent data with regard to the given stop (e.g. stop name, GPS coordinates) and lists the routes this stop is part of. Furthermore, the *trip ID* of every active trip is listed in the route node.

GTFS Emulator, Analytics Module

The GTFS Emulator provides the static timetable information, if it is available, as the initial service. It basically uses the officially distributed GTFS database of the public transport operator of the given city. However, it also relies on another data source, which is OpenStreetMap (OSM) (Haklay & Weber 2008), a crowdsourcing based mapping service. In OSM maps, users have the possibility to define terminals, public transportation stops or even public transportation routes. Thus, the OSM based information is used to extend and clean the information coming from the GTFS source. The resulted data set reflects more accurately the actual situation in the given territory because the OSM data is updated more frequently than the GTFS data set.

The Analytics module is in charge of the business logic offered by the service, e.g. deriving crowdedness information or estimating the time of arrivals at the stops from the data collected by the crowd.

Front-end Application

The front-end application, called TrafficInfo, handles the subscription to the pub/sub channels, collects sensor readings, publishes events to and receives updates from the XMPP server, and visualizes the received information.

TrafficInfo

TrafficInfo has four main functions, such as visualization, information sharing, sensing and stop event detection. These functions are discussed below.

Visualization

Most of the users benefit from the visualization capability of TrafficInfo that visualizes public transport vehicle movements on a city map. An example of this primary function can be seen on Figure 6a displaying trams of line 1, 4, 6 and buses of line 7 and 86 on the Budapest map in Hungary. The depicted vehicles can be filtered to given routes. The icon of a vehicle reflects various attributes, such as the number, progress or crowdedness of the specific vehicle. Clicking on a vehicle's icon a popup shows all known information about that specific vehicle.

Information Sharing

The second function serves for information sharing. Passengers can share their observations regarding the vehicles they are currently riding. Figure 6b shows the feedback screen that is used to submit reports. The feedback information is spread out using the framework and displayed on the devices of other passengers, who might be interested in it. It is up to the user what information and when he wants to submit.

Sensing

The third function is collecting smartphone sensor readings without user interaction, which is almost invisible for the user. It is done automatically in the

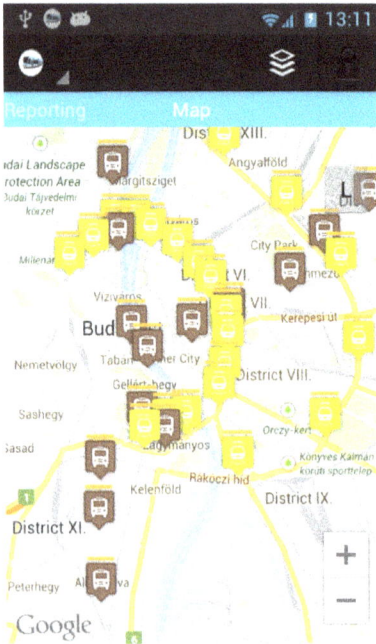

a) Vehicle visualization   (b) User feedback form

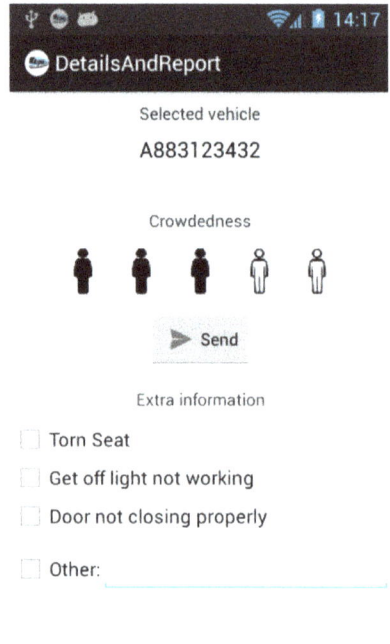

Figure 6: TrafficInfo screenshots.

background, the only thing the user has to do is to start TrafficInfo. User positions are reported periodically and are used to determine the vehicle's position the passenger is actually traveling on. In order to create the link between the passenger and the vehicle, the movement of the user is identified through his activities. To this end various sensors are used, e.g. accelerometer, and the timestamped stop events of the vehicles are deducted. The duration between the detected stops coupled with GPS coordinates identifies the route segment, which the user actually rides. With regard to the energy consumption of the sensor readings we carried out some measurements. Our results showed that in case of a normal daily scenario, such as traveling approx. 1 hour to the workplace in the morning and 1 hour back home in the afternoon, the readings and local processing consume 1.48 Wh energy on average. This is equivalent roughly to 20% of the capacity of an average smartphone battery (2,000 mAh, 3.7 V).

Besides the GPS coordinates Google also provides location information in those areas, where there is no GPS signal available. Usually this position is highly inaccurate, but the estimated accuracy is also provided. Moreover, the activity sensor, which guesses the actual activity of the user, is also used by TrafficInfo. Currently, the supported activities are: *in vehicle, on bicycle, on foot, running, still, tilting, walking* and *unknown*. The sensor monitoring part of TrafficInfo is active only if the activity recognition reports *in vehicle* status. Otherwise, sensor monitoring is suspended. Note, that the accuracy of activity recognition can be varying. However, in our experiments the *in vehicle* activity was recognized with more than 85% accuracy, which made this tool useful for our purposes.

The collected sensor readings on one hand are uploaded to the XMPP server. There the Analytics module processes and shares them among participants who are subscribers of the relevant information. On the other hand, they are used locally. For instance, based on the timestamps of the detected stop events the server side analytics estimate the upcoming arrival times of the given vehicle and disseminate live timetable updates to the subscribers.

Stop Event Detection

The fourth, most challenging function of TrafficInfo is to detect stop events of public transport vehicles without user interaction. TrafficInfo implements such a detector locally on the mobile device. When stop events are detected a summary of information is transmitted to the XMPP server. This summary consists of the location and timestamp of the event and the time elapsed since the last stop event. The final decision is made by the server based on the periodic reports from the passengers. It is a majority decision, so if the majority of the reports indicate a stop event within a given time window the detection is made otherwise not.

The stop event detection mechanism is based on features. Hence, several features were generated from the experimental usage logs collected during the measurements. In this work, we investigated our detection mechanism only on trams and left buses/trains as part of future work. The approximately 1GB measurement data were collected by 10 volunteers during the 1 month measurement period using Samsung Galaxy S3 and Nexus4 smartphones. The gathered context data included among others GPS, Wi-Fi, cellular network and acceleration sensor readings. For classification the J48 decision tree implementation of the Weka data-mining tool (Hall et al. 2009) was used. With the combination of the defined features and models the detector can detect stop events automatically with relatively high accuracy (with 0.86 AUC – Area Under the Curve) within 13 seconds after the arrival at the station. The place of the stop event is decided by investigating the GPS position and/or the Wi-Fi/cellular network fingerprint of the environment, so stops at stations can be distinguished from other stops with high probability.

Summary

In this chapter, after a short literature review a generic, XMPP based communication framework was introduced which was designed to facilitate the development of crowd assisted smart city applications. Then a prototype crowdsensing based smart timetable service was presented. Its front-end Android application, called TrafficInfo, together with an automatic stop event detector was introduced in detail. This service was implemented on top of the introduced generic framework. It updates static public transport timetables and delivers the updated information to its subscribers in real-time.

Acknowledgement

This work was partially supported by the TÁMOP-4.2.2.C-11/1/KONV-2012-0001 and the EITKIC 12-1-2012-0001 projects. The author would like to acknowledge the support and help of the participants of these projects, especially the contribution of Timon Tomás, Ádám Zsolt Nagy, Róbert Szabó, Imre Lendák, Bernát Wiandt, András Benczúr, Csaba Sidló and Gábor Fehér. Károly Farkas has been partially supported by the Hungarian Academy of Sciences through the Bolyai János Research Fellowship.

References

Bedogni, L., Di Felice, M., & Bononi, L. 2012. By Train or by Car? Detecting the User's Motion Type Through Smartphone Sensors Data. In: *Proceedings of IFIP Wireless Days Conference (WD 2012)*. Dublin, Ireland on 21–23 November 2012, pp. 1–6.

Biagioni, J., Gerlich, T., Merrifield, T., & Eriksson, J. 2011. EasyTracker: Automatic Transit Tracking, Mapping, and Arrival Time Prediction Using Smartphones. In: *Proceedings of the 9th ACM Conference on Embedded Networked Sensor Systems (SenSys 2011)*. Seattle, WA, USA on 1–4 November 2011, pp. 1–14.

Google Inc. 2006. *General Transit Feed Specification Reference*, 25 September 2006. Available at: https://developers.google.com/transit/gtfs/reference/ (Last accessed 25 June 2015).

Haklay, M. M., & Weber, P. 2008. OpenStreetMap: User-Generated Street Maps. *IEEE Pervasive Computing, 7*(4): 12–18. DOI: http://dx.doi.org/10.1109/MPRV.2008.80

Hall, M., Frank, E., Holmes, G., Pfahringer, B., Reutemann, P., & Witten, H. I. 2009. The WEKA Data Mining Software: An Update. *SIGKDD Explorations, 11*(1): 10–18. Available at: http://www.kdd.org/sites/default/files/issues/11-1-2009-07/p2V11n1.pdf.

Moovit Developers. 2014. Moovit. Available at: http://www.moovitapp.com/ (Last accessed 25 June 2015).

Saint-Andre, P. 2011. Extensible Messaging and Presence Protocol (XMPP): Core, RFC 6120 (Proposed Standard), Internet Engineering Task Force, March 2011. Available at: http://www.ietf.org/rfc/rfc6120.txt (Last accessed 25 June 2015).

Szabo, R. L., & Farkas, K. 2013. A Publish-Subscribe Scheme Based Open Architecture for Crowd-sourcing. In *Proceedings of 19th EUNICE Workshop on Advances in Communication Networking (EUNICE 2013)*. Chemnitz, Germany on 28–30 August 2013, pp. 1–5.

Thiagarajan, A., Biagioni, J., Gerlich, T., & Eriksson, J. 2010. Cooperative Transit Tracking Using Smart-phones. In: *Proceedings of the 8th ACM Conference on Embedded Networked Sensor Systems (SenSys 2010)*. Zurich, Switzerland on 3–5 November 2010, pp. 85–98.

Zhou, P., Zheng, Y., & Li, M. 2012. How Long to Wait?: Predicting Bus Arrival Time with Mobile Phone based Participatory Sensing. In: *Proceedings of the 10th International Conference on Mobile Systems, Applications, and Services (MobiSys 2012)*. Low Wood Bay, Lake District, UK on 25–29 June 2012.

CHAPTER 26

Mobile crowd-sensing in the Smart City

Imre Lendák

Faculty of technical sciences, University of Novi Sad, Serbia,
lendak@uns.ac.rs

Abstract

The Smart City connects citizens in novel ways by leveraging the latest advances in information and communication technologies (ICT). Smart citizens have various ICT solutions at their disposal, which allow them to optimize their day-to-day activities in the urban environment they live and/or work. The integration of rich sensing capabilities (e.g. camera, microphone, GPS, accelerometer, barometer) in today's mobile devices allows their users to sense their urban environment in often unforeseen ways. In mobile crowd-sensing the citizens of the Smart City collect, share and jointly use services based on the sensed data, e.g. the Waze application for optimized car-based navigation, the Smart Citizen project for collecting meteorological measurements. This paper presents the current state-of-the-art and future challenges in mobile crowd-sensing in urban environments, by focusing on sensing in the following focus areas: environment, citizen collaboration, urban traffic systems, health/fitness and social networking. From each of these areas a set of representative applications (e.g. Waze, Foursquare, Ushahidi) were selected, analyzed and compared based on the following criteria: expected social and economic impact, novelty and sophistication of system architecture, sensing methods applied, motivation techniques and user privacy.

Keywords

crowd-sensing, Smart City, mobile devices, sensors

How to cite this book chapter:
Lendák, I. 2016. Mobile crowd-sensing in the Smart City. In: Capineri, C, Haklay, M, Huang, H, Antoniou, V, Kettunen, J, Ostermann, F and Purves, R. (eds.) *European Handbook of Crowdsourced Geographic Information*, Pp. 353–369. London: Ubiquity Press. DOI: http://dx.doi.org/10.5334/bax.z. License: CC-BY 4.0.

Introduction

The rich sensing capabilities integrated into modern mobile devices allow their users to use them for novel, often unforeseen activities. The list of sensors integrated into the latest flagship mobile devices includes basic ones like the microphone necessary to record the user's voice and the touchscreen necessary for text input, through the also visible, one or more cameras used to record images from the users' surroundings and/or of the users themselves (i.e. selfies), as well as more obscure sensors, like the accelerometer for sensing acceleration, gyroscope for orientation, proximity for distance, compass for spatial bearing, GPS for geographic location and barometer for atmospheric pressure. The latest offerings (e.g. the Samsung Galaxy 6 Edge in early 2015) might offer personal health related sensing as well in the form of heart-rate and oxygen saturation sensing. Devices might identify their users with built-in fingerprint sensors, or via scanning and recognizing their fingerprints via their touchscreens.

The microphone, touchscreen and camera form a sufficient subset of sensors for the majority of use cases. The rest of the sensors might be used by mobile device producers to develop more user friendly behavior, e.g. automatically detecting the tilt of the mobile device with the gyroscope in order to rotate the screen accordingly.

Mobile crowd-sensing (MCS) is a relatively new discipline, in which the users of modern smart phones use the rich sensing capabilities of their devices to collect and share information while on the move, as well as to form micro-crowds around a certain crowd-sensing activity (Cardone et al. 2013). Current state and future MCS challenges were discussed in Ganti, Ye and Lei (2011), while Zambonelli (2011) dissects a more general theme, namely urban crowd-sourcing. Goodchild (2007) was one of the first identifying the crowd as a possible sensing 'tool'. The efficiency and efficacy of mobile crowd-sensing is discussed in Ma, Zhao and Yuan (2014).

The Smart City is the future city which leverages the latest advances in information and communication technologies (ICT) in order to optimize its operations and the everyday processes in which the smart citizens take part. Mobile crowd-sensing is one ICT tool which might be leveraged in Smart Cities, as it reaches the smart citizens and involve them in the optimization of the city's processes.

Crowd-sensing simulation efforts (Farkas & Lendák 2015; Lendák & Farkas 2015; Tanas & Herrera-Joancomart 2013) aim to simulate crowd behavior, forecast sensing patterns and help researchers and solution developers to choose what to sense as well as to identify the minimum user threshold necessary for an application to collect sufficiently 'big' data, which the algorithms can crunch in order to produce useful information. Trustworthiness of the data sensed by the crowd is also relevant and analyzed in Tanas & Herrera-Joancomart (2015).

Both mobile crowd-sensing and the Smart City are intriguing novel research and development domains, with numerous magazine and journal special issues devoted to their analysis, e.g. the August 2015, June 2013 and June 2011 issues of the IEEE Communications Magazine, the June 2013 issue of the Journal of Knowledge Economy, etc. This paper builds on those results and discusses the latest mobile crowd-sensing efforts in the Smart City setting, with a special focus on the following areas: environment, citizen collaboration, urban traffic systems, health/fitness and social networking. From each of these areas one or two representative applications were selected based on their technical sophistication and size of user base, as an easy measure of success. The solutions chosen were analyzed by trying to answer the following questions:

- How do they impact society at large (i.e. societal impact)?
- What is their expected economic impact?
- What is their system architecture like? Is it sophisticated and does it contain novel solutions?
- Which sensors and how do they employ towards reaching their goals?
- How do they motivate their users to contribute and use the application?
- How do they address the sensitive question of user privacy?

Apart from this introduction, the paper contains five sections discussing mobile crowd-sensing based applications from the above identified five focus groups. Their descriptions are followed by their comparative analysis in section seven.

Urban environment

The latest offerings in the smartphone arena come equipped with a limited set of meteorological sensors, e.g. barometer, thermometer. Apart from the obvious meteorological sensors, the integrated microphone can be used for sensing noise levels, and the camera for recording specific meteorological or other phenomena. Noise level sensing can be automated, while using the camera requires human interaction. In general, modern mobile devices are still lacking in sensing capabilities focused on collecting information about our (natural) environment. These limitations might be mitigated by purpose-built sensing hardware.

The Smart Citizen (SC) project is a mobile crowd-sensing based project whose goal is to build a platform for collecting environmental measurements in urban settings. Its website[1] is pictured in Figure 1. SC is an open-source platform consisting of a hardware device (the Smart Citizen Kit), an application programming interface utilizing RESTful web services, a mobile application, a website and a web based community of volunteers, i.e. the 'crowd'. The hardware kit is equipped with sensors which measure air composition (CO and NO2), temperature, light intensity, sound levels, and humidity. It is able

---

[1] Smart Citizen project's website, https://smartcitizen.me/

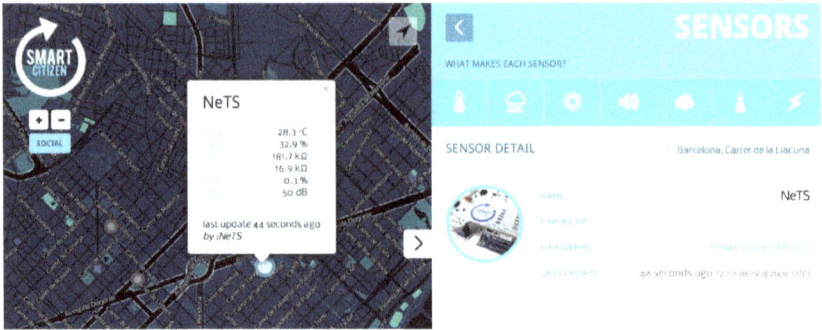

Figure 1: The Smart Citizen website.

to stream data measured by the sensors over Wi-Fi. Power to the device can be provided by a solar panel and/or battery, and it can be placed on balconies or window sills.

SC does not employ gamification or other motivation mechanisms, which might increase public interest towards this solution. It has a detailed privacy policy available via the web-based interface. Username and geographic location can be grabbed from the screen – which negatively impact smart citizen privacy. By removing the username from the web interface, the privacy level of the application could be significantly improved.

Another interesting project dealing with environmental issues in and near urban environments is Danger Maps[2], a crowdsourced, web-based environment monitoring solution originating from China. Its primary goal is to collect and share the locations of the various sources of pollution, e.g. garbage dumps, toxic-waste treatment facilities, oil refineries and power plants. The application is popular in China where pollution is a serious issue. It was created after its founder learned that the Shanghai apartment he bought in 2007 was near a landfill – something he wasn't informed of when negotiating the purchase. Originally, the 'old' Danger Maps contained official data and maps released by the Chinese Environmental Protection Agency, but since 2013[3] the crowd is allowed to create detailed custom maps themselves via a web based interface. Danger Maps relies on social sensors (i.e. human users) as the reports are posted in textual format. The camera might be used as well for taking pictures of the pollution sources. Unfortunately, the website is available only in Chinese – or at least the author failed to find the link to an English language version.

Efforts similar to the Smart Citizen or the Danger Maps projects have significant societal impact, as they allow the crowd to collect and share information

---

[2]   Danger Maps official website, http://www.epmap.org/ngo
[3]   Custer, C., 'Danger Maps' Invites You to Map China's Polluted Areas via New Open-Platform Maps, 2013, https://www.techinasia.com/danger-maps-invites-map-chinas-polluted-areas-openplatform-maps/

about both major sources of pollution and the quality of the air we breathe, which might not have been mapped otherwise. Solutions similar to Danger Maps might be especially interesting in the developing world, where laws regulate the protection of our immediate environment to a limited extent, or where modern legislation is available, but not enforced. Hopefully the 'power of the crowd' exercised via solutions similar to the projects discussed in this section, might put additional pressure on both legislative and administrative bodies in the environmental protection domain and force them to act more quickly and decisively. The immediate economic impact of these two solutions is limited at the moment, but might rise with the wider adoption of crowd-sensing, i.e. when the user bases of these projects become larger and the societies built around them gain more lobbying power.

Citizen collaboration

Crowd-sensing applications give a powerful tool into the hands of human societies, e.g. they allow citizens to reach their governments about non-essential issues they detect within their communities, like issues reported in the streets with FixMyStreet[4] (see Figure 2) and similar solutions (e.g. SeeClickFix[5] in the USA). These applications usually consist of a mobile application which is used for sensing and a website which displays the sensed events in near real-time (see Figure 2). FixMyStreet (FMS) is a crowd-sensing platform which can be customized for any urban area. In FMS the issue reports are linked to the reporter's email address, but FMS ensures its users that only the representatives

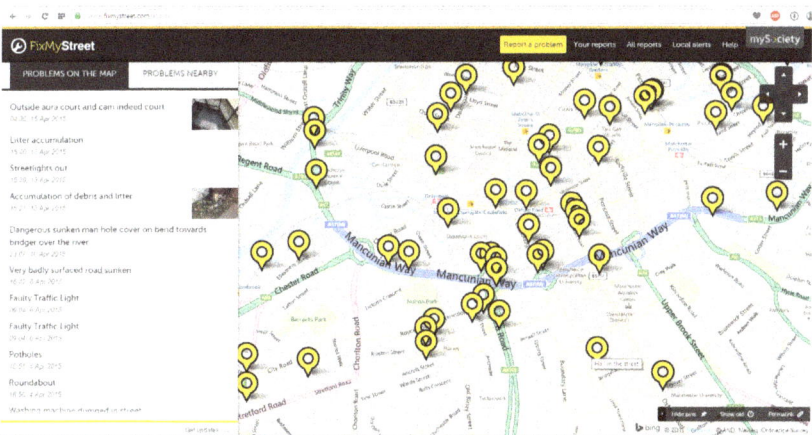

Figure 2: FixMyStreet issue reports in Manchester, UK (April 19th, 2015).

---

[4] FixMyStreet issues in Manchester, UK: https://www.fixmystreet.com/around?pc=manchester
[5] SeeClickFix website, http://en.seeclickfix.com

of the council who will address the report, and FMS' administrative staff might be allowed to see the users' email addresses. The camera is the most important physical sensor used to take pictures of the issues the users are reporting. It does not contain motivation tools which would possibly allow it to build more effective human sensor networks, by allowing them to compete and take part in in-application games.

Crowd-sensing based citizen collaboration efforts like FixMyStreet or See-ClickFix might allow smart citizens to collaborate on issues affecting local groups in an urban environment. The loosely coupled social networks formed around these solutions might more easily obtain the attention of local admin-istration and coax them into taking corrective action in the areas of interest, e.g. fix a pothole, or clean up an unplanned garbage dump. These solutions might have a measurable economic impact as well, mainly for local adminis-trations, which might find out about issues sooner, fix them and spend less on paid inspectors who would travel around the urban areas and look for potholes, garbage and similar.

Crowd-sensing might allow wider collaboration during disaster relief and during political turmoil, e.g. the Ushahidi[6] (Swahili for 'testimony' or 'witness') website started after Kenya's disputed presidential elections in 2007. In Usha-hidi, when an event occurs a volunteer sends a brief report from a smartphone, via the Web or text message, and the software annotates it with time and loca-tion information. Such information can then be visualized and 'mined'. It relies on social sensors and on the camera, as the reports are written by people in natural language and might be accompanied by a photograph or a video. The human sensors are usually motivated by our built-in altruism, i.e. our urge to help others or contribute towards a greater good. Ushahidi itself does not con-tain a motivation scheme which would reward its users for sharing informa-tion. User privacy is quite important in Ushahidi, especially in countries where people might get into trouble for sharing negative views about the govern-ment or other bodies. The Ushahidi privacy policy claims that only aggregated and non-personally identifiable information is shared with third parties. The reports shared via the Ushahidi application for Android devices do not contain personally identifiable information.

Ushahidi allows the crowd of its users to report issues affecting large groups or societies, e.g. the attempted rigging of elections in Kenya[7] or the 2010 earth-quake in Haiti[8]. Therefore, it has a quite significant societal impact. Its immedi-ate economic impact is limited, or at least it is very hard to measure.

---

[6]   Ushahidi official website, http://www.ushahidi.com/

[7]   Beyond Voting on the Ushahid official website, http://www.ushahidi.com/2015/05/21/beyond-voting-using-ushahidi-to-help-citizens-protect-their-elections/

[8]   Ushahidi Haiti Project – Evaluation Final Report, http://www.ushahidi.com/2011/04/19/ushahidi-haiti-project-evaluation-final-report/

Traffic

Modern vehicles have rich sensing and computing capabilities, which might be used to sense and share information, e.g. they might detect when a parking spot is taken and share the information automatically. Smartphones might also detect and share certain events automatically, e.g. the Google Activity Recognition library can discern whether the device holder is walking, driving a car or running, based on the accelerometer's measurements. As it will be shown below, the most important sensors used in traffic system related crowd-sensing applications are the GPS sensor and the accelerometer.

Waze[9] is arguably the most successful crowd-sensing based application in the traffic systems domain. Its primary function is point-to-point navigation, but it performs this function with a twist: it allows drivers and other participants (e.g. co-driver) to share roadside events, e.g. road works, accidents, police presence, traffic jams. These event reports are then aggregated, shown on the map and used by the navigation algorithm, which might help drivers to avoid roadside events leading to traffic jams, e.g. a collision during rush hour. Waze is not limited to urban environments, i.e. it is not a strictly a Smart City application.

Waze consists of a mobile application used for navigation and issue reporting, a big data storage, a service for running the data analysis algorithms, and a web-based live map showing the latest events. It applies an intricate motivation scheme, which includes user levels (ranging from baby, via warrior to 'king') based on the amount of points, which might be collected through long hours of active use and issue reports, user avatars, in-app messaging and occasional in-app games. In one such game the mobile application generated Easter eggs in the streets near the driver and awarded extra points for collecting them. Waze uses the GPS sensor to calculate the current location, the camera to take pictures of the events and the accelerometer to automate certain event detections, e.g. stop-and-go traffic. Most of the events are sensed by the social sensors, i.e. the users manually annotate a roadside event by clicking on its icon in the mobile application or choosing its type from a list.

Waze links the user's account to his/her mobile phone (number) and shows an avatar onscreen (both on the mobile and in the Web based live map) at the GPS position of the user. Other nearby users are shown onscreen, not just friends, thereby allowing to learn their whereabouts based on grabbing screenshots containing their avatars. Additionally, it is possible to report non-existing roadside events, e.g. traffic jams by sending in well-formed Waze messages from a custom-built application.[10] Such message fabrication attacks might be used to cause havoc in traffic systems.

---

[9] Waze website, https://www.waze.com
[10] T. Jeske, "Floating Car Data from Smartphones: What Google And Waze Know About You and How Hackers Can Control Traffic", https://media.blackhat.com/eu-13/briefings/Jeske/bh-eu-13-floating-car-data-jeske-slides.pdf

Apart from steering drivers clear of congestion, crowd-sensing might come in handy in solving parking problems in busy urban areas, where there are no funds to develop an advanced infrastructure of parking sensors in the streets, as done in San Francisco with the SFPark[11] system, or in the City of Westminster (London) with Smart Parking.[12] One such, crowd-sensing based solution, Google's OpenSpot[13] tried to use the power of the crowd to sense parking related events, and based on that data provide suggestions to drivers who were looking for parking. It was cancelled in 2012 due to its limitations, mainly its inability to adapt to busy urban environments where a parking spot might remain unoccupied only for a couple of seconds, as well as for the lack of user base, i.e. the size of its user base was insufficient to make it successful. Anagog[14] is a promising new player in the urban parking arena (see Figure 3). It uses Waze's data to automatically sense and share parking events – in essence it learns the habits of drivers, i.e. where and when they park their cars.

The analyzed applications address two pressing matters in the crowded urban environments of the 21st century, namely congestion and the limited availability of parking. Waze helps its users steer clear of traffic jams by utilizing the reports received from other Wazers (i.e. Waze users) who had the misfortune

Figure 3: The Anagog parking app.[13]

---

[11] SFPark website, http://sfpark.org/how-it-works/
[12] Smart Parking website, http://www.smartparking.com/about-us
[13] OpenSpot in the news, http://www.androidauthority.com/google-labs-open-spot-a-useful-application-that-no-one-uses-15186/
[14] Anagog website, http://anagog.com

of getting stuck in congestion. The parking assistance applications aim to use crowd-sensed big data in order to suggest the most likely location of an optimal parking spot, thereby reducing the time spent in 'cruising for parking', consequentially lowering petrol costs and time wasted. As both costs and time wasted might be measured in money, we conclude that these applications have a significant economic impact, especially if they reach the threshold number of active users allowing them to provide useful suggestions to the crowd.

Health and fitness

The power of the masses can contribute towards sharing information among patients suffering from specific illnesses. Apart from allowing patients to link with others who have similar health problems, the data collected and shared by patients might be used for health-care optimization, e.g. cancer survivors were planned to be brought Together in one such solution.[15] Patient networking websites like PatientsLikeMe (PLM) allow individuals with certain health conditions to share and compare their symptoms and responses to the treatments they received. PLM relies on social sensors for data collection, i.e. people themselves describe their mood and physical condition, either by answering questions asked by the application, or writing textual descriptions. PLM and other similar tools might allow healthcare professionals to create more precise measurement and assessment tools based on crowd-sensed/crowd-sourced data, or might even offer early warning in case of infectious disease outbreaks. Therefore both the societal and economic impact of these solutions are significant as they can improve the prospects of sick people via mining the information shared by them and using it to develop better medicines and procedures on one hand, and potentially lowering costs on the other hand, by allowing people to learn more about their condition even before visiting a physician, and being capable of better describing how and what they feel based on the information shared by others with similar conditions.

The above listed crowd-enabled medical solutions usually have simple architectures, e.g. a website where the users might share information about their health. The users are the sensors themselves, as they describe in text how they feel and what symptoms they have. PLM applies a simple motivation scheme in which it polls its users daily in order to remind them to share information about how they feel. It also awards users with stars for sharing information about their ailments. Some news agencies reported[16] that personal health information was possible to be collected from the PLM website by interested third

---

[15] Together by Medstartr, http://www.medstartr.com/projects/192-together
[16] CBS News, "PatiensLikeMe is more villain than victim in patient data "scraping" scandal", http://www.cbsnews.com/news/patientslikeme-is-more-villain-than-victim-in-patient-data-scraping-scandal/

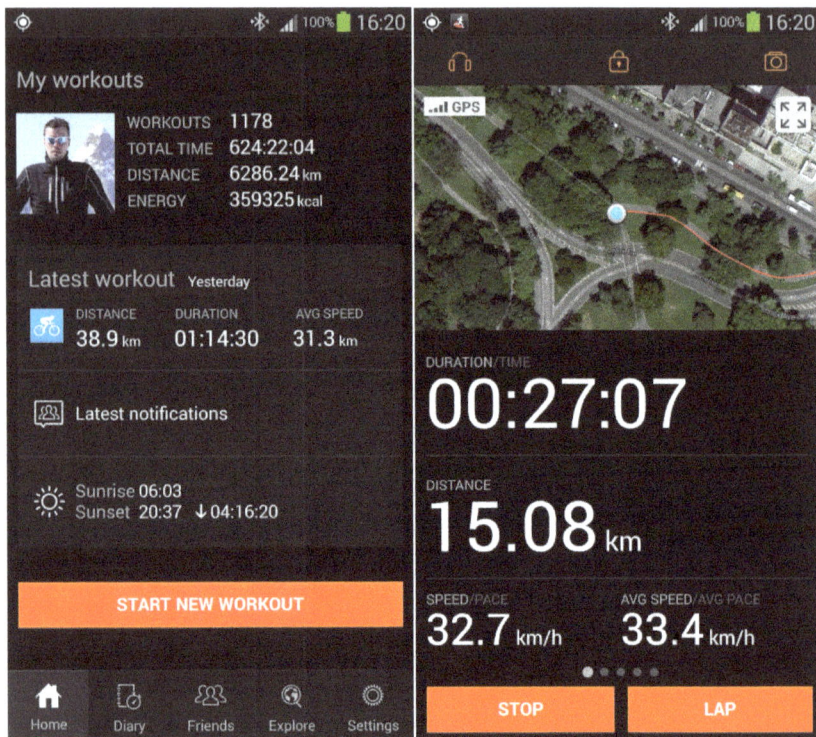

Figure 4: Sports Tracker fitness app showing fitness statistics and an on-map route.

parties, i.e. their privacy policy and its enforcement might not have been as strong as necessary.

Fitness apps (e.g. Sports Tracker[17] – see Figure 4) allow their users to measure their achievements while exercising, either automatically by reading the necessary information from the built-in sensors, or by allowing users to manually enter their results. The most important sensors in fitness applications are the GPS, accelerometer and lately the heart rate monitor and oxygen saturation sensor. Fitness solutions also tend to have relatively simple architectures, consisting of a mobile app and a (cloud based) data storage, where the results of exercises are stored. Users are motivated by allowing them to post their achievements (e.g. kilometers ran) on social networks or organizing competitions with other users. Sports Tracker also has an elaborate privacy policy clearly outlining how the system uses the data collected and shared. Crowd-sensing based fitness applications can boost people's enthusiasm towards physical exercise

---

[17]   Sports Tracker, http://www.sports-tracker.com

and thereby improve public health and lower the amount of funds spent on healthcare, i.e. they have a measurable societal and economic impact.

Social networks

Although Facebook posts and Twitter Tweets might contain descriptive information about our environment, events in the traffic system and even about our health and/or fitness, their primary aim is not mobile crowd-sensing. As opposed to the above named leaders in the social networking area, Foursquare (FS) (see Figure 5), the local search, discovery and recommendation app for mobiles has both social networking and sensing elements. FS takes into consideration where its users go and what they tell the application about those places, and advises other users where to go and what to visit near their current location, e.g. it might allow a user in a foreign country to find points of interest around him/her with only a smartphone and an internet connection.

The users 'sense' information about the points of interest (POIs) near their location by answering questions asked by the application. The social networking features contained in earlier versions (e.g. check-in at a location and sharing the event with friends also using FS) were factored out into a separate application named Swarm. This contributes towards privacy, as check-ins are not necessarily visible to followers. Personally identifiable information is visible in the application, as the users' full names and home town can be accessed via

Figure 5: Foursquare location based recommendations.

the ratings and tips they leave. Foursquare, and especially its earlier versions (prior to Swarm) employed an elaborate motivation scheme, consisting of a *point system* in which points were scored for each check-in, *badges* achieved via checking in at certain locations, and the users could become 'Mayor' of a certain venue by checking-in more often than any other user.

The architecture of Foursquare became considerably more elaborate with the introduction of Swarm, as now the solution consists of a mobile application, a backend service used for storing the big data collected from contributors and running the recommendation algorithms, as well as Swarm, which cooperates with the mobile application and adds social networking features.

Foursquare has a measurable economic impact, as it can steer visitors towards POIs (e.g. restaurant, bar) which have high ratings thereby increasing their incomes. Its societal impact is not so clear, but it can surely help its users when they are in an unknown and new environment, e.g. visiting a city in a foreign country without a local guide, by steering them towards interesting places and making them feel less out-of-place and better connected to the location they are visiting.

It is important to note that the users of other crowd-sensing based applications (e.g. Waze, FixMyStreet) might also be regarded as members of dynamic social networks, formed around certain events or activities (e.g. drivers in and around a single urban area).

Comparative analysis

The applications and solutions described in this paper were analyzed based on the following criteria: social visibility and impact, expected economic impact, sophistication of their system architectures, sensing method(s), motivation scheme and privacy level. For their comparative analysis presented in this section, from each group one or two applications were selected and marked in each of the above listed areas. The ratings were made based on publicly available material and/or the author's own experience in using the solutions. A couple of solutions were intentionally omitted from this comparative because of various reasons, e.g. Danger Maps could not be analyzed as most of the available information is in Chinese, and there was no new data available online about the Together project.

Societal impact was measured as a combination of the number of active crowd-sensors and the (expected) level of contribution of the selected application towards a greater good, e.g. improving the lives of many. The number of active crowd-sensors is not easily measured, as a varying percentage of the user base of these solutions is passive, i.e. they do not sense, just consume the services based on the sensations of others. Waze and Foursquare stand out based on the total number of their (millions of) users. Although contribution towards a greater good is even more challenging to measure, Ushahidi, FixMyStreet and PatientsLikeMe stand out as they might help people during disasters or other socially disruptive events, help local communities address their issues more effectively, as well as help inform patients and collect statistical information about their conditions, potentially leading to novel medical findings and procedures. Waze and Foursquare have global reach, but their mission is not as 'noble' as those of the above listed three applications. As none of the analyzed solutions have both a clear mission towards achieving a greater good, as well as millions of users globally, the author felt that they all have a mid-range impact.

Economic impact is a subjective area which is similarly challenging to measure as societal impact. The author made an attempt to compare the applications by taking into consideration the expected economic impact they 'should' make based on their primary goals. Waze and Anagog stand out as they might optimize urban traffic, via avoiding traffic jams and/or steering drivers more quickly to empty parking spots. The mid-range players are FixMyStreet, PatientsLikeMe and Foursquare. FMS allows its users to report issues in their neighborhoods to their councils, which might not know about those issues otherwise, or would need to employ people to manually identify them. PLM allows its users with various illnesses to learn about their condition from other people having similar conditions, which in turn might allow early diagnosis even before visiting a doctor, thereby lowering healthcare expenditure. FS might allow highly rated businesses to attract tourists and other visitors and thereby raise their revenue.

The novelty and complexity of **system architecture** was measured based on counting the total number of the following components identified during the analysis of the crowd-sensing applications: purpose-built sensing hardware, a website with live data, a mobile (sensing) application, a big data store, advanced algorithms used for analyzing the data collected and generating additional value and information, as well as social networking features. Apart from the count of various elements, another important measure was the perceived level of seamlessness of their integration into a user oriented, integrated product. Foursquare (with Swarm) and Waze are similar in not building custom sensing hardware, but having all the other components and being quite user-friendly and seamlessly integrated. Foursquare boasts an elaborate mix of information sensing and sharing mobile application combined with a social networking component (Swarm) and a powerful backend aggregating the recommendations. Waze also consists of a mobile application, a Web-based live map and backend for crunching the incoming data, creating (alternative) routes and generating various interesting in-app games with rewards to the users who participate. The Smart Citizen project on the other hand has purpose built hardware, but is lacking in the area of generating additional information and social networking features. The rest of the projects had slightly less complex system architectures, as they were either mostly web based (e.g. PLM), or seemed to be more focused on their presence on mobile devices (e.g. Sports Tracker).

The **sensing capabilities** of the applications were assessed based on the number of hardware sensors used by them, the sophistication of social sensor utilization (i.e. how well they handle the data manually entered by their users), as well as the existence of advanced signal processing algorithms applied. Waze and Sports Tracker excel in all three areas. Waze utilizes the hardware sensing features of mobile devices (e.g. GPS, accelerometer), allows its users to manually enter useful data and automates sensing via its signal processing features, e.g. the automatic detection of stop-and-go traffic. Sports Tracker also automates the collection of track length ran or cycled, and calculates the length and 'cost' (e.g. in calories) of an exercise. It also relies on the hardware sensing capabilities of mobile devices (e.g. GPS) and allows users to manually enter data about exercises whose detection it does not automate (e.g. weight lifting). The Smart Citizen project also stands out, as it is the only solution, which developed a custom-built sensing module in order to achieve its 'smart city' objective, namely to measure the various characteristics of our environment.

Crowd-sensing based solutions cannot succeed without the crowd-sensors, i.e. without their users. In today's fast-paced life it is hard to learn the habit of devoting some of our time towards collecting (i.e. sensing) data and sharing it, thereby helping others. Because of that, and in order to maintain a loyal user base, it is necessary to be innovative when trying to motivate the users. During the analysis of the crowd-sensing applications, the following **motivation mechanisms** were identified: financial motivation, appealing to our built-in altruistic nature (i.e. it is natural to people to try to help others), gamification, in-app games and social networking. While financial motivation was not present in the analyzed solutions, most relied on our altruism as a driving factor, with Sports Tracker being an exception as it is built for sensing tasks about a single user. The altruistic element of motivation was most prominent in Ushahidi. Gamification usually means that the application is assigning some reward (e.g. points) for each piece of data sensed, i.e. making the users feel that the use of the application is a game and thereby urging them to keep on sensing. In-app games keep the users locked to the application longer. The social networking features allow the users to somehow reach their peers. In the motivation area Waze has a dynamic points system with levels, which are obtainable after long hours of use and contributing information, as well as in-app games and messaging. It also plays on the natural altruism of its users, as it allows them to help others via marking a police presence or speed control. Foursquare also had a points-based system, badges awarded for certain check-ins, social networking layer for staying in contact with friends and it allowed users to become 'mayors' of the places they visited more often than others. This latter might be regarded as a hybrid form of a gamification feature and an in-app game. PatientsLikeMe has a modest motivation scheme consisting of daily polling in which it asks its users to share information about their health. Apart from that it also awards up to three stars for sharing specific information.

The following **privacy aspects** of the crowd-sensing solutions were taken into consideration: the amount of personally identifiable information shown on screen to other users, the existence of news reports or other proof about privacy breaches and the sophistication of techniques used to ensure the privacy of the crowd-sensors. In FixMyStreet the personal information collected is limited, it is not shown on screen and, according to the privacy policy, it is shared only with system administrators and council members who might be contacted to address the issue reports sent via the application. Ushahidi anonymizes the contributions of its users, there were no reports about privacy breaches at the moment of writing, and it also has an elaborate privacy policy. Sports Tracker also applies the privacy components and there were no known privacy breaches. The Smart Citizen project and Waze apply the above techniques, but show both the username and the location of the sensors on screen, which might allow malevolent third parties to misuse that information. The rest of the analyzed applications also employ techniques which raise the level of user privacy, but they either do not completely hide personally identifiable information (e.g. full name and address visible in Foursquare), or there were news reports about some form of privacy breaches (e.g. PLM).

Table 1 contains a comparative overview of the analyzed crowd-sensing based solutions. The following three types of marks were assigned based on their analysis laid out above:

- Leader/High impact – the solution applies the majority of the available techniques in the subject domain or has the highest expected social/economic impact.
- Mid-range capabilities/impact – the solution applies some of the techniques relevant in the area or has medium societal/economic impact.
- Limited capabilities/low impact – the solution has limited capabilities or its societal/economic impact is marginal.

Essentially, the analyzed applications are quite different and their comparison was done by analyzing some of their common, key aspects. The author did not

App/Solution	Societal impact	Economic impact	Architect.	Sensing	Motiv.	Privacy
Smart Citizen	••	•	••	••	•	••
FixMyStreet	••	••	•	•	•	•••
Ushahidi	••	•	•	•	•	•••
Waze + Anagog	••	•••	••	•••	•••	••
PatientsLikeMe	••	••	•	•	••	•
Sports Tracker	••	•	•	•••	•	•••
Foursquare	••	••	••	•	•••	•

Table 1: Comparative analysis of crowd-sensing based applications (••• – leader/high impact, •• – mid-range capabilities/impact, • – limited capabilities/impact).

attempt to directly compare the solutions based on the number of 'points' they were awarded in Table 1.

The author plans to extend the initial results presented in this section as part of his future research, by including a more detailed measurement of the number of total and active users of each solution, adding a more detailed econometric analysis of the diverse solutions discussed, as well as by identifying and including them additional solutions in each group and localizing the comparisons inside each group, thereby avoiding directly comparing health and traffic apps to each other.

Summary

This paper contains a comparative analysis mobile crowd-sensing solutions in the Smart City environment. The analysis was focused on crowd-sensing in the following focus areas: environment, citizen collaboration, urban traffic systems, health/fitness and social networking. From each of these focus areas up to two successful applications were selected and analyzed. The applications were analyzed (e.g. Waze, Foursquare, Ushahidi, the Smart Citizen project, PatientsLikeMe) by a compound comparison criteria consisting of the following elements: societal and economic impact, sophistication of system architecture, novel methods of sensor use, motivation scheme and steps taken towards ensuring user privacy.

As a continuation of this work, the author intends to identify additional crowdsensing-based applications in the Smart City ecosystem and do a more detailed comparative analysis, especially in the areas which might be objectively measured, e.g. number of total vs active users, economic impact via econometrics, etc. Apart from that, the author also plans to analyze crowd-sensing application use through space and time and find patterns leading to success or failure.

Acknowledgment

This work was partially supported the Hungarian National Academy of Sciences under grant Domus 105/6864/HTMT.

References

Cardone, G., Foschini, L., Bellavista, P., Corradi, A., Borcea, C., Talasila, M., & Curtmola, R. 2013. Fostering ParticipAction in smart cities: a geo-social platform. *IEEE Communication Magazine, 51*(6): 112–119.
Farkas, K., & Lendák, I. 2015. Simulation Environment for Investigating Crowd-sensing Based Urban Parking. In: *Models and Technologies for Intel-*

ligent Transportation Systems (MT-ITS 2015), Budapest, Hungary, June 2015, pp. 320–327.

Ganti, R. K., Ye F., & Lei, H. 2011. Mobile crowdsensing: current state and future challenges. *IEEE Communications Magazine, 49*(11): 32–39.

Goodchild, M. F. 2007. Citizens as sensors: the world of volunteered geography. *GeoJournal, 69*(4): 211–221.

Lendák, I., & Farkas, K. 2015. Evaluation of simulation engines for crowdsensing activities. In: *3rd International Conference & Workshop on Mechatronics in Practice and Education (Mechedu 2015)*. Subotica, Serbia, May 2015, pp. 82–87.

Ma, H., Zhao, D., & Yuan P. 2014. Opportunities in mobile crowd sensing. *IEEE Communications Magazine, 52*(8): 29–35.

Tanas, C., & Herrera-Joancomart, J. 2014. Crowdsensing simulation using ns-3. *Citizen in Sensor Networks, Lecture Notes in Computer Science*, pp. 47–58.

Tanas, C., & Herrera-Joancomart, J. 2015. When users become sensors: Can we trust their readings? *International Journal of Communication Systems, 28*(4): 601–614.

Zambonelli, F. 2011. Pervasive urban crowdsourcing: visions and challenge. In: *2011 IEEE International Conference on Pervasive Computing and Communications Workshops (PERCOM Workshops)*. Seattle, USA, March 2011, pp. 578–583.

CHAPTER 27

Mobile crowd sensing for smart urban mobility

Dragan Stojanovic*, Bratislav Predic and
Natalija Stojanovic

University of Nis, Faculty of Electronic Engineering
Aleksandra Medvedeva 14, 18000 Nis, Serbia
*dragan.stojanovic@elfak.ni.ac.rs

Abstract

In this paper, we present the application of the mobile crowd-sensing para-digm in supporting efficient, safe and green mobility in urban environments. We have developed the *CitySensing* framework demonstrating the viability of a common crowd-sourcing platform applied to various urban mobility domains. We argue that today's mobile devices, with integrated or add-on sensors, can be efficiently used to crowd source diverse information in domains that are relevant to urban life and mobility (traffic, air quality and citizens' everyday activities). This is illustrated by three distinct mobile applications, developed on top of the *CitySensing* framework, that contribute to a common goal of smarter urban mobility. Commonly integrated accelerometer and GPS are used to infer traffic events and conditions. Externally attached or integrated air qual-ity sensors enable suggestions for city areas adequate for outdoor activities on a specific day of the week or hour of the day. Mobile phone usage statistics and analysis can present valuable information to urban planning services to better adapt to citizens' habits and mobility. The analysis of this massive amount of crowd sensed data (so-called Big Data) within the cluster/cloud infrastructure enables detection of situations and events that influence human mobility, and dissemination of notifications and recommended actions.

How to cite this book chapter:
Stojanovic, D, Predic, B and Stojanovic, N. 2016. Mobile crowd sensing for smart
urban mobility. In: Capineri, C, Haklay, M, Huang, H, Antoniou, V, Kettunen, J,
Ostermann, F and Purves, R. (eds.) *European Handbook of Crowdsourced
Geographic Information*, Pp. 371–382. London: Ubiquity Press. DOI: http://dx.doi.
org/10.5334/bax.aa. License: CC-BY 4.0.

Keywords

crowd sourcing, mobile sensing, smart city, smart mobility

Introduction

With advancements and the proliferation of mobile devices with increasing computing, communication and sensing capabilities, mobile users have become important sources of sensing data in a globally spread wireless sensor network (Campbell et al. 2008). Mobile crowd sensing represents an approach for a new sensing and geo-crowdsourcing paradigm that leverages the power of various mobile devices, such as smartphones, tablets, wearable devices, smart sensors, etc. It is based on the human ability to acquire local geospatial information and knowledge through sensor-enhanced mobile devices, and the possibility to share this information/knowledge with other users and a wide community (Goodchild 2007; Kamel Boulos et al. 2011). As the world population becomes increasingly urban, cities worldwide need to leverage information and communications technologies to improve their functions, enhance efficiency, improve competitiveness and the economy and provide better environment for their citizens; to become 'smarter'. Smart mobility is one of the key characteristics of a smart city, and also one of the biggest challenges that smart cities face (Batty et al. 2012). The world population makes more than 64% of all travel kilometers within urban environments, which is expected to triple by 2050. Thus, a high priority for cities around the world is to support citizens' mobility within the urban environment based on safety, efficiency and environmental protection (Motta et al. 2015). Mobile crowd sensing paradigm will enable key methods, techniques and systems in that direction. The application of analysis, reasoning and data mining techniques on the mobile crowd sensed data provides useful insights in citizens' mobility and supports better citizen involvement in monitoring urban spaces.

   In this article, we describe main concepts, methods and technologies of mobile crowd sensing in smart cities and present the *CitySensing* framework developed to support smart urban mobility through participation and intelligence of the crowd (Section 2). Several mobile demo applications have been developed based on this framework that illustrate the use of mobile crowd sensing in different city's scenarios and provide insights how such applications could improve citizens' mobility (Section 3). Although there are still open challenges and real-life deployments of developed applications, the initial evaluations look promising in supporting smart citizens' mobility, as concluded in Section 4.

Mobile crowd sensing for smart cities

The mobile crowd sensing refers to geo-crowdsourcing and VGI (Volunteered Geographic Information) paradigm in which mobile users use their mobile computing, communication and sensing devices to collect, locally process and analyze, as well as distribute geo-referenced information (Chatzimilioudis et al. 2012). Mobile crowd sensing paradigms have become the significant source of VGI and crowdsourced geo-information owing to the large number of mobile devices carried by people worldwide to support their daily activities (Zhang et al. 2014). Such crowd sensed data is filtered, aggregated, processed and analyzed at the server(s) and appropriate information services and notifications are delivered to mobile users to support their mobility. In such scenario mobile users are both the producers of mobility sensing data, and the consumers of services, notifications and recommendations based on processing and analysis of the massive amount of crowd sensed data at server(s).

There are several sources of mobility sensing data originated at mobile devices, classified as:

• Physical sensors,
• Virtual (logical) sensors,
• Social sensors.

Physical sensors include sensors integrated in, or attached to mobile devices (smart phones, tablets, etc.), such as: GPS, microphone, camera, ambient light sensor, accelerometer, gyroscope, compass, proximity sensor and also the temperature and humidity sensors available on advanced smartphones. The development of wearable and pervasive systems, such as Sensordrone[1] and iWatch[2], provides integration of additional sophisticated sensors, worn by users and attached to their mobile devices, to measure air pollution, personal health parameters and the emotional and physiological status of users. Virtual sensors are not hardware sensors but software applications that run at user devices and collect information about users, their profile and preferences, detecting their context and situation. Such sensors detect information related to user communications (voice, SMS, etc.), user activities and interaction with devices (active applications, application in focus, the type of the interaction, etc.), user preferences and profile, user-generated content (texts, speech, videos, photos, sounds), etc. Virtually sensed information is referenced in space and time and attached to a certain location, symbolic or geographic. Social sensors detect user social status and activities, social network and social media interactions (tags, likes, Tweets, photos, etc.), currently connected friends and their status, connections in vicinity, etc. Some of such information can be detected by accessing social network/media services through appropriate APIs.

Mobile crowd sensed data collection can be performed with the participation and active user involvement, called *participatory sensing*, or without active involvement of users, i.e. *opportunistic sensing* (Campbell et al. 2008). The

---

[1] http://sensorcon.com/products/sensordrone-multisensor-tool
[2] https://www.apple.com/watch/

research of mobile crowd sensing for smart cities focuses on the collection, representation, filtering, aggregation, processing and analysis of large volumes of mobility sensing data, both on mobile devices and at central server(s). Such data describe citizens' movements and activities, environmental conditions they face (e.g. high pollution level, traffic congestion, crowded or noisy surroundings), communication and interaction they make (talking, searching on the Web, tagging a photo, connecting with friends, etc.). The analysis and mining of mobility data obtained through mobile crowd sensing provide insights into important features of the urban mobility, behavior of people moving around a city and possible prediction of future mobility. There are two approaches for collection and distribution of mobile crowd sensed data:

- Server-based approach – Sensed data in its raw form are sent to a server without processing and analysis at the mobile devices. Such an approach results in very high demands for wireless network bandwidth, as well as for storage, processing and analysis of raw mobility data at the server.
- Distributed approach – Data from physical, virtual, and social sensors are collected, processed and analyzed at the mobile devices and high level contextual and mobility information are sent to a server.

Mobile crowd sensed data and information represent the foundation for mobility information services that are delivered to the same users to support their smarter mobility (Ilarri et al. 2015). The crowd sensed data collected from a large number of mobile users/moving objects are characterized by massive volume, velocity and variety, and can be regarded as Big Data. The server applications running on a cluster/cloud that support mobile crowd sensing should provide data collection, processing, reasoning, analysis and mining over massive mobility data sets, from structured and unstructured sources. Such a collaborative crowd intelligence provides detection of aggregated mobility patterns and trajectories, group activities and behavior, as well as complex situations (e.g. interesting places, traffic congestions, popular city routes or crowded evacuation paths in an emergency situation) in smart cities (Cardone et al. 2013).

We have developed a framework, named *CitySensing*, to support the development of mobile crowd sensing applications for various smart city scenarios. The *CitySensing* framework consists of mobile application components, server components and visualization/analytics components organized in a distributed architecture given in Figure 1. It supports both opportunistic and participatory methods of crowd sensed data collection using various physical, virtual and social sensors available in today's mobile devices. It fully leverages processing, sensing and communication capabilities of mobile devices and provides distributed and scalable storage, processing, analysis and mining of crowd sensed mobility data at mobile devices within *CitySensing* mobile components. High-level mobility information generated at users' devices is further aggregated, processed and analyzed at the *CitySensing* server components running

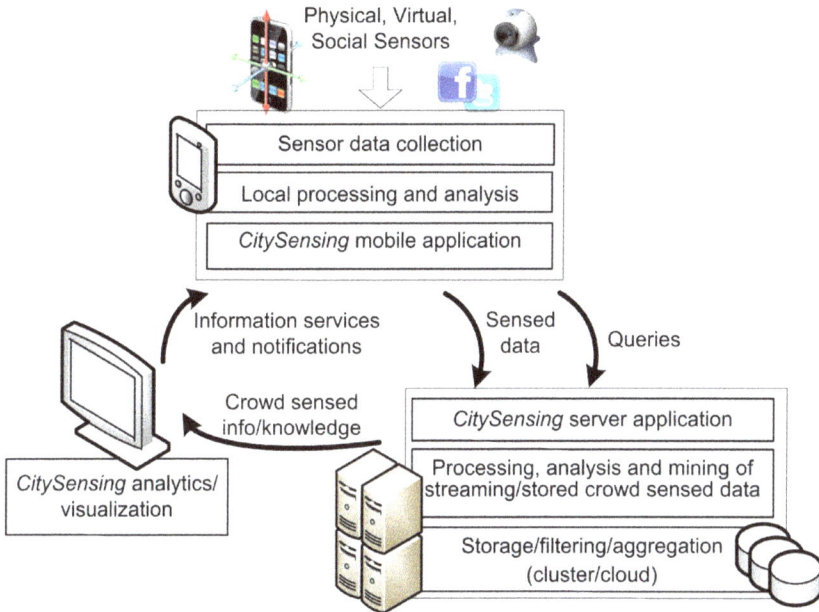

Figure 1: A general architecture of *CitySensing* framework.

on a cluster/cloud infrastructure. The *CitySensing* framework is based on the open-source sensor data collection framework, Funf,[3] and the data analysis and mining framework WEKA[4]. For processing and analysis of big spatio-temporal data, *CitySensing* server components are based on distributed processing frameworks, such as MapReduce/Hadoop[5] and Spark,[6] for processing offline crowd-sensed data stored in a distributed file system over a cloud/cluster. For real-time processing and analysis of large crowd sensed data streams appropriate data stream processing frameworks are employed, such as Apache Storm[7] and Spark Streaming.

Dynamic crowd-sensed information for smart mobility

Smart mobility in urban environments is based on integrated mobile information services to support efficient, safe and green transport of people and goods and user activities in smart cities. We are developing several demo smart

---

[3] http://www.funf.org/
[4] http://www.cs.waikato.ac.nz/ml/weka/
[5] https://hadoop.apache.org/
[6] https://spark.apache.org/
[7] https://storm.apache.org/

mobility applications, namely *DriveSense*, *ExposureSense* and *UrbanSense*, based on *CitySensing* framework, to support different urban mobility scenarios.

Crowd-sensing traffic information

Recent research on smart mobility has started to leverage the principle of mobile crowd sensing by employing drivers/passengers and pedestrians equipped with mobile devices as real-time sources of navigation, environmental and traffic information. The basic idea implemented in *DriveSense* application is to use drivers/passengers and pedestrians as moving sensor platforms that can collect traffic, driver and road related conditions, events and information and send them to a central server, or to mobile devices in the vicinity, using appropriate wireless communication mechanism (3G, Wi-Fi, Bluetooth, etc.). Such dynamic information should support citizens in their everyday mobility, both indoors and outdoors.

Regarding vehicle transport, mobile crowd sensing systems could detect and report traffic events and conditions, the state of road infrastructure, driver behavior and activities, as well as accidental events:

- Traffic events and condition
  - Travel times over the street segments.
  - Slowed down traffic (traffic jams).
  - Congestion points (start-stop locations)
  - Parking places.
  - Traffic stops.
- Road infrastructure state
  - Bumpy road.
  - Slippery road surface.
  - Damaged road surface location (potholes).
- Dynamic traffic events and driver behavior (accidents and situations that could potentially cause accidents)
  - Sudden breaking, decelerations and acceleration.
  - Lateral skidding.
  - Violent (sudden) change of direction and traffic lane at a high speed.
  - A vehicle moved out of a road; an accident or a collision happened, etc.

The *DriveSense* application is used as a tool for crowd sensing of dynamic traffic information, and also as dynamic navigation service based on current traffic conditions. Both participatory and opportunistic modes can be used for collection of information on traffic condition and events (Figure 2a). During driving, relevant traffic events (e.g. sudden lane changes, numerous hard breaking events, etc.) are reported by moving users and delivered to all drivers in affected street segments. As a result, a service also proposes re-routing to drivers to avoid an intersection that is probably heavily congested, or potentially unsafe for driving. Spatio-temporal aggregation and analysis of crowd sensed traffic information

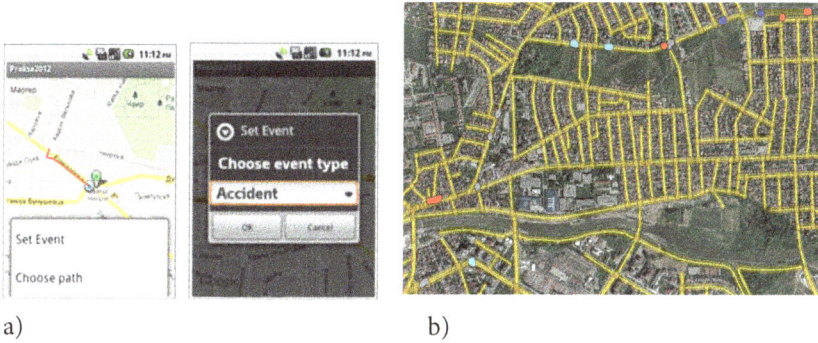

Figure 2: *DriveSense* application a) Mobility information service b) Space-time clustering of reported traffic events.

collected over longer time periods can provide valuable information to city traffic service and identify parts of city's street network that could affect safe and efficient traffic. The results of this analysis are shown in Figure 2b showing parts of a street network with detected potholes (blue dots), areas with excessive harsh breaking (red dots) and frequent violent lane changes (green dots).

Air quality sensing and reporting

Air pollution is identified as a major health concern in urban areas. The citizens are also exposed to other types of urban 'pollutions', such as noise and electromagnetic waves. By applying crowd sensing concepts to users with mobile devices, both citizens and city authorities can detect and be aware of pollution exposure during their everyday activities in urban areas, especially for pedestrians, as demonstrated by *ExposureSense* application (Figure 3). As with previous *DriveSense* example, users of *ExposureSense* act as both sources of information on air quality and consumers of such information which support planning their daily physical activities in city areas with less exposure to pollution. By cross-correlating their physical activities with pollution information, users can receive information and services regarding their pollution exposure and recommendation for activities, as well as navigation instructions to 'cleaner' areas (Predic et al. 2013).

The *ExposureSense* application can act as a personal diary of physical activities giving both map view (Figure 3a) and timeline view (Figure 3b). This information can be combined with sensed air-quality parameters (like temperature, concentration of different particles or gasses in the air). Figure 3c shows map visualization of NO_2 concentration sensed using sensors attached to a mobile device during typical recreation activity. Summarized information is shown in a calendar type view in Figure 3d.

The drivers/passengers and pedestrians are also the consumers of smart mobility information services generated by such crowd intelligence approach.

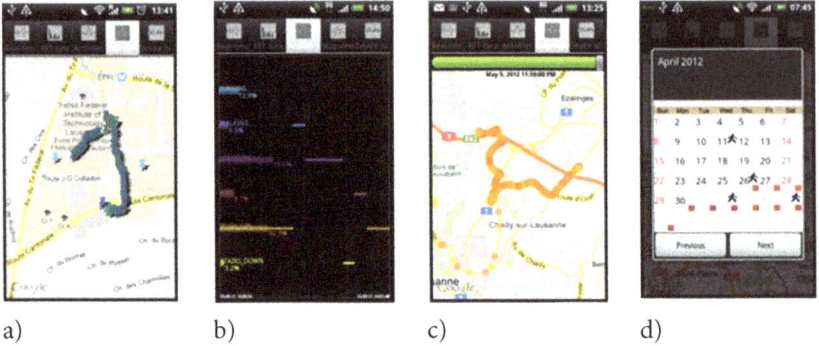

a)   b)   c)   d)

Figure 3: *ExposureSense* mobile crowd sensing application a) Map view of activities b) Timeline view of activities c) Exposure to NO_2 during recreation activity, d) Summary information of activities and exposures to pollutants.

They receive real-time traffic information, navigation instructions and notifications, recommended activities and environmental conditions and actuate upon them (Predic & Stojanovic 2015). City authorities can use such crowd intelligence to improve urban mobility by detecting mobility behaviors and patterns of citizens/tourists movement. Such behavior and movement patterns, related to background geographic information (POI, road network data, indoor maps), time context (time of the day, day of the week, month, season), weather conditions, means of transport, environmental conditions, social/cultural events in the city or indoors, can enable better understanding of the city dynamics.

Behavior patterns sensing using mobile activity collection

Ubiquitous and transparent collection of mobile usage statistics as geo-referenced mobility data can provide valuable information to urban planning city services. Crowd sensed information showing urban areas where mobile users tend to spend their leisure/break time, types of physical activities during parts of a day, and their physical environment during these activities can allow city planners to adapt city's business and recreational areas according to citizens' mobility in urban environment. The *UrbanSense* mobile application has been developed as a mobile diary service that detects and stores information about users' environment and mobile interaction by collecting following data:

- Physical environment: temperature, air pressure, air humidity, light intensity, etc.
- Phone communication usage: SMS sending/receiving and phone calls.
- User's physical activities.
- Wireless communication devices in surrounding, e.g. WiFi access points, cellular network base stations, Bluetooth devices, other mobile devices, etc.

- Active usage of a phone based on a screen on/off state.
- Usage of certain types of mobile applications especially social networking ones.
- Recording of photos/videos using phone cameras.
- Playback of sound/video files.
- Battery level / charging data.

The *UrbanSense* demo application consists of a mobile service tasked with collection of sensing data, a mobile application used for service configuration and access to mobility information services, as well as a server(s) which receives, stores, aggregates, processes and analyses crowd sensed data. The mobile service collects data in a local mobile database in predefined time intervals (usually daily) and sends the collected data to the server. A service configuration allows the user to filter which types of activity information can be collected during specific periods of a day/week, for the reason of privacy preservation (Figure 4a). All sensitive personal information in collected data is hashed in order to preserve data topology and semantics while maintaining anonymity and privacy. This includes phone numbers and SMS messages' content. Prior to submitting collected data to the server user can view and filter it in the mobile application, as shown in Figure 4b. At the server, collected data are processed and analyzed by applying spatio-temporal clustering techniques and social connections based on SMS and voice calls history are detected. After clustering, urban areas with intensive physical and communication activities can be visualized on a map and filtered according to the type and the period of day/week (Figure 4c).

The main goal of *UrbanSense* application is to infer social network connections and detect urban areas with intensive social interaction, or which contain a specific mobile usage pattern. The identification of such areas, along with time periods when certain patterns of social activities occurred, can be used for urban planning, coverage of city areas with marketing, sports or leisure events and supporting infrastructure.

a)   b)   c)

Figure 4: *UrbanSense* application a) Data collection configuration b) Overview and filtering of collected data c) Spatio-temporal clustering of collected communication data.

Server-side processing techniques for crowd-sourced spatial data

Using their mobile devices while walking and driving, and participating in social networks and media, citizens provide a wealth of information related to their mobility in urban environments and can support numerous applications for the benefit of the urban community. Smart mobility crowd sensing applications presented so far are based on local processing and analysis of mobile sensor data to detect high level, semantic mobility information at mobile devices. Such information is transferred to central server(s) that perform aggregation of semantic mobility data collected from a large number of mobile users and provide query processing, reasoning, analysis and mining, over big mobility data sets. The server performs detection of aggregated mobility patterns and trajectories, collective activities and behavior, as well as complex mobility situations (e.g. traffic congestions, risky and dangerous events, frequent city routes, crowded evacuation paths in an emergency situation, popular places and orders of their visit, etc.) (Ilarri et al. 2015). Mobile crowd sensed data processing, analysis and visualization, either for single citizen or aggregated set of citizens through time, enables early detection of situations and events that affect urban mobility, as well as the estimation of their impact to the community. Also, they provide dissemination of proactive location-based and context-aware services, notifications and alert messages, as well as recommendations for the actions to citizens on the move.

Crowd sensed data collected from *UrbanSense* mobile application can be analyzed and visualized to detect user activities on the move. The timeline and a map overlay visualization of per-user activity data collected by *UrbanSense* application is shown in Figure 5a.

Crowd sensed data from numerous mobile citizens are streamed to server applications based on *big data* processing and analysis frameworks deployed over distributed computing infrastructure, such as MapReduce/Hadoop, Apache Spark and Apache Storm and Web-based visual analytics components and technologies. Such architecture allows building of flexible and scalable crowd-sourcing software services that can effectively analyze citizens' mobility patterns in modern cities. In Figure 5b, a heavy traffic on major city streets (different levels of red color) detected by crowd-sourcing traffic information is co-occurred with bursts of 'traffic' on Twitter around the same locations (blue colored circles). The contents of Tweets could offer a clear and fast explanation of the traffic jam, e.g. a protest of local farmers.

Conclusions

Presented demo mobile applications confirm that mobile crowd sensing represents a very promising approach for mobilization of citizens in order to improve and adapt their urban mobility. Demonstrated and other freely available mobile crowd sensing applications suggest that crowd-sourcing dynamic

a)   b)

Figure 5: Visualization and analysis of crowd sensed mobility data a) Citizen's activities visualized on a timeline and a city map b) Traffic congestion and Twitter activities in Nis based on Storm server application.

geographic information has potential to improve various aspects of urban life and urban mobility. In this paper we focused on probably the most prominent domains such as traffic, air quality and citizens' activities on the move that have the great impact on citizens' mobility and city dynamics. Presented research work demonstrates that other aspects of urban life and planning can also benefit from crowd-sourcing geographic information.

There are still a number of open issues and research challenges in mobile crowd sensing approach applied to smart cities. To ensure the citizens' participation, reliability, security and privacy must be covered in the future work. Approproate methods and tools should be investigated and developed to ensure the accuracy of the information collected, an adequate level of privacy to citizens, as well as citizens motivation through gamification and micro payments. Also, the future research must take into account the high volume, velocity, variety and veracity of data provided by the citizens, and collected from other structured and unstructured data sources (e.g. sensor networks, Internet of Things, etc.). Accordingly, advanced methods and techniques for processing, analysis and visualization of such Big mobility data should be developed, that will unearth new information and knowledge by exploring the correlation and patterns across multi-sources of crowd sensed data. It will provide more effective services and solutions for citizens and decision makers in urban mobility.

References

Batty, M., Axhausen, K. W., Giannotti, F., Pozdnoukhov, A., Bazzani, A., Wachowicz, M., Ouzounis, G., & Portugali, Y. 2012. Smart cities of the future. *European Physical Journal: Special Topics, 214*: 481–518. DOI: http://dx.doi.org/10.1140/epjst/e2012-01703-3

Campbell, A. T., Eisenman, S. B., Lane, N. D., Miluzzo, E., Peterson, R. A., Lu, H., Zheng, X., Musolesi, M., Fodor, K., & Ahn, G.-S. 2008. The Rise of People-Centric Sensing. *IEEE Internet Computing, 12*(4): 12–21. DOI: http://dx.doi.org/10.1109/MIC.2008.90

Cardone, G., Foschini, L., Bellavista, P., Corradi, A., Borcea, C., Talasila, M., & Curtmola R. 2013. Fostering participaction in smart cities: A geo-social crowdsensing platform. *IEEE Communications Magazine, 51*(6): 112–119. DOI: http://dx.doi.org/10.1109/MCOM.2013.6525603

Chatzimilioudis, G., Konstantinidis, A., Laoudias, C., & Zeinalipour-Yazti, D. 2012. Crowdsourcing with Smartphones. *IEEE Internet Computing, 16*(5): 36–44. DOI: http://dx.doi.org/10.1109/MIC.2012.70

Goodchild, M. F. 2007. Citizens as sensors: the world of volunteered geography. *GeoJournal, 69*(4): 211–221. DOI: http://dx.doi.org/10.1007/s10708-007-9111-y

Ilarri, S., Stojanovic, D., & Ray, C. 2015. Semantic management of moving objects: A vision towards smart mobility. *Expert Systems with Applications, 42*(3): 1418–1435. DOI: http://dx.doi.org/10.1016/j.eswa.2014.08.057

Kamel Boulos, M. N., Resch, B., Crowley, D. N., Breslin, J. G., Sohn, G., Burtner, R., Pike, W. A., Jezierski, E., & Chuang, K-Y. 2011. Crowdsourcing, citizen sensing and sensor web technologies for public and environmental health surveillance and crisis management: trends, OGC standards and application examples. *International Journal of Health Geographics, 10*(67): 1–29. DOI: http://dx.doi.org/10.1186/1476-072X-10-67

Motta, G., Sacco, D., Ma, T., You, L., & Liu, K. 2015. Personal Mobility Service System in Urban Areas : the IRMA Project. In: *IEEE Symposium on Service-Oriented System Engineering*, pp. 1–10.

Predic, B., & Stojanovic, D. 2015. Enhancing driver situational awareness through crowd intelligence. *Expert Systems with Applications, 42*(11): 4892–4909. DOI: http://dx.doi.org/10.1016/j.eswa.2015.02.013

Predic, B., Yan, Z., Eberle, J., Stojanovic, D., & Aberer, K. 2013. ExposureSense: Integrating daily activities with air quality using mobile participatory sensing. In: *2013 IEEE International Conference on Pervasive Computing and Communications Workshops (PERCOM Workshops)*, pp. 303–305.

Zhang, D., Wang, L., Xiong, H., & Guo, B. 2014. 4W1H in Mobile Crowd Sensing. *Communications Magazine, IEEE, 52*(8): 42–48. DOI: http://dx.doi.org/10.1109/MCOM.2014.6871668

PART VI

VGI in spatial planning

Using mobile crowdsourcing and geotagged social media data to study people's affective responses to environments

Haosheng Huang* and Georg Gartner

Department of Geodesy and Geoinformation, Vienna University of Technology, Vienna, Austria
*haosheng.huang@tuwien.ac.at, huanghaosheng@gmail.com

Abstract

When travelling in space, humans perceive the environment and evaluate it affectively. This chapter illustrates how mobile crowdsourcing and social media data can be used to study people's affective responses to different environments. It also showcases how these affective responses can be used to provide a better understanding of human–environment interaction, as well as to enable smart geospatial applications (particularly navigation systems). This chapter also discusses some essential challenges that need further investigations when crowdsourcing people's affective responses. Some of these challenges are participation motivating, data quality and privacy.

Keywords

Affective responses to environments, subjective geo-information, crowdsourcing, geotagged social media data

How to cite this book chapter:
Huang, H and Gartner, G. 2016. Using mobile crowdsourcing and geotagged social media data to study people's affective responses to environments. In: Capineri, C, Haklay, M, Huang, H, Antoniou, V, Kettunen, J, Ostermann, F and Purves, R. (eds.) *European Handbook of Crowdsourced Geographic Information*, Pp. 385–399. London: Ubiquity Press. DOI: http://dx.doi.org/10.5334/bax.ab. License: CC-BY 4.0.

Introduction

When travelling in space, people perceive the environment and interpret it affectively. A lot of our daily behaviour and decision making is influenced by this kind of interpretation and affective evaluation of space (Kaplan & Kaplan 1989; Borst et al. 2009). For example, during wayfinding, which route to choose and which one to avoid may not only depend on distance, but also be influenced by people's affective evaluation of the environment, such as safety and attractiveness. Studying people's affective responses to environment contributes to a better understanding of people's spatial experiences and behaviour, as well as enables many applications, such as navigation systems and urban planning.

Conventionally, affect has been a central research topic within modern psychology. There were also many geographers, especially human geographers, approaching affective responses to environments when studying place. These conventional studies often employed empirical experiments in labs or in fields, which are often very expensive and time consuming, and hard to apply for large-scale studies. In recent years, with the increasing availability of smartphones, researchers start to apply the principle of 'citizens as sensors' (Goodchild 2007), and explore the potential of crowdsourcing affective experiences from a large group of people. Furthermore, with the impetus of social networking services (e.g. Facebook, Foursquare, Flickr, and Twitter), large volumes of social media data have been continually created. These data, especially geotagged ones, contain lots of information about people's experiences and activities in various environments, which is a new and significant source for studying people's spatial experiences at different environments.

This chapter presents our on-going research efforts towards these trends. We illustrate how mobile crowdsourcing and social media data analysis can be used to collect people's affective responses to space, as well as the potential applications of these affective data.

Affective responses to environments

According to American Psychological Association (2006), affect is the experience of feeling or emotion. It is a key part of the process of an organism's interaction with stimuli (e.g. the environment). Russell (2003) argues that some affective response is always present within a person, as neutral, moderate or extreme.

There are many studies in literature focusing on structuring or modelling people's affective responses (Ekman and Friesen 1971, Russell 2003, Barrett 2006). The first group of studies tries to define basic distinct affective responses such as happiness, anger, fear, disgust or sadness (e.g., Ekman & Friesen 1971). This approach was widely applied in the field of 'Affect Computing', which uses these basic affective responses to label facial expressions, and adapts services according to these labelled affective responses. Others adopt a dimensional approach to study people's affective responses, and argue that affective responses can be represented using some principal dimensions. For example, Russell and Barrett (1999) and Russell (2003) vary affective responses on the dimensions of valence (i.e. 'the subjective positive-to-negative evaluation', e.g. pleasant-unpleasant, comfortable-uncomfortable), and arousal (i.e. 'a sense of mobilization or energy', e.g. activating–deactivating). There are other studies proposing to include one more dimension – motivational intensity (e.g. the strength of an urge to move toward or away from a stimulus) (Harmon-Jones et al. 2013). Barrett (2006) stated that valence is a basic component of affective responses (all people seem to be able to differentiate between pleasant and unpleasant affective states), while the degree of arousal may not be always experienced.

In this paper, we are particularly interested in affective responses evoked by environments. According to Russell (2003) and Barrett et al. (2007), environments are perceived not only according to their physical features, but also affectively. Some places are experienced as pleasing, while some others as disgusting and unsafe. These affective responses to environments are experienced as attributes or qualities of environments, which are commonly described with affect-related adjectives such as boring, safe, and beautiful.

Acquisition of affective responses

Affective responses to environments can be collected through various approaches, among which self reports and physiological recordings are the most prominent conventional methods for collecting such data. However, these approaches were traditionally used in laboratories with highly controlled conditions. Recently, with the rapid development of new technologies, novel methods and sensors have become available for collecting affective responses to environments (Huang et al. 2014; Resch et al. 2015). In the following, we highlight two of these current approaches.

Mobile crowdsourcing

Self-reports are the most direct way to gather information about people's affective responses (Barrett et al. 2007). Recently, the ubiquitous availability and use of smartphones, and the rapid spread of Web 2.0 potentially enable researchers to collect self-reported affective responses from a large group of people. In the EmoMap project (Klettner et al. 2013), we applied this mobile crowdsourcing approach to acquire people's affective responses via GPS-enabled smartphones.

To help people report their affective responses via smartphones, we developed an Affect-Space-Model. Based on the dimensional approach described above, the model adopts a two-level structure. Firstly, users are asked to rate their 'level of comfort' (i.e. the valence dimension) in their current environment on a 7-point Likert scale, from uncomfortable ('1') to comfortable ('7'). Secondly, users can optionally provide further details about their affective responses, particularly on the aspects of safety, attractiveness, diversity, and relaxation. These four aspects were obtained from an empirical study (Huang et al. 2014). In summary, each affective response to environments is collected as ratings on these five aspects (i.e. level of comfort, safety, attractiveness, diversity, and relaxation). Each affective response is then annotated with the GPS location obtained from user's smartphone.

The Affect-Space-Model was implemented in an Android mobile application (Figure 1) to enable people to report their affective responses to environments anytime and anywhere. To help users locate themselves, this application shows the current location (obtained from GPS on smartphones) as a marker in an OpenStreetMap. Following the suggestions by Russell (2003), it particularly asks the question 'Here it is …' to explicitly make users focus on the environment. To better understand users' affective responses, some contextual information was also collected, i.e. company, familiarity with the place, time and weather (obtained from the Web). To encourage people's active contributions, this application was promoted to students, as well as to an urban walking community.[1] Users were asked to contribute their ratings anywhere and anytime they want in their daily life.

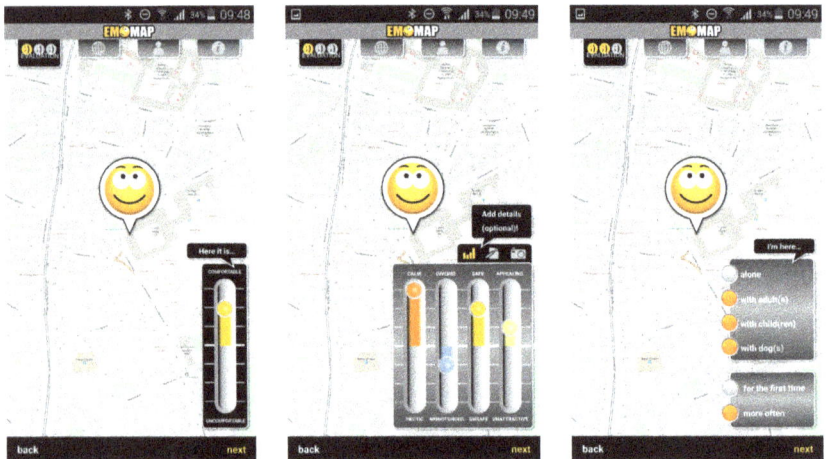

Figure 1: Main screenshots of the mobile application for collecting people's affective responses to environments (Map data: OpenStreetMap and Contributors, CC-BY-SA).

---

[1] http://www.wildurb.at/

Experiences within the EmoMap project show that with this mobile crowd-sourcing approach, we are able to acquire affective data, which are evoked by realistic scenarios, directly from a large number of people. Until December 2013, more than 3,500 contributions were collected from more than 200 people. The contributions were distributed at different places of the world, with 98% of them for the city of Vienna (Austria). As an example, we visualize part of these contributions for the city of Vienna in Figure 2.

Extracting affective responses from social media data

Recently, the increasing availability of social media (e.g. Foursquare and Flickr) has led to the accumulation of large volumes of open social media data. Recent research has started to use social media data for studying people's affective responses (Mislove et al. 2010; Hauthal & Burghardt 2013). However, these approaches extract people's affective responses *in* various environments, which might *not necessary be caused by* or *towards* these environments. In the following, we extend existing research, and illustrate how social media data can be harnessed to extract people's affective responses to environments. Particularly, we focus on geotagged photos in Flickr.

For extracting affective responses from social media data, we apply sentiment analysis technique. Sentiment analysis (or opinion mining) is a natural language processing (NLP) technique, and aims to determine an author's attitudes, opinions or sentiments with respect to the topic written about. Different methods have been proposed for sentiment analysis, among which lexicon-based method is one of the most popular ones. Lexicon-based sentiment analysis employs NLP techniques to tokenize, split, and lemmatize text into a list words, and use word lexicons to determine the affective valence of these words as well as the valence of the text. Several affective word lexicons exist for this purpose, such as Affective Norms for English Words (ANEW) (Bradley & Lang 1999) and Finn Arup Nielsen's word list (AFINN) (Nielsen 2011). These word lexicons contain lists of English words with a valence score attached to each. For example, in AFINN, on a scale of -5 (negative, unpleasant) and +5 (positive, pleasant), "nice" is rated as +3, and terrible as -3. In this research, we use AFINN for the sentiment analysis, this is mainly because AFINN is particularly designed for microblogs and social media, and has been used by many other researchers (See http://neuro.imm.dtu.dk/wiki/AFINN for a list of studies using AFINN).

Due to the different natures of different languages, we only focus on Flickr photos with title/description in English language. To achieve this aim, we used an open language detection library provided by Cybozu Labs, Inc., which has a precision of 99% for 53 languages.[2] According to the observation of Kisilevich et al. (2010), the language pattern *adjective–noun* is the simplest and

---

[2] https://github.com/shuyo/language-detection.

Figure 2: The affective ratings in the city of Vienna (Map data: OpenStreetMap and Contributors, CCBY-SA). Colours of the markers indicate values of the ratings, with green being comfortable, grey being neutral, and red/yellow being uncomfortable.

most popular example in English language to describe the characteristics of an object. For example, 'a beautiful place' and 'a dirty street'. Therefore, to extract affective responses to environments from geotagged photos in Flickr, we propose a lexicon-based sentiment analysis method, as depicted in Figure 3.

1) Firstly, for each geotagged photo, we use Stanford CoreNLP 1.3.4 library (http://stanfordnlp.github.io/CoreNLP/) to tokenize, split and lemmatise its title and description, and clean the sentences by removing English stop words (e.g. 'a', 'by' and 'since'). A part-of-speech (POS) tagging process is also applied. Results of these steps are a list of words and their lexical category (e.g. noun, verb and adjective).

2) Secondly, we extract adjective–noun sets, e.g. 'interesting building'. Again, Stanford CoreNLP library is applied. Results of this step is a list of adjective–noun sets.

3) Thirdly, we filter out adjective–noun sets that are not place-related. To achieve this aim, a list of place nouns is created, which consists of English place nouns (e.g. 'building', 'street', 'restaurant', 'park', 'museum', 'opera'…) and study-area specific place-names from GeoNames (http://www.geonames.org/) (e.g. 'Stephansdom', 'Karlskirche'). If the noun part of an adjective–noun set is not in the list, we consider the adjective–noun set as non-place-related.

4) Finally, for each adjective within the remaining adjective–noun sets, we check whether it is in the AFFIN affective lexicons. If yes, we assign the valence value to the photo. Otherwise, we use Java WordNet Library to get synonyms of the adjective, and check whether one of the synonyms is in the AFFIN affective lexicons. If yes, the valence value is also assigned to the photo. For each photo, we then average all the valence values of its title and description, and assign the result as the valence value of this photo.

This lexicon-based sentiment analysis method was applied to the Flickr photos uploaded for the city of Vienna (Austria) in the period of January 2007 and October 2011, which contain 107,353 data rows. Figure 4 shows the results. As can be seen from the results, different places are connected with different affective responses. However, to interpret these results, one should consider the bias brought by Flick, especially the contributing inequality.

Several limitations of the proposed sentiment analysis method should be also pointed out. Firstly, it ignores slang and incomplete words, which often exist

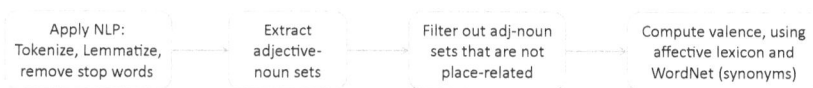

Apply NLP: Tokenize, Lemmatize, remove stop words	Extract adjective-noun sets	Filter out adj-noun sets that are not place-related	Compute valence, using affective lexicon and WordNet (synonyms)

Figure 3: Workflow of extracting affective responses to environments.

Figure 4: Affective responses to environments extracted from geotagged photos in Flickr (Map data: OpenStreetMap and Contributors, CCBY-SA). Colours of the markers indicate valence values of the responses, with green being very positive, grey being neutral, yellow being a bit negative and red being very negative.

in Flickr photo title or description. Secondly, this method uses a very simple language pattern (i.e. adjective–noun set) to identify affective responses caused by the environment. It should be improved to consider more language patterns. Furthermore, the method should be improved to check whether the geolocation of the photo actually matches the place the user is referring to in the title/description of the phone. This will help to address the problem of location accuracy, as well as discover photos that are taken for remote environment objects ('pictures of the Eiffel tower might not be taken directly at the Eiffel tower').

Selected applications of affective responses

In general, collecting affective responses from a large number of people enables many innovative applications. They can not only be used to understand people's behaviour at different environments, as studied in environmental psychology and GIScience, but also bring benefits to other disciplines and application fields, such as Information and Communication Technology, urban planning, architecture and policy making. In the following, we illustrate two different case studies.

The first case study used the collected affective data (see Section 3.1) to study the impact of environmental characteristics on people's affective responses. We focused on the area surrounding the Vienna University of Technology (Austria), mainly due to the diverse environmental settings within this area. We subdivided the area into three distinctive urban scenes according to their level of traffic and vegetation: A) green urban area (urban-green), B) urban area with light or no traffic (pedestrian lanes and one-lane street, urban-light traffic) and C) urban area with heavy traffic (roads ranging from two to three-lanes, urban-heavy traffic). These three urban scenes are compared according to the level of comfort ratings reported by the participants (Figure 5). The results suggest that the level of comfort ratings significantly differ between the three environmental settings ($H(2)= 103.4$, $P < 0.001$). For more details on the data analysis, refer to Klettner et al. (2013).

As can be seen from Figure 5, urban green areas show the most positive ratings among the three urban scenes, followed by areas of urban-light traffic. Urban areas with heavy traffic, on the other hand, show highly negative ratings. However, we argue that in order to draw a clearer conclusion, more research should be done on this aspect, e.g. a further classification of the study area, consideration of other contextual factors (such as time), validation of the quality of the affective data collected, and investigation of the bias caused by the 'participation inequality' (see Section 5 for more discussions).

Our second case study focused on how the collected affective data can be used to design smart human-centred geospatial applications, particularly navigation systems. Navigation systems aim to provide users with navigation guidance when visiting unfamiliar environments (Gartner et al. 2011). Route planning is a key component in navigation systems, and aims to compute an optimal route

Figure 5: Level of comfort according to three urban scenes (Map data: OpenStreetMap and Contributors, CCBY-SA).

from an origin to a destination. Current route planning algorithms often fail to provide other routes aside from shortest routes or fastest routes. However, research in geography and environmental psychology has shown that humans may prefer many other different route qualities when choosing 'which way to take', such as safety and attractiveness (Golledge 1995; Zacharias 2001).

We proposed that incorporating people's affective responses into route planning can help to provide routes with other qualities and characteristics. The basic idea is to aggregate affective ratings of similar users to model/approximate the current user's perception of different street segments. With this, a street network, in which each segment is encoded with a collective rating, can be generated. Based on this kind of street networks, we can compute routes with different qualities and characteristics, such as the most comfortable route (i.e. route avoiding uncomfortable areas) and the safest route. For more details on the algorithm, refer to Huang et al. (2014).

Figure 6 shows an example of the route computed by considering the 'level of comfort' ratings, comparing to its shortest counterpart. An online empirical study with 64 human subjects was designed to evaluate the proposed routing

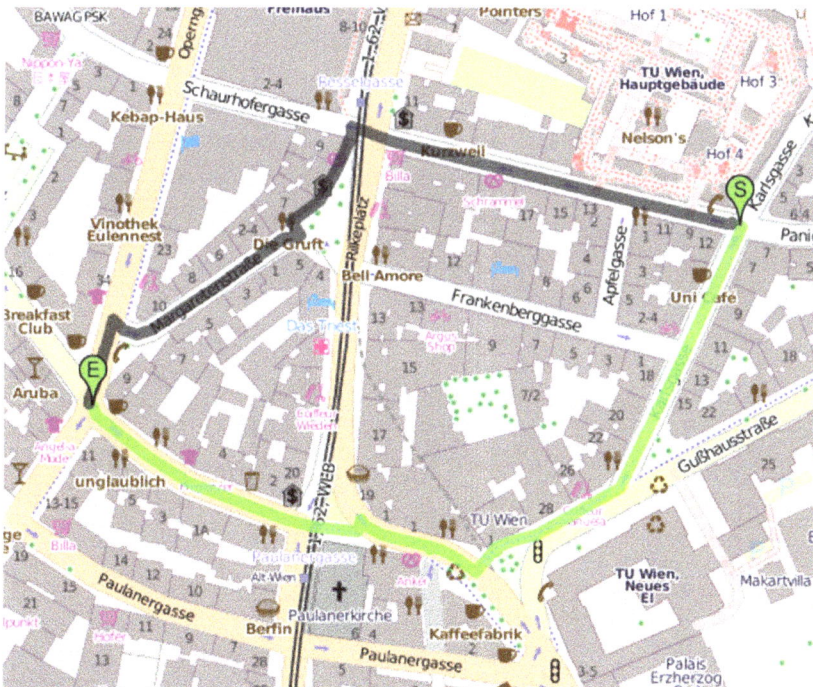

Figure 6: The most comfortable route (green one, 478 meters) computed by considering people's affective responses (solely the 'level of comfort' ratings), comparing to its shortest counterpart (grey one, 442 meters).

algorithm using a video task (participants were asked to choose the walk they prefer to take after watching the videos, which were taped along the Affec-tRoute and the shortest route) and two map tasks (given a start and an end, participants were asked to draw their preferred route). In the map task, a chi-square goodness-of-fit test showed that the routes computed by considering affective responses were significantly preferred over the shortest routes [$P = 0.041$]. Similar significant results were found in the map tasks. In conclusion, considering people's affective responses to environments leads to more satisfy-ing routing results.

In the current research, we aggregated affective ratings from a large number of other users to provide route recommendations for the current user. How-ever, it is still unclear how the routing results are influenced by the size/density of the affective ratings. More studies should be done on this aspect. On the other hand, the current experiment was implemented via an online question-naire. Further research should be done to explore more realistic evaluation in more diverse environments.

Challenges and Lesson Learned

While mobile crowdsourcing and social media data are promising for study-ing people's affective responses to environments, and these affective responses enable many innovative applications, many challenges also exist. In the follow-ing, we highlight some of the essential challenges, and describe the lesson we learned from our current research, and discuss some hints for addressing these challenges.

Motivating participation. Motivating people to participate is key to crowd-sourcing projects. This is even more important for crowdsourcing people's affective responses, which are subjective. Without enough contributions, it is impossible to get a comprehensive understanding about how environments are perceived by different groups of people. From our research, we learned that in order to motivate people's participation, it is essential to provide real or perceived 'benefits' for the contributors, and as well as to provide simple and intuitive ('easy-to-use') interface/tool to facilitate people's participation and contribution.

Data quality. Data quality is another big challenge when using mobile crowdsourcing and social media data to study people's affective responses. It is influenced by the 'participation inequality' of the contributors. Some of the examples of the aspect are: contributors might not be representative samples of the population; contributors might prefer to contribute at some places or some occasions/time, and avoid some other places or occasions/time. All these factors might strongly influence the quality of the data collected. Furthermore, compared to other data (such as those crowdsourced in OpenStreetMap), people's affective responses are subjective in nature, and no reference data are available for cross-checking. How to deal with the data quality of the crowd-sourced affective data is still an open research question. A promising solution is to combine self-reports, social media data analysis and technical sensors (such as Galvanic skin response GSR, and Electrocardiogram ECG) when studying people's affective responses. With this, we can cross-check the data we collect.

Privacy. People's affective responses might contain lots of sensitive personal information. Therefore, when studying people's affective responses, especially at an individual level, it is important to develop methods to address the privacy concern. Anonymization technique, which was often employed in traditional empirical studies, might not work well for crowdsourcing projects and social media data analysis. For example, if we know an anonymous user often contributes at a particular place in the early morning (e.g. at 6:00) and at another place in the afternoon (e.g. at 14:00), we are potentially able to re-identify who the user is, as these two places might correspond to his/her home and office (Huang et al. 2013). More research should be done to develop privacy-preserving techniques for mobile crowdsourcing and social media data analysis.

Summary and Outlook

The literature has shown that humans perceive and evaluate environments affectively, and these affective responses to environments influence our daily behaviour and decision-making in space. This paper presented our recent efforts towards these aspects. We illustrated two approaches to collect and acquire affective responses from a large number of people. Particularly, mobile crowdsourcing and social media data analysis were introduced. This paper also presented two different case studies to illustrate the potential applications of affective data.

Currently, we are exploring hybrid methods to study people's affective responses to environments, combining mobile crowdsourcing (self-reports), social media data analysis, and mobile sensing with sensors like Galvanic Skin Response. The privacy concerns of these affective contributions will also be addressed. We are also applying the affective data to other disciplines, such as urban planning and transportation.

Acknowledgments

The authors would like to thank the supports from Austrian Ministry of Transportation, Innovation and Technology (BMVIT), via the research project EmoMap.

References

APA. 2006. APA dictionary of psychology. Washington, DC: American Psychological Association.

Barrett, L. F. 2006. Valence is a basic building block of emotional life. *Journal of Research in Personality, 40*(1): 35–55. DOI: http://dx.doi.org/10.1016/j.jrp.2005.08.006

Barrett, L. F., et al. 2007. The experience of emotion. *Annual Review of Psychology, 58*: 373–403. DOI: http://dx.doi.org/10.1146/annurev.psych.58.110405.085709

Bradley, M., & Lang, P. 1999. Affective norms for English words (ANEW): Instruction manual and affective ratings. Technical Report C-1, The Center for Research in Psychophysiology, University of Florida.

Ekman, P., & Friesen, W. V. 1971. Constants across cultures in the face and emotion. *Journal of Personality and Social Psychology, 17*(2): 124–129. DOI: http://dx.doi.org/10.1037/h0030377

Gartner, G., Huang, H., Millonig, A., Schmidt, M., Ortag, F. (2011). Human-centred mobile pedestrian navigation system. *Mitteilungen der Österreichischen Geographischen Gesellschaft, 153*: 237–250. doiDOI: http://dx.doi.org/10.1553/moegg153s237

Golledge, R. G. 1995. Path selection and route preference in human navigation: a progress report. In: Frank, A. U., & Kuhn, W. (Eds.) *Spatial information theory A theoretical basis for GIS*. Berlin: Springer, pp. 207–222.

Goodchild, M. F. 2007. Citizens as sensors: the world of volunteered geography. *Geojournal, 69*(4): 211–221. DOI: http://dx.doi.org/10.1007/s10708-007-9111-y

Harmon-Jones, E., Gable, P., & Price, T. 2013. Does Negative Affect Always Narrow and Positive Affect Always Broaden the Mind? Considering the Influence of Motivational Intensity on Cognitive Scope. *Current Directions in Psychological Science, 22*(4): 301–307.

Hauthal, E., & Burghardt, D. 2013. Extraction of location-based emotions from photo platforms. In: Krisp, J. M. (Ed.) *Progress in location-based services*. Berlin: Springer, pp. 3–28.

Huang, H., Gartner, G., & Turdean, T. 2013. Social media data as a source for studying people's perception and knowledge of environments. Mitteilungen der Österreichischen Geographischen Gesellschaft, pp. 155, 291–302.

Huang, H., Klettner, S., Schmidt, M., Gartner, G., Leitinger, S., Wagner, A., & Steinmann, R. 2014. AffectRoute – Considering people's affective responses to environments for enhancing route planning services. *International Journal of Geographical Information Science, 28*(12): 2456–2473.

Kaplan, R., & Kaplan, S. 1989. The experience of nature: a psychological perspective. Cambridge University Press.

Kisilevich, S., Rohrdantz, C., & Keim, D. 2010. Beautiful picture of an ugly place. Exploring photo collections using opinion and sentiment analysis of user comments. Computer Science and Information Technology (IMCSIT), pp. 419–428.

Klettner, S., Huang, H., Schmidt, M., & Gartner, G. 2013. Crowdsourcing affective responses to space. KN Kartographische Nachrichten. *Journal of Cartography and Geographic Information, 2 & 3*: 66–72.

Mislove, A., et al. 2010. Pulse of the Nation: U.S. mood throughout the day inferred from Twitter [online]. Available from: http://www.ccs.neu.edu/home/amislove/twittermood/.

Nielsen, F. 2011. A new ANEW: Evaluation of a word list for sentiment analysis in microblogs. In: *Proceedings of the ESWC2011 Workshop on 'Making Sense of Microposts': Big things come in small packages 718 in CEUR Workshop Proceedings*, pp. 93–98. Available at: http://arxiv.org/abs/1103.2903

Resch, B., Summa, A., Sagl, G., Zeile, P., & Exner, J. P. 2015. Urban emotions—geo-semantic emotion extraction from technical sensors, human sensors and crowdsourced data. In Gartner, G., & Huang, H. (Eds.) *Progress in location-based services 2014*. Springer, pp. 199–212.

Russell, J. 2003. Core Affect and the Psychological Construction of Emotion. *Psychological Review, 110*(1): 145–172.

Russell, J. A., & Barrett, L. F. 1999. Core affect, prototypical emotional episodes, and other things called emotion: dissecting the elephant. *Journal of Personality and Social Psychology, 76*(5): 805–819. DOI: http://dx.doi.org/10.1037/0022-3514.76.5.805

Zacharias, J. 2001. Path Choice and Visual Stimuli: Signs of Human Activity and Architecture. *Journal of Environmental Psychology, 21*: 341–352.

CHAPTER 29

Integrating Authoritative and Volunteered Geographic Information for spatial planning

Pierangelo Massa* and Michele Campagna

University of Cagliari, DICAAR, Via Marengo 2, Cagliari 09123, Italy,
*pmassa@unica.it

Abstract

This contribution concerns ongoing research by the authors on the integrated use of Social Media Geographic Information (SMGI) and Authoritative Geographic Information (A-GI) as a support in urban and regional planning. Advances in Information and Communication Technologies (ICT) are fostering the production and the sharing of georeferenced user-generated contents, namely Volunteered Geographic Information (VGI) and SMGI, which may complement traditional spatial data sources. VGI is a voluntary contribution by users in order to collect or to disseminate geographic knowledge, while SMGI may be considered a deviation from VGI nature, due to the implicit and passive mode in disseminating geographic information, which is exclusively one embedded attribute of the main shared information. However, SMGI may offer unprecedented opportunities to investigate users' needs, opinions, behaviors and movements, thus representing a potential support for analysis and decision-making in spatial planning. In this respect, the authors present an original tool called Spatext, which allows collection, management and analysis of SMGI in GIS environment, easing the integration of SMGI with official information. Afterwards, the opportunities for spatial planning arising from

How to cite this book chapter:
Massa, P and Campagna, M. 2016. Integrating Authoritative and Volunteered
  Geographic Information for spatial planning. In: Capineri, C, Haklay, M, Huang, H,
  Antoniou, V, Kettunen, J, Ostermann, F and Purves, R. (eds.) *European Handbook of
  Crowdsourced Geographic Information*, Pp. 401–418. London: Ubiquity Press. DOI:
  http://dx.doi.org/10.5334/bax.ac. License: CC-BY 4.0.

SMGI are demonstrated through a case study where this type of information is used for investigating the geography of places. An original methodology is developed applying clustering techniques on the spatial and the temporal components of SMGI collected from Instagram. The applied methodology enables the identification of residential buildings that are not mapped in available official datasets. The results demonstrate how SMGI may be proficiently used to integrate and update A-GI, as well as to investigate the users' behaviors and movements in an urban environment.

Keywords

SMGI, VGI, Urban planning, Spatial planning, Instagram

Introduction

In recent years, continuous advances in Information and Communication Technologies (ICT), the internet and Web 2.0 technologies are strengthening the production, the sharing and the access of user-generated contents among millions of users worldwide. The availability of user-generated contents may represent a novel source of geographic information (Elwood, Goodchild & Sui 2012), inasmuch as most of these contents may embed a spatial reference, thanks to the availability of GPS and sensors in handheld devices and smartphones, as well as geo-browsers or location-based social networks, which are used for production.

This novel type of geographic information is commonly referred to as Volunteered Geographic Information, emphasizing the role of users which act as volunteer sensors to collect and contribute to this data (Goodchild 2007). However, the information produced and shared through social networks, namely Social Media Geographic Information (SMGI) (Campagna 2014), may be considered a deviation from the traditional VGI nature, since the collaborative collection and the diffusion of geographic information are not the main purposes of users (Stefanidis, Crooks & Radzikowski 2013). Despite an implicit nature of SMGI for the geographic dissemination, this kind of information, coupled with traditional VGI, proved to be useful in different application domains such as environmental monitoring, crisis management (Poser & Dransch 2010), as well as urban planning (Frias-Martinez et al. 2012). Indeed, VGI and SMGI may represent a valuable complement to traditional official information, supplying insights on users' perceptions and needs, opinions on places, as well as information about daily events, in (near) real-time, so potentially contributing to faster decisions.

In the urban and regional planning domain, VGI and SMGI may play an important role to support (1) analysis, (2) design and (3) decision-making, fostering innovations in planning methodologies, inasmuch the majority of information required and used in practices is mainly spatial. As a matter of fact, this innovative wealth of GI may integrate the current availability of official digital information with pluralist knowledge from local communities that is usually neglected in practice, paving the way for innovative analytic scenarios.

Currently in Europe, a wealth of official digital geographic information was made available since 2007 by the implementation of the Directive 2007/02/CE (INSPIRE), which fostered developments in Spatial Data Infrastructures (SDI) among European member states. This process is favoring the access and the reuse of available official digital information, namely A-GI, produced by Public Authorities. This way, planners, analysts and professionals may access A-GI according to common technology, data formats, and policy standards. The integration of available official information with SMGI may further improve this potential, enriching the official datasets with information regarding not only geographic facts but also insights on people's perceptions and feelings both in space and time.

Nevertheless, the opportunities for the use of SMGI in spatial planning as affordable and potentially boundless sources of information have to deal with diverse challenges related to data management, data quality and data analysis. Indeed, the traditional spatial analysis methodologies and techniques may not be fully suitable to tackle the complexity of this information that exhibits Big Data nature for its modes of production and consumption (Caverlee 2010). Hence, in spite of several applications, related to different application domains and built upon the integration of SMGI and VGI in recent years, the lack of common methods to deal with these issues still requires the development of novel analytical frameworks in order to fully exploit the SMGI potential for analysis, design and decision making.

In the light of the above premises, this contribution presents a review of the authors' research results on the integration and use of A-GI and SMGI, aiming at developing valuable tools and methodologies for spatial planning. The remainder of the contribution is articulated as follows. The next section briefly discusses the distinctive features of SMGI, focusing on its main issues and opportunities for analysis. Section 3 introduces an original tool, developed by the authors and called Spatext, which enables the seamless collection, management and analysis of SMGI from multiple social media in a GIS environment. In section 4 a novel approach to SMGI analytics is proposed concerning a case study related to urban planning. Finally, section 5 draws conclusions and summarizes the discussion about the opportunities and the open issues of SMGI for urban and regional planning.

Issues and opportunities of SMGI

The wealth of georeferenced user-generated contents regarding facts, opinions, and concerns of users, freely accessible through the internet by social media

Application Programming Interfaces (APIs), may affect current practices in urban and regional planning, giving opportunities for real-time monitoring of needs, thoughts and trends of local communities. However, the current public accessibility to SMGI is rather limited (Lazer et al. 2009), and common methods to manage, process and exploit these resources in practices still lack. The main hurdles limiting a wider use of SMGI may be found both in the shortage of user-friendly tools to collect and to manage huge data volumes and in the particular data structure of this information, which is burdensome to analyze by traditional methods. While the former challenge is starting to be addressed by new approaches typical of computational social science, an emerging field that aims to develop methodologies to address the complexity of Big Data (Lazer et al. ibidem), the latter challenge might require a tuning of analytical methodologies to deal with the several facets of SMGI.

First of all, although SMGI may be potentially available through the internet from any social media APIs, each platform features specific characteristics for contents production and sharing; hence SMGI from different social media may embed different attributes, causing difficulties for data integration and analysis. Moreover, SMGI is usually broadcasted through the internet by coupling alphanumeric data and multimedia clips, which complicate the analysis by means of traditional query languages. Secondly, SMGI, as user-generated contents with an associated geospatial component, combines the spatial and the temporal dimension of geographic information with a third dimension, namely the user itself, therefore extending the range of available analytical methods with further opportunities, such as users' behavioral analysis, users' interests investigation, land segmentation and potentially any analysis based on space, time and user (Campagna ibidem). These analytical methods may represent a novel way to investigate facets of the social and cultural habits of local communities, but their implementation may represent a challenge, which requires the integration of traditional spatial analysis methods with expertise and contributions from various disciplines such as social sciences, linguistic, psychology and computer science (Stefanidis, Crooks & Radzikowski ibidem).

The requirements for new analytical tools to deal with SMGI, and the opportunities resulting from the inherent nature of this information, guide the development of an original user-friendly tool, called Spatext, which eases the collection of information from multiple social networks and the integration of the data in a GIS environment for analysis.

Spatext: the SPAtial-TEmporal-teXTual Suite

Spatext is an add-in for the commercial software ESRI ArcMap© implemented in Python 2.7. It enables the contextual social media data collection, management, geocoding, as well as the spatial, temporal and textual analysis of SMGI. This SMGI Analytics suite includes a number of tools, which can be used mainly to (1) retrieve social media data from social media (including Twitter, YouTube, Wikimapia, Instagram, Instagram Places, Foursquare and Panoramio); (2) geocode or georeference data; and (3) carry out integrated spatial, temporal and textual analyses. In addition, the number of analytical methods available in the tool is steadily increasing in order to include several clustering algorithms to enable user profiling, user movement analysis, user behavioral analysis and land use detection, to name a few. Indeed, the collection, management and geocoding functionalities may turn any social media content into a workable SMGI dataset, which may then be directly integrated with other spatial data and analyzed in a GIS environment with off-the-shelf instruments.

Spatext takes advantage of the available social media APIs to perform queries directly from the GIS interface, enabling the collection of multimedia information regarding different topics, time periods and geographic areas. This way, the extension of traditional GIS tools with Spatext tools may ease the integration of SMGI with authoritative data, in order to support analysis, design and decision-making in urban and regional planning. The tools included in Spatext are developed in order to deal with the aforementioned issues regarding the access, management and analysis of 'big data' and can be categorized in three different classes according to the specific function: (1) data collection, (2) data management and (3) data analysis. The first class includes user-friendly tools that enable the harvesting of information from several social networks through spatial, temporal or textual queries. These tools can facilitate the direct access to social networks APIs avoiding programming efforts. The second class provides tools developed to ease the management, the integration and the successive analysis in GIS environment of SMGI extracted from different sources. These tools aim to limit the issues regarding the management and conversion of SMGI originated from different sources, which may present different data structures and information. Finally, the third class contains tools designed for analyzing the spatial, temporal and user dimensions of this information, as well as, for enabling the investigation of embedded textual contents. At the time being, the Spatext suite is not available for download due to minor technical revisions ongoing on APIs access. An overview of the Spatext functionalities is presented in Table 1, where the main tools are classified and briefly described according to the specific class, while the Spatext architecture is shown in Figure 1.

In the next section, functionalities of the Spatext tool for SMGI analytics are demonstrated through a case study related to urban planning in the municipality of Iglesias in Sardinia, Italy. The case study proposes the analysis of Instagram SMGI coupled with A-GI from Sardinian Spatial Data Infrastructures (i.e. Sardegna Geoportale http://www.sardegnageoportale.it) to investigate the geography of the municipality and to debate the potential opportunities emerging from the integration of implicit experiential knowledge with official information for urban planning practices.

SPATEXT TOOLS				
CLASS 1 - 'DATA COLLECTION'		**query parameters**		
		space	**time**	**keyword**
Instagram extractor	Extracts Instagram SMGI to shapefile	✓	✓	✓
YouTube extractor	Extracts YouTube SMGI to shapefile	✓	✓	✓
Instagram Places extractor	Extracts Instagram Places SMGI to shapefile	✓		
Twitter extractor	Extracts Twitter SMGI to shapefile	✓	✓	✓
WikiMapia extractor	Extracts Wikimapia SMGI to shapefile	✓		✓
Foursquare extractor	Extracts Foursquare SMGI to shapefile	✓		
Panoramio extractor	Extracts Panoramio SMGI to shapefile	✓		
CLASS 2 – 'DATA MANAGEMENT'		**function activation**		
		manual	**automatic**	
Geocoding tools	Geocode places/addresses from attribute strings	✓	✓	
Conversion tools	Convert SMGI attributes to table	✓	✓	
Decomposition tools	Slice SMGI dataset in multiple datasets	✓	✓	
CLASS 3 – 'DATA ANALYSIS'				
Temporal Reference	Enriches SMGI dataset by adding information on temporal periods (season, month, day, hour) for further investigation			
Temporal Trend	Analyzes the temporal trend of SMGI contribution providing a graphic report			
Clustering functions (DBSCAN)	Detect clusters of high density by running the Density-Based Spatial Clustering of Applications with Noise algorithm on SMGI dataset			
Textual tag-cloud	Performs a simple tag-cloud analysis on textual contents of SMGI dataset to detect topic of interest			

Table 1: Main functionalities available in Spatext tool.

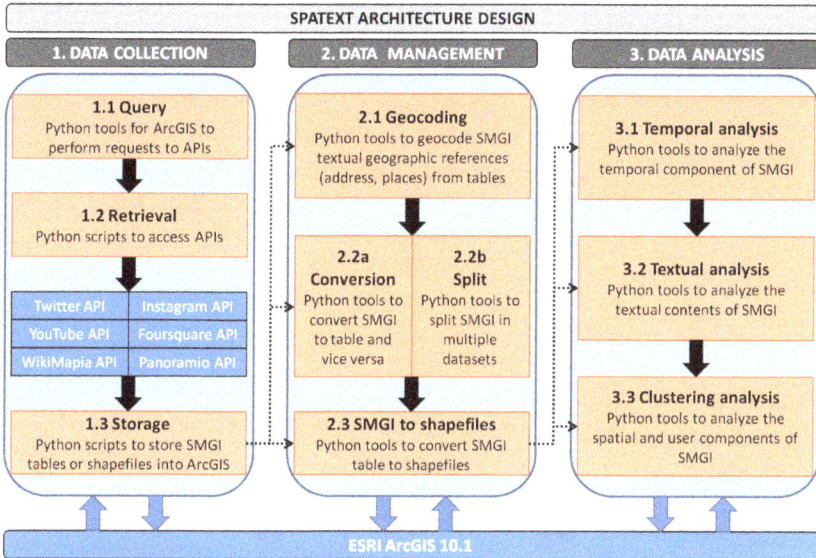

Figure 1: Spatext architecture design.

Instagram SMGI analytics: an application in urban planning

In this section, an application of SMGI analytics is proposed through the analysis of Instagram contents in the urban environment of the Iglesias municipality in Sardinia, Italy. Nowadays, Instagram is one of the most popular online social networks worldwide, and it enables users to take, upload, edit and share photos with other members of the service through the platform itself, or other social media such as Facebook, Twitter, Tumblr, Foursquare and Flickr. Approximately 20 percent of the internet users aged 16 to 64 have an account on the service, and the trend is growing over last years. In addition, demographics of active Instagram users (GlobalWebIndex 2014) show a balanced percentage between male users (51%) and female users (49%), with a high percentage (41%) of users aged 16 to 24 that prevail over users aged 25 to 34 (35%), 35 to 44 (17%), 45 to 54 (6%) and 55 to 64 (2%). Statistics on the service stress also how a major part of active users (56%) appear to be into the middle quartile (33%) or top quartile (23%) of income. Among the features offered by Instagram, the geotag allows users to embed latitude and longitude of the place with the taken photos, therefore allowing to share the contents and the geographic reference through the internet according to own privacy settings. This capability plays a central role in considering Instagram contents as SMGI and permits the development of analysis to investigate spatial and temporal patterns within any geographic area where the service is available.

The case study concerning the Iglesias municipality (Italy) took advantage of the Instagram SMGI for a twofold purpose: (1) to explore the geography of the place through spatial and temporal patterns of the contributions, investigating trends and areas of interest within the municipality, and (2) to identify and classify SMGI clusters, relying on the inherent spatial and temporal components, as well as by means of the integration with A-GI, in order to detect potential missing buildings in official datasets. The operational application of SMGI analytics on the case study of Iglesias municipality was carried out according to the following three main steps: (1) data collection, (2) analysis of spatial and temporal components, and (3) detection and classification of SMGI clusters, as explained in detail in the remainder of the contribution.

Data collection

The data collection of SMGI from Instagram was conducted through the Spatext Instagram extractor tool by setting the spatial query parameter on the municipality of Iglesias and the temporal query parameter on a one year period (from 1 August 2013 to 1 August 2014). The extraction resulted in the collection of a one year sample of approximately 14,000 geotagged photos from 1.243 users for the study area. The tool automatically generated a point feature dataset, georeferencing each photo according to the geographic reference (latitude and longitude) embedded in the spatial metadata of the content, namely the geotag. Commonly, the geotag refers the GPS position of camera when the photo was taken; however, issues in connectivity may lead toward the lack of this information. In these cases, the Instagram service sets the geographic coordinates of the contents using the user's position during the upload. In addition to the geographic coordinates, the dataset includes several attributes, such as name of the place, if set by the user during upload, user name, user id, user picture URL, media URL, date of creation, number of comments, number of likes, tags and captions. These attributes are made available for any Instagram content if the user's privacy settings are public, offering opportunities for the development of several analysis in combination with other spatial data layers. Even though these pieces of information are publicly available, data were anonymized for privacy issues before any storage or processing for the study.

An exploratory analysis of the SMGI dataset showed a mean value of 11.22 photo/user, a modal value of 1.0 photo/user and a median value of 2.0 photo/user. Indeed, the 39.82% of users contributed with only 1 photo per year, the 32.74% contributed with 5 photo or more, while only the 4.34% of users posted 50 photos or more. Despite a different degree of participation by users, the dataset was investigated in order to identify potential commonalities among contributions in terms of areas of interest and urban dynamics.

Analysis of spatial and temporal components

After the data collection, the spatial and temporal components of the SMGI dataset were investigated directly in GIS environment, in order to explore potential patterns of interest in the area and local community dynamics. At this stage, the SMGI dataset was integrated with several official datasets from the regional spatial data infrastructure of Sardinia related to the Iglesias municipality such as settlements, roads network and buildings.

A simple investigation of the dataset spatial distribution showed a high concentration of the placemarks within the built environment, with approximately the 89% of the contents taken in residential or commercial and service areas. This value may depict the users' preference to employ the Instagram service in situations strictly related to their daily life within the city and might be considered a good starting point to investigate the dynamics in the municipality. The spatial distribution of the SMGI dataset is shown in Figure 2.

With the above considerations in mind, the temporal component of the SMGI dataset was investigated for different periods by searching potential peaks of interest, trends and dissimilarities in the use of Instagram by the users in Iglesias. The temporal analysis was performed investigating seasons, months, days of the week and hours of the day, disclosing interesting patterns. The results of temporal analysis showed how SMGI was increasingly produced and shared by users during the spring (30.9%) and summer (33.3%) in opposition to winter (19.1%) and autumn (16.7%); and this phenomenon was also evident in month distribution where July presented the highest percentage of produced contents (13%) and November the lowest one (5%).

The analysis of daily distribution provided more balanced results, with a slightly higher percentage of contents produced during weekends (Saturday and Sunday). Finally, the analysis of daily hours trend allowed identifying two main peaks of interest for both workdays (Monday to Friday) and weekends (Saturday to Sunday). The peaks were identified during the periods 14:00-15:00 and 21:00-22:00 for workdays, and the periods 14:00-15:00 and 20:00-21:00 for weekends, probably identifying meals or pause times. In contrast, the period 05:00-06:00 showed the lowest percentage of produced contents both for the workdays and the weekends. In spite of similar temporal peaks, the workdays and weekends trends exposed a few differences, which might be considered to be a descriptor of the typical cultural behaviors of inhabitants or a sort of cultural footprint of the place. This assumption may be corroborated by the results of a similar study conducted on Instagram datasets by Silva et al. (2013), which demonstrated how workdays and weekends temporal patterns were similar for cities of the same country, but showed major differences among cities in different countries. The results of temporal analysis for the different periods are provided in Figure 3.

Detection and classification of SMGI clusters

The results of spatial and temporal investigations led towards the development of further analysis to investigate the geography and the urban dynamics of the municipality. Especially the major density of SMGI in the built environment

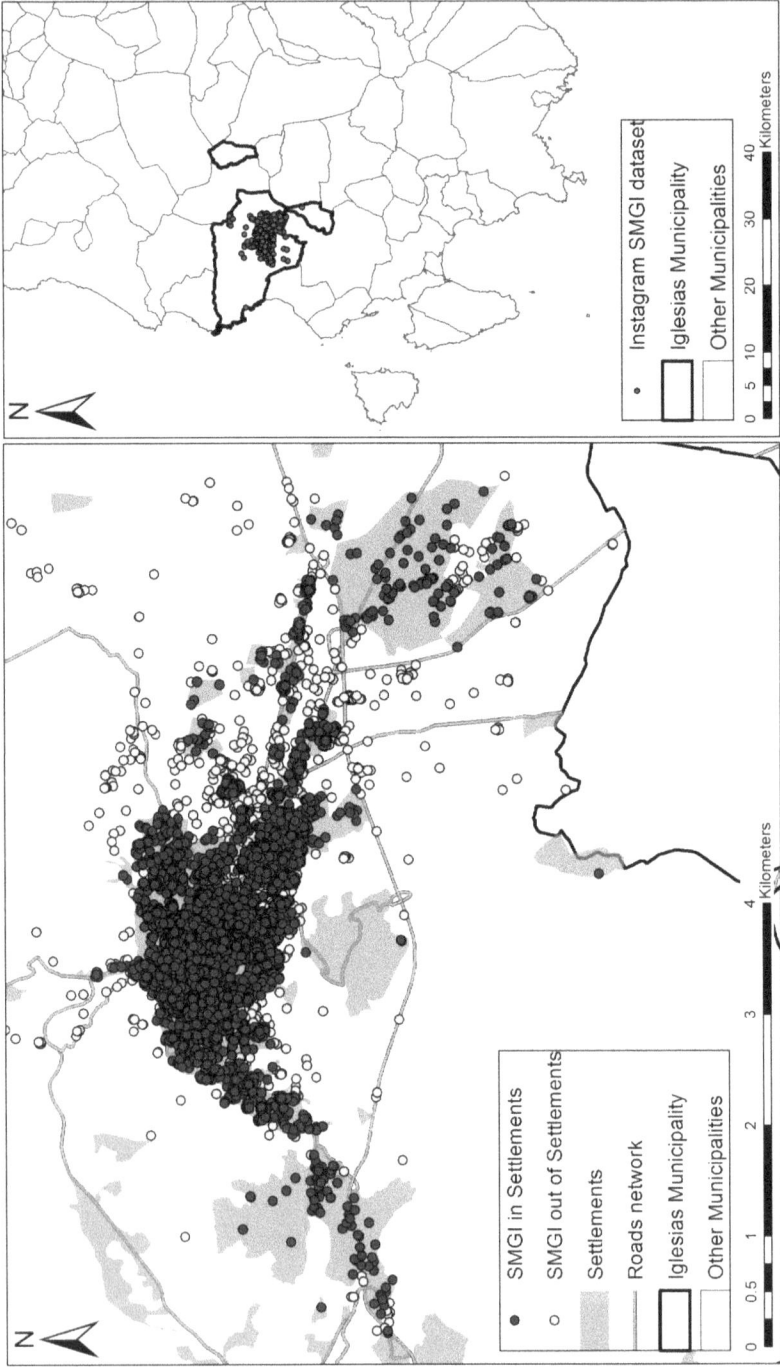

Figure 2: Spatial distribution of Instagram SMGI in Iglesias municipality.

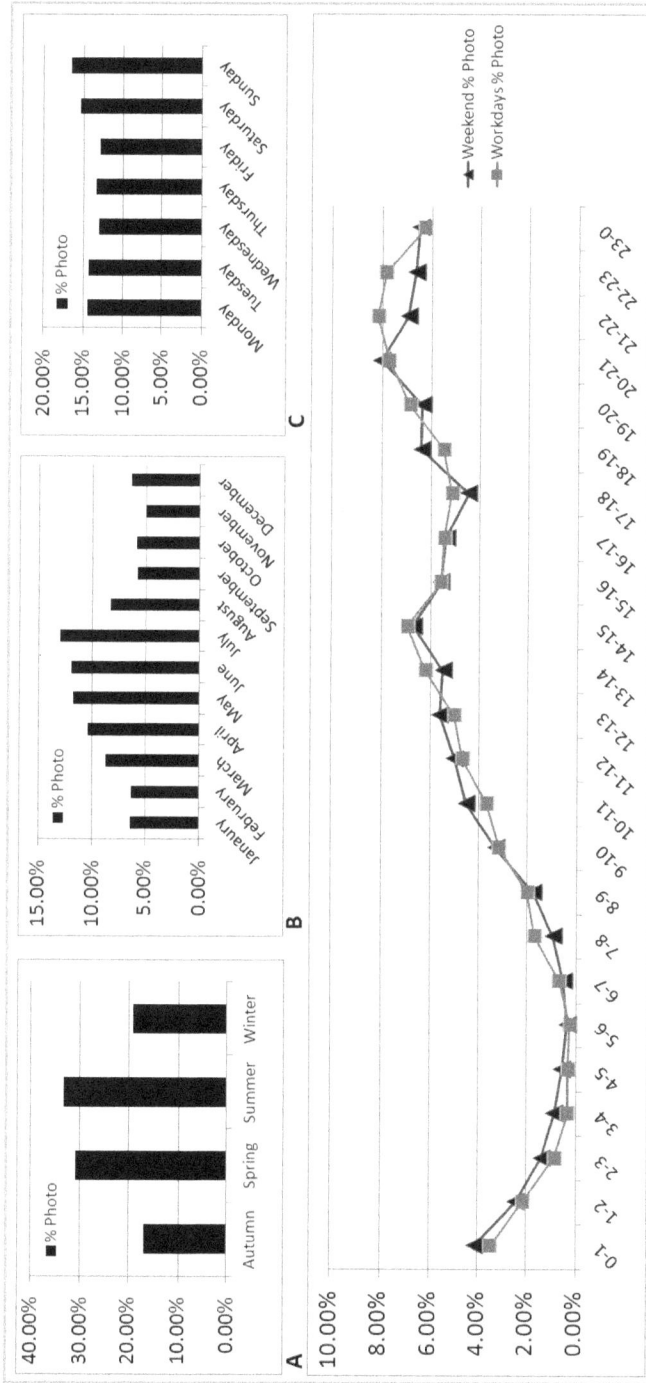

Figure 3: Temporal distribution of Instagram SMGI dataset: (A) season, (B) month, (C) day of the week, (D) hours trend during weekends and workdays.

fostered the development of analytical methods to identify, classify and interpret the users' interest toward certain specific spaces. For this purpose, the Density-Based Spatial Clustering of Applications with Noise algorithm or DBSCAN (Ester et al. 1996) and a slightly modified version called Feature-Based DBSCAN (FB-DBSCAN) were integrated in Spatext, and were used to compute clusters based on the spatial density of points.

The DBSCAN algorithm offers major advantages with respect to other clustering algorithms; firstly it is not necessary to know a priori the number of clusters, which also may differ in size and shape. Secondly, it works using two parameters exclusively: the epsilon (eps) that is the maximum threshold distance for including points in the same cluster, and the minimum number of points (min_pts) that is required to define a cluster. In the study, the goal of the clustering analysis was the identification of the places that attracted the interest of the local community, which may be measured in terms of high density of contributions. Nevertheless, operatively there was no opportunity to establish the preferable value of eps and min_pts before the computation, therefore the DBSCAN tool was applied iteratively on the SMGI dataset for different measures of the parameters in order to evaluate different results of the clustering. The assessment of clustering results led toward the identification of the following values, which proved to be the most suitable for the purpose of the study: eps = 20 meters and min_pts = 5. Indeed, this eps value, or threshold distance, was able to cover the dimension of a medium-sized fabric, while the min_pts value was set to 5 as a compromise value to avoid false positive in clusters detection and, at the same time, to prevent the dismissal of clusters with a modest participation of users. The results of clustering analysis with the above set of values enabled the identification of 290 clusters within the urban area of Iglesias, with a major concentration near the city center. In addition, two large clusters with an area greater than 50,000 square meters emerged from the analysis, identifying the areas attracting the highest interest by users within the urban context. These areas concerned both the historic centre of Iglesias and several service and public space areas. A closer look to the clusters showed that the top cluster contained the historic Cathedral of Santa Chiara, the main avenue for leisure and night life of the municipality, two of the main squares of Iglesias, as well as the train station area. At the same time, the bottom cluster contained several areas related to medical services, leisure, nightlife as well as the public park of the municipality.

Along the same vein, the FB-DBSCAN tool was used on the SMGI dataset in order to detect the places of major interest for each user. In fact, the FB-DBSCAN algorithm processes the dataset after performing a selection for attribute on the sample, in this case the users. This way, the algorithm computes clusters by processing only points related to a specific user for each iteration, offering opportunities to develop more specific analysis on the users' behavior. The analysis through FB-DBSCAN with the parameters eps = 20 meters and min_pts = 5 identified 368 clusters concerning 266 users. In this case the number of identified clusters was higher than the one in the previous analysis, but the clusters' sizes were notably smaller, identifying specific places or fabrics within the municipality. The results of the clustering analysis performed by DBSCAN and FB-DBSCAN are shown in Figure 4.

Each cluster identified through the FB-DBSCAN tool belonged to the contributions of a single user, and could be considered representative of a specific use regarding residence, work or leisure activities. The current use of a cluster may be discovered by analyzing several parameters related to spatial and temporal characteristics, as well as by integrating further spatial information. The aim of the study was the identification of not mapped buildings in the official information; therefore the latest official buildings dataset from the Regional SDI was integrated. This official dataset was selected in order to check the consistency of the clusters' location with the urban fabrics and to ease the identification of suitable parameters to detect residential clusters. As a matter of fact, the clusters related to a specific land use, in this case residential use, may expose similar patterns for certain characteristics such as number of intersections among clusters, temporal span among contributions, number of contributions and density of contributions, to name few, paving the way to the identification of common patterns for classification. In the study, six different parameters were selected with regards to the cluster itself and to the contributions, as described in Table 2.

The values of the six parameters were estimated for each cluster, and several combinations of the values were iteratively evaluated to identify exclusively the residential clusters. The following set of values resulted as the most suitable to classify a cluster as residential in the study area: *Cluster Centroid* and *Contributions Centroid* had to be 1 (yes), while *Number of Contributions* and *Time Span Among Photos* had to present the highest values among clusters of the same user, or the values had to be higher than 10 and 30, respectively. Finally, *Cluster Intersections* had to be equal or lower than 2, while *Cluster Density* had to be higher than 4. The above parameters allowed the identification of 47 residential clusters, which were confirmed by an overlay analysis with satellite imagery in GIS environment. Furthermore, the used parameters avoided potential biases caused by temporary phenomena such as massive tourists' presence or extremely popular events thanks to the threshold interval set for the parameter *Time Span Among Photos*, which considered only time periods equal or higher than 30 days to classify a cluster as residential. Afterwards, the same set of parameters was used to identify potential missing buildings in the official dataset by setting to 0 (no) the values of *Cluster Centroid* and *Contributions Centroid*, while leaving unchanged the other parameters values. Indeed, the values of *Number of Contributions, Cluster Intersections, Time Span Among Photos* and *Cluster Density* were considered as a sort of residential parcels footprint among clusters and were used for the investigation.

The analysis identified 40 clusters, which were then visually assessed through satellite imagery to confirm the presence of not mapped buildings in A-GI. The visual assessment allowed the detection of 9 not mapped buildings; at the same

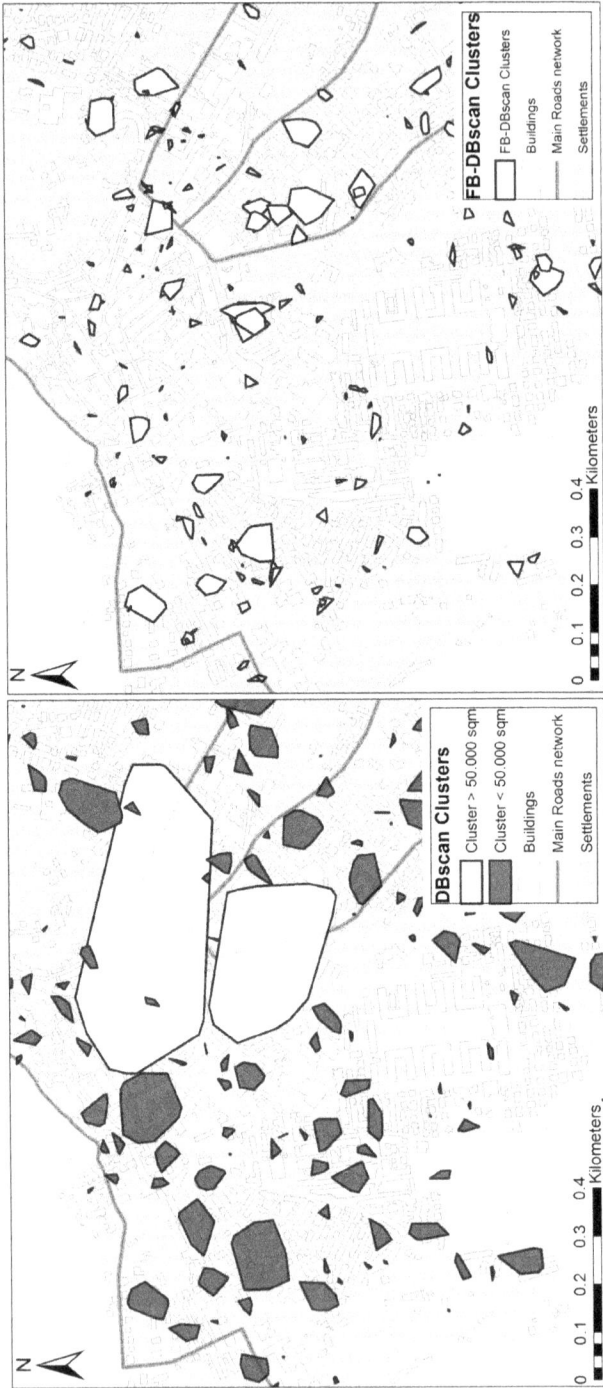

Figure 4: Clustering analysis of SMGI dataset: DBSCAN results (left), FB-DBSCAN results (right).

Parameters	Description	Units of measure
Cluster Centroid	The overlap of the cluster's centroid with an official building footprint is estimated	Boolean
Contributions Centroid	The overlap of the cluster's contributions centroid with an official building footprint is estimated	Boolean
Number of Contributions	The total number of contributions contained in the cluster is estimated	Number of contributions
Cluster Intersections	The total number of intersections between the cluster's shape with other clusters	Number of intersections
Cluster Density	The ratio between the cluster's area and the number of contained contributions	Square meters
Time Span Among Photos	The time passed between the first contribution and the last one in the cluster	Days

Table 2: Parameters used to identify residential clusters.

time the other 31 clusters were confirmed as residential areas, but the buildings were already mapped in A-GI. This issue can be explained by the lack of tolerance during the estimation of *Cluster Centroid* and *Contributions Centroid* values with the official buildings dataset. An example of the analysis results is provided in Figure 5, where six different clusters (i.e. A, B, C, D, E and F), their barycenter, the existing buildings footprints from the official dataset, the main roads network, and the Instagram SMGI dataset are shown.

In this example, the manual investigation through the Google Maps satellite image enabled the detection of two buildings which were not mapped in official dataset, namely cluster B and D. At the same time, the visual assessment confirmed the building presence in cluster A, C, E and F.

This example demonstrates the potentialities of Instagram SMGI to elicit information related to geography of places, and also shows how this information may be potentially used as a support for the update and the integration of official datasets.

Conclusion

The results of the proposed study offer an overview of potential uses of SMGI for integrating and updating the available official information, as well as for obtaining information about the physical geography of places in the domain of spatial planning analysis. Currently, the wealth of information enclosed in SMGI may be used to investigate the concerns and the attentions of people toward places and also their behaviors and movements in space and time. These opportunities arise from the increasing availability of SMGI produced

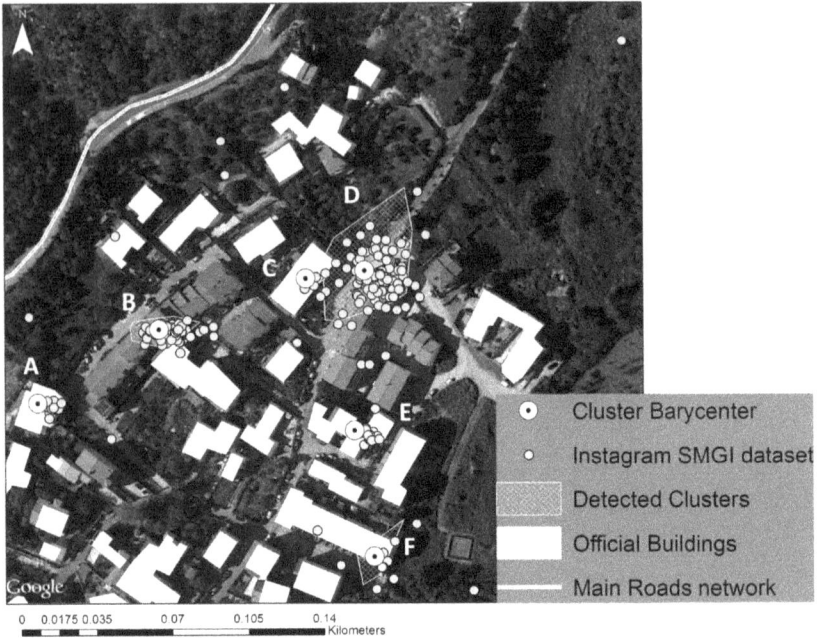

Figure 5: Results of clusters investigation in Iglesias municipality.

through several social networks, which may be considered as affordable and potentially boundless sources of near real-time information about any topic. Hence, the collection of SMGI and the integration with official dataset may represent a valid support for analysis, design and decision-making, offering a pluralist perspective from local communities to enhance methodologies and practices in urban and regional planning.

Nevertheless, despite the several opportunities for analysis, it is important to be aware that the SMGI datasets should be not considered representative of the whole local community. The social network services are used differently by diverse segments of the population, that are the users of the service itself, and the preferences and cultural biases of these groups highly affect the phenomena under observation in SMGI, raising issues about the data representativeness. Furthermore, as for the subject of the proposed case study, namely the study of geography of a place by Instagram, the social platforms used to collect SMGI suffer of a different degree of penetration worldwide according to users' pref- erence, limiting de facto the analysis opportunities only for areas where the services are available. In the future, a wider diffusion may occur to this respect as suggested by the current social network growth trends, but definitely for the time being both SMGI and A-GI show different diffusion rates in diverse regions and countries worldwide. Therefore, different analytical approaches based on several platforms may be required in order to investigate the local contexts appropriately. Much more research is needed to assess the full potential of SMGI and several issues should be addressed regarding data quality and representativeness; however, current results disclose challenging research opportunities, which may lead to advances in spatial planning methodologies and practices, as well as in other domains.

Acknowledgments

The work presented in this chapter was developed by the authors within the research project "Efficacia ed efficienza della governance paesaggistica e territoriale in Sardegna: il ruolo della VAS e delle IDT" [Efficacy and efficiency of landscape and environmental management in Sardinia: the role of SEA and of SDI] CUP: J81J11001420007 funded by the Autonomous Region of Sardinia under the Regional Law n° 7/2007 "Promozione della ricerca scientifica e dell'innovazione tecnologica in Sardegna".

References

Campagna, M. 2014. The geographic turn in Social Media: opportunities for spatial planning and Geodesign. In: Murgante, B et al *Computational Science and Its Applications – ICCSA 2014*. Springer International Publishing. pp. 598–610. DOI: http://dx.doi.org/10.1007/978-3-319-09129-7_43

Caverlee, J. 2010. A few thoughts on the computational perspective. In: *Presented during the Specialist Meeting on Spatio-Temporal Constraints on Social Networks*. Santa Barbara, CA, USA on December 2010.

Elwood, S., Goodchild, M. F., & Sui, D. Z. 2012. Researching Volunteered Geographic Information: Spatial Data, Geographic Research, and New Social Practice. *Annals of the Association of American Geographers, 102*(3): 571–590. DOI: http://dx.doi.org/10.1080/00045608.2011.595657

Ester, M., Kriegel, H. P., Sander, J., & Xu, X. 1996. A density-based algorithm for discovering clusters in large spatial databases with noise. In: Simoudis, E., et al (Eds.) *Proceedings of the Second International Conference on Knowledge Discovery and Data Mining (KDD-96)*. AAAI Press, pp. 226–231. ISBN 1-57735-004-9.

Frias-Martinez, V., Soto, V., Hohwald, H., & Frias-Martinez, E. 2012. Characterizing urban landscapes using geolocated Tweets. In: *Privacy, Security, Risk and Trust (PASSAT)*. 2012 International Conference on and 2012 International Confernece on Social Computing (SocialCom) IEEE, pp. 239–248.

GlobalWebIndex. 2014. GWI Infographic: Instagram Users. Available at: https://www.globalwebindex.net/blog/instagram-infographic (Last accessed 08 April 2015).

Goodchild, M. F. 2007 Citizens as Voluntary Sensors: Spatial Data Infrastructure in the World of Web 2.0. *International Journal of Spatial Data Infrastructures Research*, 2: 24–32.

Lazer, D., Pentland, A., Adamic, L., Aral, S., Barabasi, A. L., Brewer, D., & Van Alstyne, M. 2009. Life in the network: the coming age of computational social science. *Science*, *323*(5915): 721.

Poser, K., & Dransch, D. 2010. Volunteered Geographic Information for disaster management with application to rapid flood damage estimation. *Geomatica*, *64*(1): 89–98.

Silva, T. H., de Melo, P. O. V., Almeida, J. M., Salles, J., & Loureiro, A. A. 2013. A picture of Instagram is worth more than a thousand words: Workload characterization and application. In: *International Conference on Distributed Computing in Sensor Systems (DCOSS) 2013*. IEEE, pp. 123–132.

Stefanidis, A., Crooks, A., & Radzikowski, J. 2013. Harvesting ambient geospatial information from social media feeds. *GeoJournal*, *78*(2): 319–338. DOI: http://dx.doi.org/10.1007/s10708-011-9438-2

A Proposed Crowdsourcing Cadastral Model: Taking Advantage of Previous Experience and Innovative Techniques

Sofia Basiouka* and Chryssy Potsiou†

*National Technical University of Athens, s.basiouka@gmail.com
†National Technical University of Athens

Abstract

Are Land Owners able to participate in official Cadastral Surveys? Can the official procedures be modified in order for crowdsourcing techniques to be incorporated in them? How many different stakeholders can actively get involved and which is the optimum workflow that could be adopted? This chapter addresses the main question of whether the spatial and attribute data that is collected by volunteers – land owners – can be used in official Land Administration Systems (LAS) and explores the potential introduction of crowdsourcing techniques into the official cadastral surveys as a simplified and transparent procedure with the aid of citizens.

As the title reveals, this chapter aims to propose a crowdsourcing cadastral model as an alternative option to the official cadastral procedures. The research investigates a voluntary model and documents it in terms of participants and structure according to various international case studies that were analyzed in depth within the research that was carried out by Haklay et al. (2014). The main lessons that were collected by each separate crowdsourced case study were used as a guideline for the suggested cadastral model. The opportunities and the weaknesses are isolated and explored in terms of a cadastral survey. The main advantage of the adopted case studies is the opportunity for a simulation of

How to cite this book chapter:
Basiouka, S and Potsiou, C. 2016. A Proposed Crowdsourcing Cadastral Model: Taking Advantage of Previous Experience and Innovative Techniques. In: Capineri, C, Haklay, M, Huang, H, Antoniou, V, Kettunen, J, Ostermann, F and Purves, R. (eds.) *European Handbook of Crowdsourced Geographic Information*, Pp. 419–433. London: Ubiquity Press. DOI: http://dx.doi.org/10.5334/bax.ad. License: CC-BY 4.0.

real circumstances in cadastral surveys and not a simplistic logic. The main innovation is focused on the design of the model in an a priori approach with well-defined and already tested successful case studies.

The model is clarified in technical and societal aspects while it sheds light on the proposed process. The workflow, the stakeholders and the adopted main lessons of the crowdsourced case studies are the key components that are explored and analyzed in this chapter.

Keywords

Volunteered Geographic Information, Crowdsourcing, Cadastre, Land Administration & Management, Spatial Data Infrastructure.

Introduction

The great revolution that has taken place in Geographic Information Science (GIS) has led to dramatic changes in the source, use and manipulation of spatial information during the last decade. Seeger (2008) and others name it 'the digital spatial data which is collected and edited not by data producers but by citizens who are not experts but willing to disseminate their spatial knowledge and observations' without any special invitation. Blaut et al. (2003) had earlier noted the specific predisposition of all human beings to map by underlining that all people have natural mapping abilities. According to Kingsley (2007), *civil society* shares the same goals and has created a non-hierarchical network of self-organized individuals who participate in it. In fact he predicted the evolution of mapping, the involvement of amateurs with the aid of web tools and the alteration of role distribution between mapping agencies and users (Budhathoki et al. 2008). Three principle definitions were introduced to give a general title to the phenomenon; *Neogeography, Volunteered Geographic Information* and *Crowdsourcing*. Turner (2006) refers to Neogeography as a set of techniques and tools that fall outside the realm of traditional GIS. Goodchild (2007) coined the term Volunteered Geographic Information (VGI) which is one of the most widely deployed and disseminated terms and Sieber (2007) refers to Crowdsourcing in the way that it is currently used. All terms investigate GI and find innovative adjectives to shed light on different aspects of the phenomenon, which can be simply understood as voluntary manipulation of spatial data by citizens.

Within the new web-based era and the great possibilities offered to users via dynamic maps and new technologies to manipulate data, another field presents great interest due to its socio-economic perspectives. *Land Administration* is a relatively new term, which was introduced in the 1990s. Theoretically, Land Administration is defined as the process of determining, recording and disseminating information about the tenure, value and use of land when implementing land management policies (UNECE 1996). Many academics have investigated Land Administration and have contributed to various aspects of the field. Dale and McLaughlin (1999), Williamson (2001) and Bogaerts et al. (2002) are just a few of the academics contributing to this field.

Cadastre is a significant part of Land Administration and has various definitions according to different scientific sources. The United Nations (in 1985) and the International Federation of Surveyors (FIG) (in 1995) gave the predominant definitions. The former underlines that 'the cadastre is a methodically arranged public inventory of data on the properties within a certain country or district based on a survey of their boundaries; such properties are systematically identified by means of some separate designation. The outlines of the property and the parcel identifier are normally shown on large-scale maps' (UN, 1985 In: Steudler, 2004: 13). The latter notes, 'A cadastre is normally a parcel based, and up-to-date land information system containing a record of interests in land (e.g. rights, restrictions and responsibilities). It usually includes a geometric description of land parcels linked to other records describing the nature of the interests, the ownership or control of those interests, and often the value of the parcel and its improvements'" (FIG 1995 In: Steudler, 2004: 14). The need for an accurate and up-to-date cadastral system is so vital that Kaufmann and Steudler (1998), in their publication 'Cadastre 2014', support the inclusion of public and traditional law aspects in the definition given by FIG in 1995.

Cadastral survey according to Steudler (2004: 14) is 'simply defined as a survey of boundaries of land units'. Generally, cadastre is an essential tool for land management and administration as it records the land parcels which constitute a part of a country's spatial information infrastructure. The Bogor Declaration on Cadastral Reform (UN-FIG 1996) in other words declared that the development of modern cadastral infrastructures facilitate efficient land and property markets, protect the land rights of all, and support long term sustainable development and land management. It also facilitates the planning and development of national cadastral infrastructures so that they may fully service the escalating needs of greatly increased urban populations.

The main question that is posed in this chapter, taking into consideration all the above perspectives on technology and potential scenarios, is whether VGI and crowdsourcing techniques more generally can be incorporated in cadastral surveys and to what extent. The testing of the results is done through the Hellenic Cadastre project, which is a well-known, long-lasting project. The idea of incorporating VGI into the cadastral procedure is based on the power of locality and the participation of citizens in land planning as active parts of society.

The official cadastral procedure

The HC Project started in 1995 and cadastral surveys have been carried out in 340 regions all over the country while 106 cadastral offices have already begun operations in these regions. The responsible agency for the HC project is the National Cadastre and Mapping Agency S.A. (NCMA S.A.). The Hellenic Cadastre is a uniform, public, systematic and on-going title registration information system in fully digital form with spatial and attribute records of each land parcel. Before investigating on the alternative crowdsourced proposal, an overview of the official procedure is thoroughly presented.

The main cadastral survey includes the following stages (Basiouka & Potsiou 2012) (Figure 1):

- Declarations are submitted to the cadastral survey offices by the right holders and the registration of the declared rights is added to a digital database.
- Interim cadastral tables and diagrams are formed based on the data from the submitted declarations, which has been processed by lawyers and surveyors.
- Interim cadastral maps and data are published for a two-month period and extracts are sent to the rights holders for their information and acceptance.

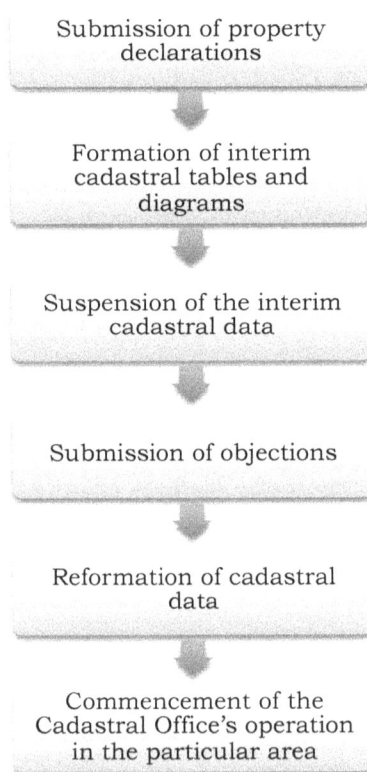

Figure 1: Main Cadastral Stages.

- Objections or applications for correction of a cadastral registration are submitted and forwarded to independent administrative committees, depending on the case, by whoever has a legal right.
- The cadastral data is reformed after examination of the objections and the correction claims and the final cadastral tables and diagrams are formed. These registrations are called Initial Registrations and they constitute the first registration in the Hellenic Cadastre.
- The cadastral office in operation in the particular area replaces the old Land Registry Office (NCMA 2011).

Given that the official process, according to Greek legislation, is based on a core scientific number of stages where surveyors, lawyers and IT technicians are the ones to collect, edit, store and handle data, the proposed idea has proven somewhat innovative. To date, citizens are excluded at all stages, except from the declaration of their private property at the beginning of the process.

The Proposed Crowdsourcing Cadastral Model

The proposed crowdsourced model for cadastral surveys is the result of the main lessons that have been collected from a series of successful governmental crowdsourced projects that were analyzed in depth (Haklay et al. 2014); the requirements posed by the nature of a cadastral project; the innovative ideas the researcher has decided to introduce; and the necessity for coordination with formal cadastral procedures and realities. As previous research carried out by Haklay et al. (2014) indicated, the key factors for a successful crowdsourcing experiment include building on previous experience, leveraging existing technology and having the support of key partners such as governments or other authorities. The proposed model aims to take advantage of previous experience of official procedures and to introduce innovative techniques that will facilitate them.

The structure of the proposed model is divided into four main sections: the adopted main lessons identified from the successful case studies, which will be incorporated in various parts of the model: the proposed workflow, which will replace the official procedure; and finally the participants and stakeholders of the project. The aim of the research is to launch a model that may be applied and used either at a local or a national level; and it may have a wider application in many countries and communities that face land issues.

The successful crowdsourced case studies

The research on the successful case studies was carried out by Haklay et al. (2014). An investigation was made of the incentives, scope and aims behind the practical experiments; their participants and stakeholders; their relationships and the modes of engagement. The research also shed light on technical aspects, successful factors and the problems that were encountered in the evaluation of the examples.

Each case study is intended to provide an example of the use of VGI by government or by the public, and summarizes the context, positive and negative outcomes and main lessons. The focus of 6 case studies is divided into three main categories: land management, public administration and disaster response. Although the case studies are differentiated by content and scope, all of them are governmental projects that incorporate voluntary and crowd-sourced data collection, so their study can isolate those components that are crucial for the success of the current project. The main lessons derived from the case studies and the outcomes are described briefly below, with text taken from Haklay et al (2014, CC-BY):

Mapping of South Sudan This was launched because of the need for a temporally accurate and up-to-date map when the new nation was created. Google Map Maker, the Sudanese diaspora and various organizations carried out workshops to train people to work separately on the digitization of aerial imagery. A significant amount of work was completed in a very short amount of time by adopting local knowledge and providing technical tools. Those who experienced the training sessions were inspired to transmit the experience and recruit new volunteers.

Main lessons:
- Crowdsourcing projects can be coordinated and implemented from a distance.
- Great participation of volunteers and transmission of motivation to others are key factors in terms of participation in crowdsourcing applications.

Informal settlement mapping, MapKibera, Nairobi, Kenya. Map Kibera was carried out in the most crowded slum in Nairobi, Kenya, in an effort to improve its reputation and offer an accurate picture of the area, which is quite dynamic due to the movement of the population. Local people collected and edited GPS tracks. Innovative techniques such as SMS, video and voice reporting were also launched and a small amount of compensation offered to participants.

Main lessons:
- Mapping can be achieved by young local people relatively quickly.
- Basic topographic maps can be enhanced with essential thematic layers.
- A combination of open source and conventional software can facilitate VGI projects.
- Compensation may be needed to improve participation in locations where participants suffer great survival issues.
- Innovative methods such as SMS, voice and video reporting can support the appeal of mapping projects.

Crowdsourcing The National Map, National Map Corps, US. National Map Corps has given volunteers the choice to collect and edit data about ten different human-made structures in fifty states in an effort to provide accurate and authoritative spatial data. The methodology includes various steps such as adding new features, removing obsolete points and correcting existing data. A pilot test in Colorado showed that the VGI was satisfactory in its accuracy.

Main lessons:
- Adoption of challenging techniques such as gamification has been successful and attracts volunteer interest.
- Evaluation of the quality indicated that the participation improves accuracy and reduces the errors.
- Organizational resistance to accepting data from volunteers is one of the major challenges for VGI projects of this kind.
- Key factors to successful crowdsourcing include building on past experience, leveraging existing technology and having the support of key individuals within the organization.

iCitizen, mapping service delivery, South Africa. This project is at the design phase and aims to involve the public at a local level to collect data points via mobile phones and adopt different ways for geotagging of photos in real time or via SMS and email. The purpose is to report infrastructure issues.

Main lessons:
- Projects can be used for a variety of tasks at local level, not just that for which they were designed.
- Using a range of software, programming languages and platforms can broaden a project's horizons.
- VGI applications face financial issues due to their nature and the resources of the organizations involved.
- Concerns from agencies about the quality of generated data sets and improving public adoption of mobile applications are common challenges.

Haiti disaster response. One of the most well-known crowdsourcing applications, developed after the earthquake hit Haiti in 2010. Within 48 hours, the capital was mapped by volunteers who contributed from every part of the world to create an up-to-date topographic map to fill the gap left by the official mapping agency. The maps were used by various organizations to allocate supplies and medicine.

Main lessons:
- An integrated methodology of this kind follows four steps: spatial data contributed by official providers, supplemented with GPS tracks, integrated

into OpenStreetMap and updated by a great number of volunteers from each part of the world.
- Time, cost and official trust of data by Non-Governmental Organizations (NGOs) and other official partners are key to success.
- Lack of coordination and experience between different actors can lead to duplication of data and waste of resource.
- Differentiation between conventional and governmental data in terms of engagement to the project did not prevent success.

Community Mapping for Exposure in Indonesia. The goal of this project is to reduce vulnerability to natural disasters. Young people have successfully collected spatial and attribute data, and traced them in the OSM platform so that thematic maps can be created to show potential damage in case of physical disasters.

Main lessons:
- Interaction between official providers and VGI is a parameter of success not only for the beginning of the project but also for its continuity.
- Open source data can be reliable for scenario building but its quality can vary, especially in terms of attribute data.
- The coordination of participating organizations and volunteers is important to take full advantage of human resources and technical innovations.

The main lessons adopted

Within the specific alternative proposed model for cadastral surveys, a series of main lessons, technical aspects and successful factors are adopted as vital for the viability of the model. The majority of them serves the nature of the research and plays a crucial role in its design. The predominant key factors are given below and are explained in terms of the proposed cadastral model.

- *Training and workshops* should take place at the beginning of each application. Most of the practical applications that constitute successful crowdsourced paradigms – such as Community Mapping for Exposure in Indonesia, MapKibera in Kenya and mapping of South Sudan – introduced workshops as a key factor for their success. This is of vital importance due to the technical requirements of a cadastral survey and the quality controls that should be satisfied at the end of any project. The workshops have three different targets: (a) to inform locals about the necessity and benefits of the project, (b) to recruit them and (c) to train them to properly manipulate spatial and attribute data.
- *Recruitment of volunteers.* Except for local people – land owners who will play a vital role in data collection – the proposed model is based

on non-governmental organizations and undergraduate students of the schools of Surveying, Geography or any other relevant field. Both will keep the cost low and the students will support the cadastral surveys both practically and technically. The undergraduate students will participate actively in the data collection and editing. The idea for this came accidentally due to the nature of the study, however, Community Mapping for Exposure in Indonesia had already recruited undergraduate students to collect a huge amount of data in a relatively short time and scholarships were offered in exchange. The idea behind the recruitment of non-governmental organizations is to carry out all workshops at the beginning of the process. Students should also deal with the great amount of data that should be collected and manipulated. The main question over the participation of land owners as volunteers is in terms of their motivations. Land owners may be more easily recruited if cadastral fees were eliminated, or similar taxation rates lowered, as a result of their participation. Experience from other countries has indicated that the incentives that lead volunteers to participate in crowdsourcing projects are a mixture of various parameters. The land owners will participate voluntarily for altruistic reasons if they recognize the necessity of the project, how their lives will be affected by its implementation and how their properties will be protected. This is especially the case for those who face difficulties in land transactions, use and development. However, the majority will be motivated if a compensation rate is offered to them.

- *Partnership with scientific organizations.* Collaboration between the various organizations is proposed within a well-defined and compact pattern where the roles and duties are made clear from the beginning of the procedure. To avoid the lack of coordination and duplication of data noted in Haiti disaster response, each stakeholder should be responsible for a specific part of the cadastral survey and all the participants should work in cooperation in order to produce the final result. Supervision and quality controls should also be carried out by experts. The main innovation of the project, which has a national application, is based on the participation of various NGOs at local level. The NGOs that take action at local level should be responsible and should participate actively in the cadastral surveys in these areas.

- *Crowdsourcing projects can be coordinated and implemented from a distance.* The importance of this keynote is crucial for the success of the first phase of the cadastral survey where citizens should digitally declare their ownership from a distance instead of hard copy declarations and indicating their ownership on digital orthophotos at the cadastral offices. The phase may be carried out in digital form and implemented from a distance. The attribute data declaration may be replaced by an online declaration with the aid of Apps or online tools and the field work for spatial data collection in urban areas can be replaced by the online digitization of land parcels on orthophotos provided by the website of the official mapping agency or on OSM. This specific strategy worked efficiently in the mapping of South Sudan.

- *Open – source and commercial software* should be used because experience has indicated that their combination offers the required freedom and openness for the project. The informal settlement mapping in Nairobi, Kenya, flourished with the aid of open-source and conventional software. Open-source software may be used in data collection while conventional software may be used in data manipulation. The contribution of volunteers in data collection requires the use of flexible and easy-to-use tools. The editing of data demands advanced functionalities that may be available only in conventional GIS software packages. Previous experience has indicated that using a range of software, programming languages and platforms may broaden a project's horizons.

- *Innovative techniques.* In an effort to meet the requirements of the project, especially in terms of quality, a series of innovative techniques have been adopted in data collection and manipulation. For the first time, OSM is tested for use in data manipulation and storage in cadastral surveys. Also, smartphones and handheld GPS gradually replace the expensive equipment that is needed for data collection. All these innovations were introduced to propose an alternative, viable solution to the official procedure.

The Workflow

The workflow of the proposed model follows a general pattern that includes all different occasions of mapping in rural or urban areas, and it constitutes of four main steps. The proposed model may be applied at the beginning of a cadastral survey and it may replace the first phase of official cadastral mapping. The hard copy declaration of ownership, the identification on orthophotos of properties in the rural areas and the acceptance of the produced result are only a few stages of the official process that may be modified. Although the procedure may be categorized in these four stages, it may include further parameters that are identified within practical experiments (Figure 2).

Training of volunteers is the first step of the process. The NGOs and the undergraduate students will train the land owners both theoretically and practically. The award for the participation of students will be in the form of scholarships and work experience.

Data collection as the second step of the workflow differs depending on the nature of the area and the kind of data that should be collected. Rural areas require a different approach in comparison to urban ones while spatial data collection requires different tools to attribute data. Data in rural areas may be collected by using handheld GPS devises, tablets or smartphones while in

Training

Spatial &
Attribute Data
collection

Data Evaluation

Maintenance &
Storage of Data

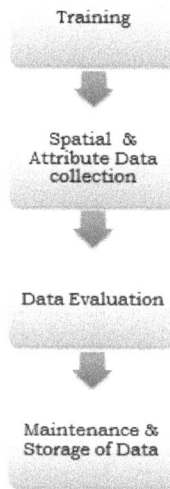

Figure 2: The workflow of the proposed Cadastral Model.

urban areas by using accurate orthophotos provided via the website of the official mapping agency or various online maps such as OSM in case of a lack of other accurate basemaps.

In rural areas, the citizens may either declare their ownership by giving the point of its centroid (Figure 3, right) in a quick and inexpensive approach or collect their parcel boundaries (Figure 3, left) with expert support after been trained. Both approaches fit in interim cadastral mapping. The methodology may vary depending on parcels' shape.

In urban areas, citizens can simply declare their ownerships by using online tools and orthophotos as basemaps which are provided via the website of the official mapping agency or online dynamic maps such as OSM. The choice depends on the availability of the recourses. The experience has indicated that OSM can be adopted for cadastral purposes, taking into consideration specific generalizations and rules (Basiouka et al. 2015).

The attribute data collection can be also implemented with the aid of online databases, which can store and maintain the saved information so that the hard copy declaration to be replaced.

Data evaluation, as the third step of the workflow, should be guaranteed by the national mapping agency, which will also be responsible for the maintenance and storage of the data in its servers. Thus, one of the most important concerns in terms of viability – the difficulty of keeping data up to date – can be bypassed.

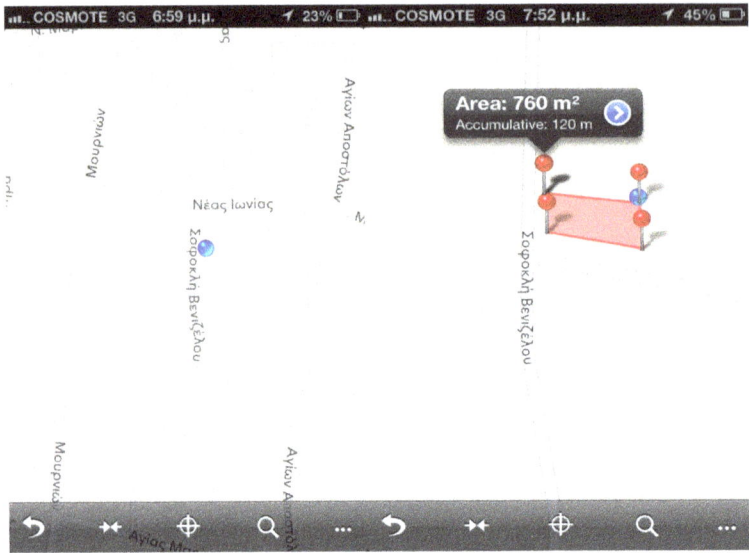

Figure 3: Different approaches in spatial data collection.

The Stakeholders of the Model

The model is based on the participation of citizens in a hybrid approach where citizens and property owners participate as volunteers and experts work as team leaders supervising the whole process. Thus, the hybrid approach involves both amateurs and experts for its implementation. In order to be successful, it requires the participation of locals, experts and NGOs, while the coordination of these parties should be carried out by the official mapping agency. Specific parts of the public sector should also participate in certain parts of the procedure (Figure 4).

The term 'locals' refers to the land owners and volunteers who will participate in the cadastral surveys. Their participation should be based on data collection and editing during the first steps of the process and the acceptance of the result at the end of the cadastral survey. Citizens may collect not only spatial but also attribute data. The locals constitute the base of the pyramid that is given in Figure 4.

The next level of the pyramid refers to the NGOs who will train volunteers in data collection and editing. Their coordination highlights another important parameter: the cooperation of participating organizations and volunteers is crucial for the full advantage of human resources and technical innovation. Their interaction is also a parameter of success required not only at the beginning of the project but throughout.

Furthermore, the public sector should act to supplement these actors and provide the required equipment such as GPS and total stations, computers,

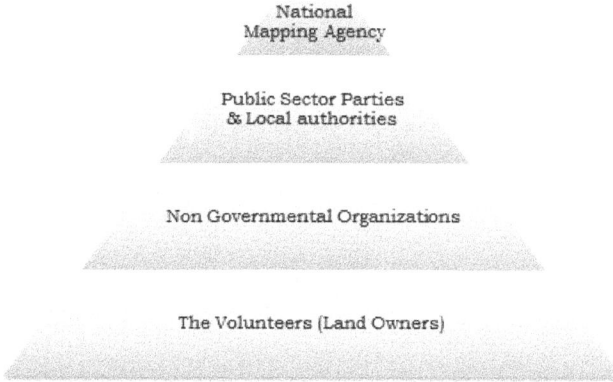

Figure 4: The stakeholders of the Project.

vehicles and facilities. The local authorities should also play a critical role in data collection and they will contribute effectively to the collaboration among teams of local people at the stage of data collection, depending on their knowledge. This ensures that participants can contribute to specific tasks and stages of the data collection.

The national mapping agency should keep a supervising role and should be actively involved in all stages of the process. Coordination of the procedure, evaluation of the result and maintenance of the data in their server are their predominant tasks.

Concluding Remarks

The model proposed here follows the trend that has been adopted during the last decade by the governmental bodies to open, inexpensive, quick and transparent processes in Land Administration Systems with the participation of citizens and the use of new technologies. The proposed model constitutes a general process that may be applied or modified according to the special needs of the official mapping agency and can be adopted in various land administration issues in various countries. Its main innovation is based on the combination of current trends and previous experience in similar case studies. Its main advantage is stated not only through the participation of citizens in important decision making policies but also through its flexibility. There are still many concerns and parameters that have not yet been identified, quality and privacy queries that a conservative part of the society poses, or problems of credibility that make governmental bodies reluctant to use it. However, it is clear that the new crowdsourcing trend is not temporary and will continue to play a fundamental role in this recently begun revolution. The general outcome is condensed not in a unique solution but in a general idea that should be flexible and can be easily modified based on the needs of society and the available tools and funds.

Acknowledgments

Special thanks to Professor Muki Haklay, coordinator of the World Bank project titled "Crowdsourced geographic information use in government". The findings of the previous research inspired this publication and evolved the findings in a cadastral project.

Sofia's Basiouka PhD studies are funded by the Greek State Scholarships Foundation. This support is greatly acknowledged.

References

Basiouka, S., & Potsiou, C. 2012. VGI in Cadastre: a Greek experiment to investigate the potential of crowd sourcing techniques in Cadastral Mapping. *Survey Review*, *44*(325): 153–161(9).

Basiouka, S., Potsiou, C., & Bakogiannis, E. 2015. OpenStreetMap for cadastral purposes: an application using VGI for official processes in urban areas. *Survey Review*, DOI: http://dx.doi.org/1752270615Y-0000000011

Blaut, J. M., Stea, D., Spencer, C., & Blades, M. 2003. Mapping as a cultural and cognitive universal. *Annals of the Association of American Geographers*, *93*(1): 165–185.

Bogaerts, T., Williamson, I. P., & Fendel, E. M. 2002. The Role of Land Administration in the Accession of Central European Countries to the European Union. *Journal of Land Use Policy*, *19*(1): 29–46.

Budhathoki, N., Bruce, B., & Nedovic-Budic, Z. 2008. Reconceptualising the role of the user of spatial data infrastructure. *GeoJournal*, *72*(3): 149–160.

Dale, P., & McLaughlin, J. D. 1999. *Land Administration Systems* Oxford University Press.

FIG. 1995. *Statement on the Cadastre*. Report by Commission 7 (Cadastre and Land Management), No. 11, FIG Publication.

Goodchild, M. F. 2007. Citizens as sensors: the world of volunteered geography. *GeoJournal*, *69*(4): 211–221.

Haklay, M., Antoniou, V., Basiouka, S., Soden, R., & Mooney, P. 2014. *Crowdsourced geographic information use in government*. Report to GFDRR (World Bank). London, UK.

Kaufmann, J., & Steudler, D. 1998. *Cadastre 2014 – A Vision for a Future Cadastral System*. Report by Workgroup 1 of FIG Commission 7, Rüdlingen and Bern, Switzerland, FIG Publication.

Kingsley, D. 2007. Technologies of civil society; communication, participation and mobilization. *Innovation, 20*(1): 19–33.

National Mapping Agency. 2011. Cadastral Survey Available at: http://www.ktimatologio.gr/ktima/EN/index.php?ID=MpuNFMi98NlPs1QZ_EN [accessed 15 August 2011].

Seeger, C. 2008. The role of facilitated volunteered geographic information in the landscape planning and site design process". *GeoJournal, 72*(3): 199–213.

Sieber, R. 2007. Geoweb for social change. Available at: http://www.ncgia.ucsb.edu/projects/vgi/supp.html [Accessed 12 September 2012].

Steudler, D. 2004. A framework for the evaluation of land administration systems. Unpublished thesis (PhD), The Department of Geomatics, The University of Melbourne.

UNECE. 1996. *Land Administration Guidelines.* ECE/HBP/96 Sales No.E.96. II.E.7, New York and Geneva, Switzerland, United Nations publications.

UN-FIG. 1996. *Bogor Declaration on Cadastral Reform.* UN-FIG. United Nations Interregional Meeting of Experts on the Cadastre. Bogor, Indonesia, FIG publication No. 13A, In: Steudler, D. (Ed.) *2004 A framework for the evaluation of land administration systems.* Unpublished thesis (PhD), The Department of Geomatics, The University of Melbourne.

United Nations. 1985. *Conventional and Digital Cadastral Mapping.* Report of the Meeting of the Ad Hoc Group of experts on Cadastral Surveying and Land Information Systems. Economic and Social Council E/CONF (Vol. 77). In: Steudler, D. (Ed.) *2014 A framework for the evaluation of land administration systems.* Unpublished thesis (PhD), The Department of Geomatics, The University of Melbourne.

Williamson, I. P. 2001. Land administration "best practice" providing the infrastructure for land policy implementation. *Journal for Land Use Policy, 18*(4): 297–307.

CHAPTER 31

Modelling the world in 3D from VGI/ Crowdsourced data

Hongchao Fan* and Alexander Zipf

Chair of GIScience, Heidelberg University,
Berlinerstr. 48, 69120 Heidelberg, Germany
*hongchao.fan@geog.uni-heidelberg.de

Abstract

Within the last ten years, Volunteered Geographic Information (VGI) has developed rapidly and significantly influenced the world of GIScience. Most prominently, the OpenStreetMap (OSM) project maps our world in a detail never seen before in user-generated maps on the one hand. On the other hand, most of the urban area on our planet has been covered by hundreds of millions of photos uploaded on social media platforms, such as Flickr, Instagram, Mapillary, Weibo and others. These data are directly or indirectly geo-referenced and can be used to extract 3D information and model our world in the 3D environment. At the current stage, several approaches have been made available to visualise and generate the 3D world mainly using OpenStreetMap data. The 3D buildings are unfortunately restricted to a coarse level of detail, since no further information about facade structure is available on OSM. In this paper, the current work of reconstruction of 3D buildings by using the data both on OSM and Flickr is presented, whereby facade structure could be extracted from Flickr images and OSM footprints can be used to accelerate the process of dense image matching and to improve the accuracy of geo-referencing.

How to cite this book chapter:
Fan, H and Zipf, A. 2016. Modelling the world in 3D from VGI/Crowdsourced data.
In: Capineri, C, Haklay, M, Huang, H, Antoniou, V, Kettunen, J, Ostermann, F and
Purves, R. (eds.) *European Handbook of Crowdsourced Geographic Information*,
Pp. 435–446. London: Ubiquity Press. DOI: http://dx.doi.org/10.5334/bax.ae.
License: CC-BY 4.0.

Keywords

VGI, 3D city model, crowd sourcing, OpenStreetMap

Introduction

In recent years, the term Volunteered Geographic Information (VGI) became popular, whereat VGI describes how an ever-expanding range of users collaboratively collects geographic data (Goodchild 2007a). That is, hobbyists create geographic data based on personal measurements (via GPS etc.) and share those in a Web 2.0 community, resulting in a comprehensive data source with humans acting as remote sensors (Goodchild 2007b). With a global cast of volunteers, OpenStreetMap (OSM) is considered as one of the most successful and popular Volunteered Geographic Information (VGI) projects. In its current state, there are more than two million registered members (OSM 2015) who contribute to the rapid growth of OSM. Recent investigations on its completeness and quality have shown that urban areas in Central Europe in particular have already been mapped with an impressive level of detail (Neis et al. 2012). In those areas, OSM is well ahead of only mapping the street network. For the continuous improvement of OSM it is crucial to enable the mapping of even more detailed, three-dimensional spatial information.

Several years ago, Over et al. (2010) investigated the possibility of creating a 3D virtual world by using OSM data for different applications, and drew the conclusion that OSM has huge potential for fulfilling the requirements of CityGML LOD1 (Gröger et al. 2008), which is modelled as block models regionally. With the rapid development of OSM in recent years, especially, sparked by the availability of high-resolution imagery from Bing since 2010, there has been an increase in building information in OSM, proving that volunteers do not only contribute roads or points of interest (POIs) to the database. According to the latest statistics (the values are derived from our internal OSM database, which is updated daily), the number of buildings in OSM is above 200 million, thereof 18.4 million building footprints in Germany. The research of Fan et al. (2014) demonstrated that the building footprints data on OSM has a high degree completeness and semantic accuracy. There is an offset of about four meters on average in terms of position accuracy. With respect to shape, OSM building footprints have high similarity to those in authority data. Moreover, there is more and more information about building height and roof structures, which is required for the 3D reconstruction. From this point of view, one can say that it is possible to model the virtual world in 3D from OSM data. The 3D data could be further enriched when introducing related information from other VGI projects, such as Flickr, WikiMapia, Panoramio, Instagram and Dronestagram.

This paper provides a detailed perspective on the generation of 3D city models by using VGI data that is mainly based on OSM data. First, it gives an overview of the data sources that could be used for the 3D model generation, then the mechanism to generate 3D city models will be described. Many of the proposed algorithms have been implemented within the OSM-3D and Open-BuildingModels projects, which will also be introduced. Finally, this paper will give conclusions, a discussion and suggestions for future work.

VGI as a data source for 3D reconstruction

The earliest approach for sharing 3D models using the principle of 'everyone for everyone' is Google 3D Warehouse launched on April 24, 2006. This shared repository contains user-generated 3D models of both geo-referenced real-world objects, such as churches or stadiums, and non-geo-referenced prototypical objects, such as trees, light posts or interior objects like furniture. The former also appear in Google Earth. In order to voluntarily contribute, users have to have a certain level of 3D modelling skill. The main focus of this repository does not lie in assembling 3D city models as the non-geo-referenced objects seem to be more important in related work. They are, for example, used to improve methods of automatic object recognition in the field of laser scan classification or robotic vision. In addition, the 3D warehouse models are being integrated in several commercial systems, such as design tools or simulation software. However, the number of 3D buildings and many 3D city facilities, such as bridges, bus stations and fuel stations, has increased in recent years, thanks to the development of 3D modelling computer programs such as SketchUp and ESRI CityEngine, which make 3D editing more easy and effective.

In 2007, VGI was introduced by Goodchild (2007a,b) to describe the recent revolution of collaboratively created spatial information on Web 2.0. Almost at the same time, Microsoft Virtual Earth and Google Earth launched their pioneer projects in the way of VGI or crowdsourcing. The projects are called 3DVIA (Virtual Earth) and Building Maker (Google Earth). Both of them provide a model kit to create buildings, deriving the 3D geometry from a set of oblique (and proprietary) birds-eye images of the same object from different perspectives. In contrast to the 3D Warehouse, this tool specifically aims at geo-referenced 3D building models only. It is intended for people who do not have knowledge in 3D modelling, but still want to contribute.

In addition to 3D Warehouse, there are several free-to-use 3D object repositories on the internet, for example OpenSceneryX6, Archive3D7 or Shapeways8. These projects emerged from entirely different communities with interest in, for example, flight simulators or 3D printing. The contents usually lack connection to the real world but can nonetheless be useful to enrich real 3D city model visualisations.

The above mentioned projects serve to directly share and collect 3D models by means of crowdsourcing. The data used for 3D reconstruction might be commercial or authority data. In fact, 3D city models can also be reconstructed using 2D vector or image data contributed by crowdsourcing. The OSM community has not only captured roads and paths, but also more and more POIs, land use areas and even buildings. The latter can be extracted and extruded into 3D. At present, there are several projects that generate and visualise 3D buildings from OSM: OSM-3D, OSM Buildings, Glosm, OSM2World, etc. The major limitation of these projects is that the majority of buildings are only modelled at coarse level of detail. When applying the concept of levels level of details (LoDs) introduced in CityGML (Gröger et al. 2008), these buildings are actually LoD1, i.e. they are reconstructed by extruding footprints with flat roofs. In OSM-3D, a number of buildings are modelled in LOD2 in cases where there are indications for their roof types. Further, it is possible to integrate more detail, however, usually manually generated buildings (LOD3 or LOD4) from other sources via OpenBuildingModels (Uden & Zipf 2012).

Flickr is another VGI project that is often used for the reconstruction of 3D buildings. Preliminary experiments on reconstructing 3D scenes from Flickr imagery have been made available by Snavely et al. (2006; 2008) and Agarwal et al. (2011). However, Flickr imagery is almost untapped and unexploited by computer vision researchers, in particular when it comes to deriving representations suitable for GIS. A major reason is that the imagery is not in a form that is amenable to processing (Snavely et al. 2008). The photos are unstructured – they are taken in no particular order, and have uncontrolled distribution of the camera viewpoints und unclear positional accuracy. In addition, they are uncalibrated (Argarwal et al. 2011), and with widely variable illumination, resolution, and image quality. All this increases the difficulty in image registration and sparse 3D reconstruction. Furthermore, the existing approaches are developed based on dense image matching which leads to high computation costs.

Generating 3D city models from OSM data

A city normally consists of street network, land uses, buildings, point features and others. These would be handled separately for 3D modelling and visualisation. It should be pointed out that a digital terrain model (DTM) is required from other sources (i.e. open data) for the 3D visualisation, because OSM does not contain any information about terrain.

Integrating OSM land uses within 3D Terrain Surface

In fact, it is hard to integrate OSM data within 3D terrain surface because OSM data is recorded in 2D. The problem can be solved by overlapping OSM data as a liquid net over the terrain surface as a solid object and preserving the characteristics of OSM features (i.e. a football ground should be flat) during the process. In principle, there are three alternative ways to display OSM land-use data in 3D. This data can be displayed by mapping raster images onto a digital elevation model (DEM), by overlaying vector data on the DEM or by combining the vector data and the DEM in an integrated triangulated irregular network (TIN). Schilling et al. (2007) proposed an approach to integrate the road surface into the triangulation of the DTM, which is represented by a set of Triangulated Irregular Networks (TINs). This means that the road surface becomes a part of the TIN. The street network is treated as a layer consisting of a collection of polygons representing all the individual network segments. The borders of the polygons are integrated into the TIN as fully topological edges so that we can distinguish between triangles that are part of the street surface and the remaining triangles.

The resulting triangles within the polygon receive the attributes of the source features and can be coloured for visualisation. Another advantage of this approach is that all layers can be styled by the user on demand via a 3D styled layer descriptor (3D-SLD), which is an enhancement of the OGC SLD standard (Neubauer & Zipf 2007).

After integrating the street network with the surface layer, the street surfaces within the DTM have to be smoothed and corrected, because linear features like ditches, smaller dikes, walls, the rims of terraces and especially the hard border edges of roads can be only represented insufficiently due to the low resolution of the DTM. An example can be seen in Figure 1. Sometimes the road sidelines seem to be frayed. At steep hillsides, the road surface is inclined sideways. The situation is of course even worse with lower-resolution DTM data sources.

Another common way to support linear features is to include break lines during the terrain triangulation, e.g. using the Constrained Delaunay Triangulation

Figure 1: Comparison between the original terrain surface (left) and with the flattened road segment (right).

(CDT). However, break lines are seldom available. However, one can correct the parts of surfaces representing areas that should be actually more or less flat. A comparison between the situation before the correction and afterwards is shown in Figure 1. It is much more likely that the middle line takes a smooth course between the river and the hillside with approximately the same height, and that the profile is nearly horizontal, as can be seen on the right side, instead of being very bumpy and uneven.

3D building objects

In OSM, building footprints are modelled as closed polygons. For creating 3D models, the height must be derived from other OSM attributes (called tags). The key height, as well as the key building:height, ought to contain information about the height of a building. If such information is not available, as an alternative, the keys levels, building:levels and building:levels:aboveground can be utilised for an approximation of the building height (by multiplying the number of levels with an average level height of 3.5 meters). The key building:min_levels also needs to be considered because it describes the individual elevation of a building, thus the space between the ground and the building (part).

When computing building geometries, it is also interesting to generate proper roof geometries. The keys building:roof:shape, building:roof:style and building:roof:type contain a semantic description of the roof shape, such as gabled roof or hipped roof. In contrast, the key building:roof is supposed to contain information about the material of the roof, although it often also contains roof shape information. Similar to this key, building:roof:material can contain information about the roof material. Besides those keys, there are also some other relevant keys for roof generation. Building:roof:extent describes the extent of the roof, thus the actual distance between the roof edge and the building facade. For describing the orientation of the roof, the key building:roof:orientation is applied: if the roof ridge is parallel to the longer roof side, the value is along; otherwise the value is across. The generation of roof geometries for simple building footprints, that is footprints with rectangular shape or those that only consist of four points, is straightforward and can be applied with adequate performance to the OSM dataset. For more complex roofs, such as those with holes or arbitrary shapes, the generation of roof geometries is quite challenging. Some early results have already been gained by using procedural extrusion with Skeleton computations, but until now a broad application of those algorithms for the whole OSM on the one hand is very time consuming (about factor 100) and on the other hand does not, due to special cases and exceptions, lead to satisfying results. A detailed description of the building generation process can be found in Götz and Zipf (2011).

The generation of 3D buildings from OSM data is achieved by using the algorithm *create3dBuildingModel*, as below:

Algorithm create3dBuildingModel(G, A)
Input: G = 2D Geometry (Polygon) from OSM
Input: A = Attributes as OSM key-value pairs
 1: 3dm[] <-- empty
 2: height <-- getHeight(A[height], A[building:height], A[levels],
    A[building:levels], A[building:levels:aboveground])
 3: roofShape <-- getRoofShape(A[building:roof:shape],
    A[building:roof:style], A[building:roof:type])
 4: roofAttr <-- getRoofType(A[building:roof:extent],
    A[building:roof:orientation], A[building:roof:angle],
    A[building:roof:height])
 5: roofColor <-- getRoofType(A[building:roof:colour],
    A[building:roof:color])
 6: color <-- getRoofType(A[building:colour], A[building:color],
    A[building:façade:colour], A[building: façade:color])
 7: body <-- computeBuildingBody(G, height, color)
 8: roof <-- computeRoof(G, roofShape, roofAttr, roofColor)
 9: building <-- combine(body, roof)
10: triangulate(building)
Output: 3D Building

3D indoor modelling

The 3D indoor environment of buildings could be generated by using the indoor information mapped on OSM using IndoorOSM. It is an OSM-based indoor extension proposed by Götz and Zipf (2011). The schema follows existing OSM methodologies; thus, it only uses nodes, ways, relations and key-value pairs. That is, existing OSM editors, such as JOSM7 or Potlatch, 8, are suitable for mapping IndoorOSM data. The schema is defined as follows: a whole building is represented as one OSM relation, whereas the different relation members (the children of the relation) are the different building levels (floors). A level itself consists of one or several closed way(s) for representing the shell of the level, that is the outer boundary, and several other closed ways representing the inner building parts (e.g. rooms, corridors, etc.).

3D information such as the height of a level or the height of a building part is attached as a key-value pair to the corresponding OSM feature with the key height and corresponding values (e.g. 3, 6 ft, etc., the default unit is meter). That is, for each level and its inner parts, a two-dimensional (2D) footprint geometry plus additional 3D information is available. Further semantic information, such as room names, level names, level numbers and so on are attached as key–value pairs to the corresponding OSM feature.

In IndoorOSM, information about windows is provided by adding nodes to the OSM features that represent the level shells. Thereby, the location of the node represents the 2D centre of the window (from a bird's perspective). Information about the breadth, width and height is attached via corresponding keys.

Point features

In OSM, point features have been captured for an abundance of different locations, shops, restaurants, facilities, technical installations and so on. They provide in part very deep information, which enables applications that go far beyond the static display of map content. For some categories, a tagging schema has been established for storing typically useful information about a specific type of facility. The schema for restaurants, for instance, includes name, address, opening hours, cuisine, telephone number and URL of the homepage.

The primary OSM key for this kind of node is 'amenity'. The value describes the type, which can be used to assign an icon or symbol. The generic 'name' key may be used for an additional label. The amenity types have been divided into the categories: accommodation, eating, education, enjoyment, health, money, post, public facilities, public transport, shop and traffic. Each category is provided as an individual layer through the Web3D Service.

For the 3D environment, these point features should be classified into two classes: points as additional attributes of buildings, and points as location indication of city facilities. The first class of points can be integrated to their corresponding buildings by using text-matching algorithms. The second class (Figure 2a.) stands for the objects of city facilities, such as bus station, traffic signal, post box, tree and streetlights. These objects of city facilities can be modelled in 3D by using generic 3D objects (Figure 2b.), because they have unified shapes and sizes in the city (Gröger et al. 2008).

City furniture   Vegetation objects   Signs & Signals for
                                                 Traffic and Tram

Generic objects   3D public
                                                       phone booth

Figure 2: Generic objects in a city and an example of 3D representation of a public phone cell.

The OSM-in-3D projects

The most advanced work in the context of creating 3D city models from VGI data is the OSM-3D project developed at Heidelberg University. It combines the extrusion of building footprints into the third dimension with a detailed integrated terrain model derived from SRTM height data. It provides the 3D data in a standardised manner through a Web 3D Service (W3DS). The OSM-3D W3DS supports different terrain generalisation levels and provides tiled 3D scenes, based on the requested point of view, in VRML, X3D, COLLADA or KML format. Tailored client software called XNavigator has also been developed, which automatically requests the data from the W3DS server and assembles complex 3D landscapes worldwide. This client also allows the integration of other OGC Web Services, such as a Web Feature Service (WFS), the OpenGIS Location Services or the Sensor Observation Service. Thus, for example, POIs or 3D routes can be included. The interoperability with different data sources (e.g. also CityGML), web services and targets has recently been examined within the OGC 3D Portrayal Interoperability Experiment. Figure 3 shows the user interface of XNavigator with 3D city models in Heidelberg, Germany. The wide applicability of the W3DS and XNavigator with heterogeneous data could be demonstrated.

3D buildings in OSM-3D are generated using the automated process described in Section 3.2. The drawback of this kind of automated approach is that the buildings can only be generated with coarse geometries. In other

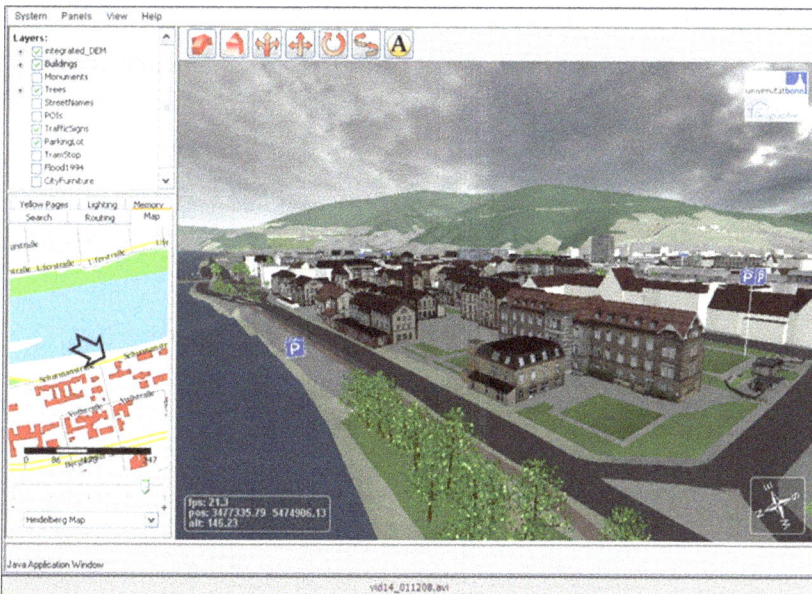

Figure 3: OSM-3D overview of Heidelberg in XNavigator.

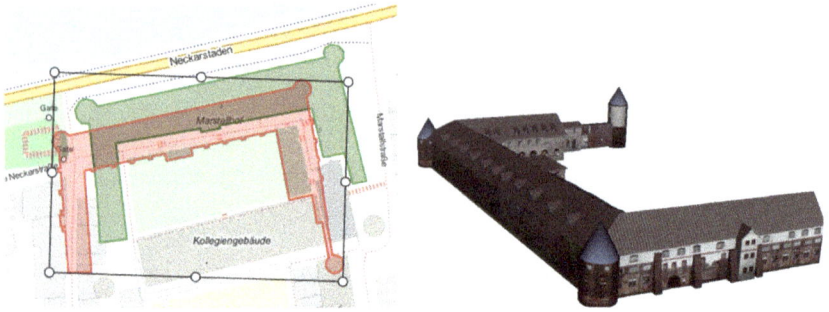

Figure 4: OpenBuildingModel.

words, buildings can only be modelled as block models (LoD1) or models with roof structures (LoD2). Architectural details on facades unfortunately cannot be modelled. Aiming to acquire 3D buildings with detailed geometries (LoD3 and LoD4), the OpenBuildingModel project was launched in 2012. It is a web-based platform for uploading and sharing entire 3D building models. In line with this project, a user-friendly web interface (see Figure 4.) has been developed, which allows: (i) uploading 3D building models (modelled by internet users) associated with a footprint in OSM (Figure 4a), and (ii) browsing, viewing and downloading existing models in the repository (Figure 4b). The processing of the OSM data and setting up of a model repository in the first prototype comprises several steps. First, building footprints have to be derived from the OSM data separately and overlaid as a vector layer by importing the OSM data into a PostgreSQL/PostGIS database with the Osmosis tool. Then building footprints can be operated interactively and individually on the web-client, whereby GeoServer is deployed for the data provision. By selecting a building footprint, one can upload his/her own 3D building (with textures) created offline. At the same time, it is also possible to add attributive information.

Future work

Collaborative mapping in 3D is more difficult than in 2D, because basic knowledge and skills about 3D modelling are essential when creating 3D buildings manually or in a semi-automated way. From this point of view, one cannot expect much contribution of 3D building models through data-sharing platforms such as 3D Warehouse and OpenBuildingModel. In order to have 3D buildings with detailed façade structures at regional, country and even global scales, an alternative solution has to be provided.

One possible solution might be the combination of the two VGI projects: OSM and Flickr. Building footprints, height information and further semantic tags given in OSM will be used as known information for extracting facade geometries from dense unorganised Flickr photos. In addition, attributive information both in OSM and in Flickr are to be integrated into the 3D building structures. Thus, the resulting city models will not only be appropriate for visualisation tasks but also usable for further analysis, e.g. urban planning, emergency management and simulations for energy consumption. To achieve this, novel intelligent modelling concepts for 3D city models that can cope with the growing needs and requirements arising in the area of geo-information science have to be developed.

However, there are still many challenges that make the 3D reconstruction from VGI data somehow difficult. First of all, the data is heterogeneous in quality, completeness and accuracy. An automated approach may achieve good results in some regions but fails in other regions. Secondly, although there are a large number of images on Flickr for landmark buildings, there are still not enough to obtain robust results for dense image matching, in order to acquire detailed geometries of 3D buildings. Images on other crowdsourcing platforms (e.g. Wikipedia, Weibo and Tweeter) may also be used for 3D reconstruction purposes. The third issue is the quality of the 3D buildings. This can be evaluated by using authority data in some regions where authority data can be made available. But the quality of 3D buildings created in this way is difficult to be controlled from the sources due to the diversity of the personal capabilities of internet contributors in terms of operating with geo-data.

References

Agarwal, S., Snavely, N., Simon, I., Seitz, S., & Szeliski, R. 2011. Building Rome in a Day. *Communications of the ACM, 54*(10): 105–112.

Fan, H., Zipf, A., Fu, Q., & Neis, P. 2014. Quality assessment for building footprints data on Open-StreetMap. *International Journal of Geographical Information Science, 4*: 700–719.

Goodchild, M. F. 2007a. Citizens as sensors: the world of volunteered geography. *GeoJournal, 69*(4): 211–221.

Goodchild, M. F. 2007b. Citizens as Voluntary Sensors: Spatial Data Infrastructure in the World of Web 2.0. *International Journal of Spatial Data Infrastructures Research, 2*: 24–32

Gröger, G., Kolbe, T. H., Czerwinski, A., & Nagel, C. 2008. OpenGIS® City Geography Markup Language (CityGML) Implementation Specification. Available at: http://www.opengeospatial.org/legal/.

Goetz, M., & Zipf, A. 2011. Extending OpenStreetMap to indoor environments: bringing volunteered geographic information to the next level. In: Rumor, M., et al. (Eds.) *Urban and Regional Data Management: Udms Annual 2011,* 28–30 September, Delft, Netherlands: CRC Press, pp. 47–58.

Neis, P., Zielstra, D., & Zipf, A. 2012. The street network evolution of crowdsourced maps: Openstreetmap in Germany 2007–2011. *Future Internet, 4*(1): 1–21.

Neubauer, S., & Zipf, A. 2007. Suggestions for extending the OGC styled layer descriptor (SLD) specification into 3D – towards visualisation rules for 3d city model. In: Urban data management symposium, UDMS 2007, Stuttgart, Germany.

OSM. (2015). Stats – OpenStreetMap Wiki. Retrieved from: http://wiki.openstreetmap.org/wiki/Statistics (18/04/2015).

Over, M., Schilling, A., Neubauer, S., & Zipf, A. 2010. Generating web-based 3D City Models from OpenStreetMap: The current situation in Germany. *Computer Environment and Urban System, 34*(6): 496–507.

Schilling, A., Basanow, J., & Zipf, A. 2007. Vector based mapping of polygons on irregular terrain meshes for Web 3D map services. In: *3rd international conference on web information systems and technologies (WEBIST),* March 2007. Barcelona, Spain.

Snavely, N., Seitz, S., & Szeliski, R. 2006. Photo tourism: exploring photo collections in 3D. *ACM Transactions on Graphics, 25*(3): 835–846.

Snavely, N., Seitz, S., & Szeliski, R. 2008. Modeling the world from internet photo collections. *International Journal of Computer Vision, 80*(2): 189–210.

Uden, M., & Zipf, A. 2012. OpenBuildingModels – Towards a platform for crowdsourcing virtual 3D cities. Progress and New Trends in 3D Geoinformation Sciences, Springer Berlin Heidelberg, pp. 299–314.

Internet links

http://osm-3d.org/home.en.htm
http://openbuildingmodels.uni-hd.de/

Glossary

The Glossary has been compiled by Linda See and Cristina Capineri and revised by Sofia Basiouka. It is a joint effort of TD1202 Mapping the citizen sensors and IC1203 Energic.

Terminology	Definition
Ambient geographic information (AGI)	AGI refers to geographic data that are passively contributed by ctizens, e.g. Twitter, which can then be used for another purpose, e.g. studying human behaviour/patterns in the data. The term first appeared in Stefanidis et al. (2013).
Augmented reality	Augmented reality (AR) refers to the ability to annotate places in the geoweb with multiple layers of information: real-world environment are supplemented (or augmented) by computer-generated sensory inputs such as sound, video, graphics, texts or GPS data. AR is indeterminate, unstable, context dependent and subjective (Graham et. al. 2012: 465; Graham, Zook & Boulton 2013).

Terminology	Definition
Citizen science	Citizen science is the engagement of the general public in scientific research including collection of data, hypothesis generation and the design of research experiments (Bonney et al. 2009a). The Christmas Bird Count is a prime example in which amateur ornithologists are enlisted to conduct a mid-winter census of bird populations. Participants require a fairly high level of skill, and over the years a number of protocols have been established to ensure that the resulting data have high quality. Citizen science is also about education, whereby volunteers learn new skills and gain a better understanding of scientific research (SOCIENTIZE, 2013).
Coded space	Coded spaces are parts of the environment in which code (formalised rules for information into other outputs and representations) or software contributes to the production or enaction of that space. In Zook and Poorthius (2014) coded space emerges when software and space become mutually constituted in this process of transduction: space produces code and code on its turn produces space (Dodge and Kitchin 2005).
Collaborative mapping	Collaborative mapping is the collective creation and annotation of web-based maps and UGC online (MacGillavry, 2003).
Collaboratively contributed geo-spatial information	First appearing in Bishr and Kuhn (2007), this term refers to user generated geospatial information and is a precursor to Volunteered Geographic Information (VGI).
Collective Intelligence	According to the collective intelligence handbook of MIT center, the idea of the collective intelligence is not new, although during the last decade its meaning has changed. The term combines by collective the individual actors such as people, computational agents, organizations and by intelligent, the collective behavior of the group exhibits characteristics. The most representative and simple definition given in the handbook underlines that coolective intelligence is about groups of people and computers, connected by the Internet, collectively doing intelligent things.
Collective Sensing	The concept has been explained by as analysing aggregated anonymised data coming from collective networks such as Twitter, Flickr, Foursquare. It differs from the idea of People as Sensors (people contributing subjective observations) and Citizen Science (exploiting and elevating expertise of citizens and their personal, local experiences) (Resch 2012).

Terminology	Definition
	Collective Sensing is an infrastructure-based approach, which tries to leverage existing ICT networks to generate contextual information. Unlike smartphone-based or specialised web apps, which examine single input data sets, Collective Sensing holistically analyses events and processes in a network. For instance, increased traffic in the mobile phone network might be an indicator for the presence of a dense crowd of people (Reades et al. 2007). This information is generated without having to use a single person's data and their personal details.
Content	Content is used to describe a particular form of information that can be presented to an audience. It is synonymous with some form of creative work. Content is understood to be found in texts, movies, music, books.
Contributed Geographic Information (CGI)	CGI was proposed by Harvey (2013) as geographic information that is contributed by users via an 'opt-out' agreement in contrast to VGI, which is collected via an 'opt-in' agreement. Opt-out agreements are more open-ended and have fewer possibilities for controlling the data collection. This has implications for quality, bias and assessing the fitness-for-use of the data.
Credibility	Flanagin & Metzger (2008) notices credibility's importance as a combination of both trust and expertise and explains that trustworthiness and expertise have both objective and subjective components. With regard to VGI, it is useful to consider credibility according to the degree to which people's spatial or geographic information is unique and situated, and the extent to which its acquisition requires specialized, formal training. If VGI refers to perception based collections, generally carried out by "locals" or "insiders" credibility rests on the extent to which a representative sample of people provide their personal input honestly and accurately. Collective intelligence can function well in many circumstances, yet it is also subject to biases through processes of bandwagon effects and groupthink.
Crowdsourcing	In 2006, the term crowdsourcing was introduced by Howe, which is literally the combination of the words 'crowd' and 'outsourcing'. Thus crowdsourcing refers to the outsourcing of micro-tasks to large volumes of people in order to get a task done that would not have been possible through more traditional means.
Crowdsourcing Geographic Information (CrGI)	This term is similar to VGI and appeared in a paper by Goodchild and Glennon (2010) in the context of disaster management.

Terminology	Definition
Digital footprint	Digital footprint refers to the traces of data that are left by users on digital services. There are two main classifications for digital footprints: passive and active. A passive digital footprint is created when data is collected about an action without any client activation (e.g. a click, the user IP address), whereas active digital footprints are created when personal data is released deliberately by a user for the purpose of sharing information about oneself (e.g. a user being logged into a site when making a post or edit).
Digiplace	Zook and Graham (2007) define digiplace as a heuristic for the subjective mixing of the code, data and material places. It encompasses the situatedness of individuals balanced between the visible and the invisible, the fixed and the fluid, the space of places and the space of flows and the blurring of the lines between material place and the digital representations of place.
Digital divide	The idea of "digital divide" has been introduced long time ago but only recently applied to geographic information by Goodchild (2009) and M.Graham (2014) who discuss the issue that remains largely; the preservation of those fortunate to have access to the Internet—and broadband access in particular and names it a digital divide. They explain that while a growing fraction of citizens in developed countries have such access, it is largely unavailable to the majority of the world's population who live in developing countries either as users or as contributors.
Extreme citizen science	Extreme citizen science can be attributed to Muki Haklay and his team at UCL (Excites). Extreme citizen science refers to the highest level of citizen participation, which includes problem definition, data collection and analysis.
Geocoding	ArcGIS Resource Center gives the definition of geocoding as the process of transforming a description of a location—such as a pair of coordinates, an address, or a name of a place—to a location on the earth's surface. Davis et al. (2003) states geocoding as the process of locating points on the surface of the Earth from alphanumeric addressing data associated to events.
Geocollaboration	Geocollaboration was first defined by MacEachren and Brewer (2004) in which technologies enable collaboration that involves geospatial information in a visually-enabled manner. Geocollaboration can be used for knowledge construction and refinement; design; decision-support; and training and education. The multidisciplinary nature of geocollaboration has been highlighted by Tomaszewski (2010), which brings together human-computer interaction, computer sicnece and psychology.

Terminology	Definition
Geographic citizen science	This refers to citizen science where the location information is a key part of the citizen science activites.
Geographic Information Science	Goodchild (1992) has first introduced the term "Geographic Information Science" for issues raised around the technology that collects and manages spatial data in a technological manner. GIS is designed for storing, managing, analysing and presenting spatial information. As reported by Longley et al. (2005) knowing where something happens turns a situation from a curiosity-driven to a practical problem-solving science.
Georeferencing	Hackeloeer et al (2014) gives the technical definition of georeferencig. According to them, to georeference means to associate something with locations in physical space. The term is commonly used in the geographic information systems field to describe the process of associating a physical map or raster image of a map with spatial locations. Georeferencing may be applied to any kind of object or structure that can be related to a geographical location, such as points of interest, roads, places, bridges, or buildings.
Geotag	Goodchild (2007) supports that a geotag is a standardized code that can be inserted into information in order to note its appropriate geographic location. Geotags have been inserted into many Wikipedia entries, when the contents relate to a specific location on the Earth's surface, and several sites allow such entries to be accessed from maps. Antoniou et al. (2010) gives a technical definition of associating images with the co-ordinates of their capture locations.
GeoWeb or GeoSpatialWeb aka Geographic World Wide Web	The GeoWeb is the merging of spatial information with attribute (i.e. non-spatial information) on the web. The organization of information is by location, allowing spatial searching of the internet. The GeoWeb 2.0 is a system of systems (i.e. GIS clients and servers, service providers, GIS portals, standards and collaboration agreements) (McGuire, 2005).
Hybrid geography	Hybrid geography refers to practice new synthesis in geographic research by challenging existing boundaries and adopting creative connections within geographies (physical and human, critical and analytical, qualitative and quantitative). it suggests the integration of perspectives on space, place, flow, and connection (Sui & DeLyser 2012; Whatmore 2002).
Involuntary geographic information (iVGI)	This refers to VGI that has not been voluntarily provided by the individual, where the data are then used in different types of applications, e.g. behavioural studies. The term first appeared in Fischer (2012) and is similar to AGI.
Map Hacking	Map hacking is the creation of clever solutions with digital maps, e.g. using Gooogle Maps.

Terminology	Definition
Map Mashup	A map mashup is the aggregation of spatial data from different sources to create an online map (Wong, 2008; Sui, 2009).
Motivation	Motivation is a fundamental aspect in crowdsourcing activities and for volunteers manipulating geographic information. Clary & Miller (1986), Clary & Orenstein (1991), Penner & Finkelstein (1998) are among the very first studies which figured out the motivations of volunteerism in general. One important factor is the self – promotion that an individual hopes to gain from the participation in the project (Coleman et al. 2009). Basiouka and Potsiou (2013), developed further research and carried out a Survey over citizens' motivations in crowdsourced governmental projects.
Neogeography	Neogeography literally means 'new geography' and consists of techniques, tools and practices of geography that have been traditionally beyond the scope of professional geographers and GIS practitioners and are now being used by citizens (Turner, 2007). Goodchild and Turner debated the relationship between VGI and neogeography in Wilson and Graham (2013b). In terms of data practices, Goodchild argues the two are identical but as a new paradigm for interaction between people and geography, it provides a much broader perspective. Turner argues that neogeography is not limited to the information but that the intelligence and complexity provided by citizens plays an important role.
Ontologies	Ontology is a branch of metaphysics and it is defined as the science of being but during the last decade GIscience has embraced ontologies as a new research direction towards a better understanding, representation and communication of geographic information. As regards user generated content ontologies and semantic representation of geographic information are relevant in order to be properly understood and used even by non expert (Kavouras and Kokla, 2011).
OpenStreetMap	The most representative example of online mapping which is based on volunteers' participation is OpenStreetMap (OSM); it is a global online, up-to-date, dynamic map having derived from civil society. OSM was launched by Steve Coast in 2004 in United Kingdom and was spread in many countries in relatively short time. As the welcome message says in the main home page of OSM: "The OSM is a free, editable map of the whole world. It is made by people like you".

Terminology	Definition
Participatory sensing	Participatory Sensing is the contribution of georeferenced data via end user devices such as smartphones (Zacharias et al. 2012). Similar to the idea of 'people as sensors' (Goodchild 2007), the definition is more restricted, focussing on input devices, data acquisition and information processing.
Prosumers	Prosumer is a portmanteau formed by contracting either the word professional or, less often, producer with the word consumer. For example, a prosumer grade digital camera is a "cross" between consumer grade and professional grade. In the 1980 book, The Third Wave, futurologist Alvin Toffler coined the term "prosumer" when he predicted that the role of producers and consumers would begin to blur and merge (even though he described it in his book Future Shock from 1970). Toffler envisioned a highly saturated marketplace as mass production of standardized products began to satisfy basic consumer demands. To continue growing profit, businesses would initiate a process of mass customization, that is the mass production of highly customized products.
Public Participation	The international association for public participation underlines the meaning of citizens' participation that affected by or interested in a decision. Public Participation is a general term that can be applied in a variety of projects with the active involvement of citizens and may involve public meetings, surveys, open houses, workshops, polling, citizen's advisory committees and other forms of direct engagement with the public.
Public participation geographic information system	Sieber (2006) was the first who coined that PPGIS pertains to the use of geographic information systems (GIS) to broaden public involvement in policymaking as well as to the value of GIS to promote the goals of non governmental organizations, grassroots groups, and community-based organizations. The idea behind PPGIS is empowerment and inclusion of marginalized populations, who have little voice in the public arena, through geographic technology education and participation. PPGIS uses and produces digital maps, satellite imagery, sketch maps, and many other spatial and visual tools, to change geographic involvement and awareness on a local level.
Public participation in scientific research (PPSR)	This term is synonomous with citizen science and denotes public involvement in all aspects of scientific research (Bonney et al., 2009b).

Terminology	Definition
Quality criteria	Quality is defined as "totality of characteristics of a product that bear on its ability to satisfy stated and implied needs" (ISO 2002, originally in ISO standard 8402). The quality of spatial data is related to the purpose of creation, to its use and to its lineage. Quality aspects that mainly affect VGI data usability are several, among which accuracy and precision of geo-location and of observations, completeness and intelligibility of contents, as well as the reliability of information and the trustworthiness of the data source. The information content validity, known also as intrinsic quality (Bordogna, et al. 2014), depends on a combination of factors such as lineage, positional accuracy, attribute accuracy, logical consistency, completeness (Moellering, 1987). These factors make data fit for a given use. It is therefore dependent on contents' inherent characteristics. According to Ciepluch et al., 2011 literature on OSM contribution apply as quality criteria: density of users contributions, spatial density of points and polygons, types of tags and metadata used, dominant contributors in an area. Various studies have been carried out to evaluate VGI's accuracy. Haklay et al. (2010) did a further research proving that a sensible number of volunteers per area can eliminate the errors in mapping.
Quality techniques	Criscuolo et al. 2013 divided the quality techniques in two main categories. Methods applicable to audit these quality features could include ex-ante combined in different ways guided such as filling of protocols, the use of web-forms with fixed fields, the use of metadata automatically created by the measuring device (i.e. GPS information associated to a picture taken with a smartphone), the use of ontologies and geographical gazetteers, of volunteer contributors; and ex-post techniques (i.e. geo-statistical filters automatically, but also by human experts, in order to detect biases and maintain the data consistency).
Relevance	According to Cosijn and Ingwersen (1999), relevance has become a major area of research in the field of Information Retrieval, despite the fact that the concept relevance is not well understood. Saracevic (1996) previously underlined that nobody has to explain to users of IR systems what relevance is, even if they struggle (sometimes in vain) to find relevant contents. People understand relevance intuitively. However, various studies on relevance view IR as a cognitive interaction between human and computer. Furthermore, there are many types of relevance: system or algorithmic; topical, pertinence; situational; motivational.

Terminology	Definition
Science 2.0	This term refers to the next generation of collaborative science enabled through IT, the internet and mobile devices, which is needed to solve complex, global interdisciplinary problems (Shneiderman, 2008).
Self-knowledge	The term was given by Pedresci (2015) in a different way than the usual philosophical. He supported engagement of people in the creation and use of big data and knowledge, by empowering individuals with self-knowledge and tools so that they may integrate their own digital breadcrumbs into meaningful knowledge and boost self-awareness of own behavioral patterns: health, mobility, consumptions etc.
Spatial Citizenship	Spatial Citizenship as a term combines the spatial awareness in a societal framework by using geo-media and focuses on the information production and communication. Elwood & Mitchell (2013) by making a further step, relate the neogeography to the political formation which leads to the spatial citizenship. The term gains particular importance through the emergence of the Geoweb.
Spatially Explicit Web 2.0 Applications	Spatially Explicit Web 2.0 Applications refer to Web applications that are designed and built according to the Web 2.0 principles (O' Reilly, 2005) but also drive their contributors to use geography and location as a motivational and organisational factor and explicitly urge them to interact with spatial entities. According to Antoniou et. al (2010) spatially explicit applications urge their contributors to interact directly with spatial features (i.e. to capture spatial entities in their photos) while at the same time encourage that photos, and thus the content, be spatially distributed.
Spatially Implicit Web 2.0 Applications.	Refer to Web applications that are designed and built according to the Web 2.0 principles (O' Reilly, 2005) but with no explicit reference to space. Space is just one of the many interesting features that the applications have but spatial information is neither one of the core features nor is it the main motivation of their users. According to Antoniou (2011), these are more socially oriented, as they are aiming to allow people to share their photo albums with no explicit reference to space, and thus are regarded as spatially implicit Web applications.
Swarm Intelligence	This terms appears in Buecheler and Sieg (2011) as a 'buzzword' for paradigms like citizen science, crowdsourcing and open innovation, among others.
Trust of VGI	Trust of VGI may be applied similarly to credibility that first explored by Flanagin & Metgzer (2008) and it is the general term where quality criteria and techniques fall under.

Terminology	Definition
	User-generated content is also associated to the user's trust in the content (a subjective concept) which leads to a connection between content quality and provider's authority. Goodchild and Li (2010) describe three approaches to quality assurance; the crowd-sourcing, social, and geographic approaches respectively: Crowdsourcing approach: How a group of people may converge on the solution to a problem that an individual, even an expert, may be unable to solve. It also refers to the ability of the crowd to converge on the truth (Linus' law). The social approach: It relies on a hierarchy of trusted individuals who act as moderators or gate-keepers. The geographical approach: It relies on a comparison of a purported geographic fact with the broad body of geographic knowledge.
Ubiquitous cartography	This is the study of how maps are created and used at any time or location in way that is much more frequent in space and time than traditional cartography (Gartner et al., 2007).
User generated content	User generated content is the general term which is applied in various fields and includes publishing of content by users in a digital format such as data, videos, blogs, maps, photographs and comments or ratings to other UGC, among others, which as been faciliated by Web 2.0 and mobile technology (Moens et al, 2014).
User Generated Spatial Content (UGSC).	User Generated Spatial Content is an alternative term to VGI as the term "Volunteered" can be misleading regarding the particularities of the generated data and the intentions of the data providers. In a sense "Volunteered" implies a noble and altruistic gesture as if the users donate the data, personal or not, to the world for any use, known or not, to the data provider. This might not be the case in any dataset termed as VGI. Acknowledging the issues raised, a more general but still precisely descriptive term is used: User Generated Spatial Content (UGSC) (Antoniou et al., 2009, Brando and Bucher, 2010).
Vernacular geography	According to Hollenstein & Purves (2010), vernacular geography encloses the sense of place that is reflected in ordinary people's language very different from that captured by standard geographical techniques. In other words, vernacular geography concentrates on how people name and delimit space in everyday use. Waters & Evans (2003), identify the origins of the phonomenon, as people are not geographers but they exist in a geographical world. So, much of human beings' behaviour is dictated by how they perceive areas. Ordnance Survey carried out a project to study the phenomenon which leads people to talk about their world and say where things are within it in an unofficial way.

Terminology	Definition
Volunteer computing	Volunteer computing, according to Anderson and Fedak, (2006) is a concept where citizens spend their spare time to solve scientific problems via the Internet. Volunteer computing is also referred to as peer-to-peer or global computing and is applied in various academis fields such as high-energy physics, molecular biology, medicine, astrophysics, climate study, and other areas. Sarmenta and Satochi (1999) refer to the idea of volunteer computing, where people form very large parallel computing networks by using ubiquitous and easy-to-use technologies such as web browsers and Java.
Volunteer thinking	According to Munyaradzi (2013), volunteer thinking is the act of using one's brain and cognitive skills to solve a problem. Volunteer / Distributed thinking (Quinn and Bederson, 2011) is the harnessing of human brain power on the Internet to solve problems that machines are not suitable to tackle. In volunteer thinking, users are tasked to solve some fundamental problem, reduced to a simplistic level that is easy to comprehend. Using their mental and cognitive abilities, volunteers actively attempt to solve the problem at hand. The types of problems vary in nature e.g. image tagging & classification, proof-reading documents and pattern recognition. The tasks are designed in such a manner that volunteers need no previous experience to solve the problem. Anderson (2006) also developed the Bossa crowd- sourcing framework that manages distributed Web-based volunteer thinking projects.
Volunteered Geographic Information (VGI)	The term VGI was first coined by Goodchild (2007) as the voluntary collection and dissemination of spatial infromation by individuals who often have little training or formal qualifications in spatial sciences.
Web mapping	Web mapping is cartographic representation on the web, with a particular focus on UGC, user-centred design and ubiquitous access (Tsou, 2011).
Wikinomics	Wikinomics embodies the idea of mass collaboration in a business environment. It is based on four principles: a) openness; b) peering; c) sharing; and d) acting globally. The book itself is meant to be a collaborative and living document that everyone can contribute to (Tapscott and Williams, 2006).

Glossary's Bibliography

Anderson, D. P., & Fedak, G. 2006. The computational and storage potential of volunteer computing. In: *Cluster Computing and the Grid, 2006. CCGRID 06.* Sixth IEEE International Symposium. Vol. 1, pp. 73–80.

Antoniou, V., Haklay, M., & Morley, J. 2009. The role of user generated spatial content in mapping agencies. In: *Proceedings of the GIS Research UK 19th Annual Conference.* Durham, pp.251–255.

Antoniou, V., Morley, J., & Haklay, M. 2010. Web 2.0 geotagged photos: Assessing the spatial dimension of the phenomenon. *Geomatica, 64*(1): 99–110.

Antoniou, V. 2011. User generated spatial content: an analysis of the phenomenon and its challenges for mapping agencies. Doctoral dissertation, UCL.

ArcGIS. 2015. ArcGIS resource center. Available at: http://help.arcgis.com/en/arcgisdesktop/10.0/help/index.html#//002500000001000000.htm (Last accessed 18 October 2015).

Basiouka, S., Potsiou, C., 2013 The Volunteered Geographic Information in Cadastre: perspectives and citizens' motivations over potential participation in mapping. GeoJournal, vol. 79(3), 343-355.

Batty, M., Hudson-Smith, A., Milton, R., & Crooks, A. (2010). Map mashups, Web 2.0 and the GIS revolution. Annals of GIS, 16(1), 1–13.

Bishr, M., & Kuhn, W.,2007. Geospatial information bottom-up: A matter of trust and semantics. In: The European information society. Springer Berlin Heidelberg, pp. 365–387.

Bonney, R. 1996. Citizen Science: A Lab Tradition. *Living Bird, 15*(4): 7–15.

Bonney, R., Cooper, C. B., Dickinson, J., Kelling, S., Phillips, T., Rosenberg, K. V., & Shirk, J. 2009a. Citizen science: A developing tool for expanding science knowledge and scientific literacy. BioScience, 59, 977–984.

Bonney, R., Ballard, H., Jordan, R., McCallie, E., Phillips, T., Shirk, J., & Wilderman, C. C. 2009b. Public Participation in Scientific Research: Defining the Field and Assessing its Potential for Informal Science Education (A CAISE Inquiry Group Report). Center for Advancement of Informal Science Education (CAISE), Washington DC.

Bordogna, G., Carrara, P., Criscuolo, L., Pepe, M., & Rampini, A. 2014. A linguistic decision making approach to assess the quality of volunteer geographic information for citizen science. *Information Sciences, 258*: 312–327.

Brando, C., & Bucher, B. 2010. Quality in User Generated Spatial Content: A Matter of Specifications. In: *Proceedings of 13th AGILE International Conference on Geographic Information Science*. Guimarães, Portugal.

Brown, M. 2006. Hacking Google Maps and Google Earth. Indianapolis, IN: Wiley Pub.

Bücheler, T., & Sieg, J. H. 2011. Understanding science 2.0: crowdsourcing and open innovation in the scientific method. *Procedia Computer Science,7*: 327–329.

Burke, J.A., Estrin, D., Hansen, M., Parker, A., Ramanathan, N., Reddy, S., & Srivastava, M. B. 2006. Participatory sensing. Center for Embedded Network Sensing.

Ciepluch, B., Mooney, P., Jacob R., Zheng, J., & Winstanely, A. C. 2011. Assessing the Quality of Open Spatial Data for Mobile Location – based Services Research and Applications. *Archives of Photogrammetry, Cartography and Remote Sensing*, (22): 105–116.

Coleman, D. J., Georgiadou, Y., & Labonte, J. 2009. Volunteered geographic information: The nature and motivation of produsers. *International Journal of Spatial Data Infrastructures Research, 4*(1): 332–358.

Cosijn, E., & Ingwersen, P. 2000. Dimensions of relevance. *Information Processing & Management, 36*(4): 533–550.

Clary, E. G., & Miller, J. 1986. Socialization and situational influences on sustained altruism. *Child Development*: 1358–1369.

Clary, E. G., & Orenstein, L. 1991. The amount and effectiveness of Help: The relationship of motives and abilities to helping behavior. *Personality and Social Psychology Bulletin, 17*(1): 58–64.

Criscuolo, L., Pepe, M., Seppi, R., Bordogna, G., Carrara, P., & Zucca, F. 2013. Alpine glaciology: An historical collaboration between volunteers and scientists and the challenge presented by an integrated approach. *ISPRS International Journal of Geo-Information, 2*(3): 680–703.

Davis, C. A., Fonseca, F. T., & Borges, K. A. 2003. A Flexible Addressing System for Approximate Geocoding. *GeoInfo*.

Elwood, S. 2008. Volunteered geographic information: key questions, concepts and methods to guide emerging research and practice. *GeoJournal, 72*(3): 133–135.

Elwood, S., 2010 Geographic Information Science: Visualization, Visual Methods, and the Geoweb. *Progress in Human Geography, 35*(3): 401–408.

Elwood, S., & Mitchell, K. 2013. Another Politics Is Possible: Neogeographies, Visual Spatial Tactics, and Political Formation 1.Cartographica: *The International Journal for Geographic Information and Geovisualization, 48*(4): 275–292.

Estellés-Arolas, E., & González-Ladrón-de-Guevara, F. 2012. Towards an integrated crowdsourcing definition. *Journal of Information science, 38*(2): 189–200.

Fischer, F. 2012. VGI as Big Data. A New but Delicate Geographic Data-Source. GeoInformatics. Available at: http://www.academia.edu/1505065/VGI_as_Big_Data._A_New_but_Delicate_Geographic_Data-Source (Accessed May 2014).

Flanagin, A., J., & Metzger, M. 2008. The credibility of volunteered geographic information. *GeoJournal, 72*(3): 137–148.

Gartner, G., Bennett, D., & Morita, T. 2007. Toward ubiquitous cartography. *Cartography and Geographic Information Science, 34*: 247–257.

Erle, S., Gibson, R., & Walsh, J. n.d. *Mapping hacks.*

Goodchild, M. F. 1992. Geographical information science. *International journal of geographical information systems, 6*(1): 31–45.

Goodchild, M. F. 2007. Citizens as sensors: the world of volunteered geography. *GeoJournal, 69*: 211–221.

Goodchild, M. F. 2008. Commentary: whither VGI? *GeoJournal, 72*(3–4): 239–244.

Goodchild, M. F. 2009. Virtual geographic environments as collective constructions. *Virtual geographic environments*: 15–24.

Goodchild, M. F., & Glennon, J. A. 2010. Crowdsourcing geographic information for disaster response: a research frontier. *International Journal of Digital Earth, 3*(3): 231–241.

Goodchild, M., F., & Li, L. 2012 Assuring the quality of volunteered geographic information. *Spatial statistics, 1*: 110–120.

Graham, M., Hogan, B., Straumann, R. K., & Medhat, A. 2014. Uneven geographies of user-generated information: patterns of increasing informational poverty. *Annals of the Association of American Geographers, 104*(4): 746–764.

Sui, D., Elwood, S., & Goodchild, M. 2013. *Crowdsourcing geographic knowledge*. Dordrecht: Springer.

Graham, M., Zook, M., & Boulton, A. 2012 Augmented reality in urban places: contested content and the duplicity of code. *Transactions of the Institute of British Geographers.* DOI: http://dx.doi.org/10.1111/j.1475-5661.2012.00539.x

Hackeloeer, A., Klasing, K., Krisp, J. M., & Meng, L. 2014. Georeferencing: a review of methods and applications. *Annals of GIS, 20*(1): 61–69.

Haklay, M., Basiouka, S., Antoniou, V., & Ather, A. 2010. How many volunteers does it take to map an area well? The Validity of Linus' Law to Volunteered Geographic Information. *The Cartographic Journal, 47*(4): 315–322(8).

Haklay, M. 2010. Geographical citizen science–Clash of cultures and new opportunities. Available at: http://web.ornl.gov/sci/gist/workshops/2010/papers/Haklay.pdf (Last accessed 16 October 2015).

Haklay, M. 2013. Citizen science and volunteered geographic information: Overview and typology of participation. In: *Crowdsourcing Geographic Knowledge*. Springer Netherlands, pp. 105–122.

Harvey, F. 2013. To volunteer or to contribute locational information? Towards truth in labeling for crowdsourced geographic information. In: *Crowdsourcing Geographic Knowledge*. Springer, pp. 31–42.

Herring, C. 1994. An architecture of cyberspace: Spatialization of the internet. US: Army Construction Engineering Research Laboratory.

Hollenstein, L., & Purves, R. 2010. Exploring place through user-generated content: Using Flickr tags to describe city cores. *Journal of Spatial Information Science*, (1): 21–48.

Howe, J., 2006. The rise of crowdsourcing. *Wired magazine, 14*(6): 1–4.

Hub, M., Valenta, Z., & Visek. O. 2008. Heuristic Evaluation of Geoweb. *Ekonomie a Management, 11*(2): 127–131.

International Association for public participation. 2016. Home Page. Available at: http://www.iap2.org/. (Last accessed 17 April 2016).

Kavouras, M., & Kokla, M. 2011. Geographic ontologies and society. *The SAGE Handbook of GIS and Society*, 46.

Keßler, C., Janowicz, K., & Bishr, M. 2009. An agenda for the next generation gazetteer: Geographic information contribution and retrieval. In:*Proceedings of the 17th ACM SIGSPATIAL international conference on advances in Geographic Information Systems*. ACM, pp. 91–100.

Kitchin, R., & Dodge, M. 2005. Code and the transduction of space. *Annals of the Association of American geographers, 95*(1): 162–180.

Krumm, J., Davies, N., & Narayanaswami. C. 2008. User-Generated Content. *IEEE Pervasive Computing, 7*(4): 10–11.

Lastowka, F. G., & Hunter, D. 2004. The laws of the virtual worlds. *California Law Review*: 1–73

Longley, P., Goodchild M., Maguire, D., & Rhind, D. 2005 *Geographic Information Systems and Science*, John Wiley and sons Ltd, USA.

MacEachren, A. M., & Brewer, I. 2004. Developing a conceptual framework for visually-enabled geocollaboration. *International Journal of Geographical Information Science, 18*(1): 1–34.

MacGillavry, E. 2003. Collaborative Mapping. Webmapper. Available at: http://www.webmapper.net/carto2003/ (Last accessed June 2013).

Maguire, D. 2005. GeoWeb 2.0: implications for ESDI. In: *Proceedings of the 12th EC-GIGIS Workshop*.

MIT center of collective intelligence. 2016. The Collective Intelligence Handbook. Available at: http://cci.mit.edu/CIchapterlinks.html (Last accessed 17 April 2016).

Moellering, H. 1987. A Draft Proposed Standard for Digital Cartographic Data; National Committee for Digital Cartographic Data Standards: Columbus, OH, USA.

Moens, M. F., Li, J., & Chua, T. S. 2014. Mining user generated content. CRC Press.

Munyaradzi, N. 2013. Transcription of the Bleek and Lloyd Collection using the Bossa Volunteer Thinking Framework. MSc Thesis, University of Cape Town.

OpenStreetMap. 2015. Home Page. Available at: https://www.openstreetmap.org (Last accessed 18 October 2015).

Ordnance Survey. 2016 Research: Vernacular Geography. Available at: http://webarchive.nationalarchives.gov.uk/20100402191611/ordnancesurvey.co.uk/oswebsite/partnerships/research/research/vernacular.html (Last accessed 17 April 2016).

Pedreschi, D. 2015 Big Data Analytics & Social Mining for Science and Society. Available at: http://www.unece.org/fileadmin/DAM/stats/documents/ece/ces/2015/presentations/DinoPedreschi_final.pdf (Last accessed 17 October 2015).

Penner, L. A., & Finkelstein, M. A. 1998. Dispositional and structural determinants of volunteerism. *Journal of Personality and Social Psychology, 74*(2): 525–537.

Quinn, A. J., & Bederson, B. B. 2011. Human computation: a survey and taxonomy of a growing field. In: *Proceedings of the SIGCHI conference on human factors in computing systems*, ACM, pp. 1403–1412.

Reades, J., Calabrese, F., Sevtsuk, A., & Ratti, C. 2007. Cellular census: Explorations in urban data collection. *Pervasive Computing, IEEE, 6*(3): 30–38.

Resch, B. 2013. People as sensors and collective sensing-contextual observations complementing geo-sensor network measurements. In: *Progress in Location-Based Services*. Springer Berlin Heidelberg, pp. 391–406.

Saracevic, T. 1996. Relevance reconsidered '96. In: Ingwersen, P., & Pors, N. O. (Eds.) *Information science: integration in perspective*. Copenhagen, Denmark: Royal School of Library and Information Science.

Sarmenta, L. F., & Hirano, S. 1999. Bayanihan: Building and studying web-based volunteer computing systems using Java. *Future Generation Computer Systems, 15*(5): 675–686.

Scharl, A. 2007. Towards the Geospatial Web: Media Platforms for Managing Geotagged Knowledge Repositories. In: *The geospatial web*. Springer London. pp. 3–14.

Schuurman, N. 2009. The New Brave NewWorld: Geography, GIS, and the Emergence of Ubiquitous Mapping and Data. *Environment and Planning D: Society and Space, 27*: 571–580.

Sczott, R. 2006. What is Neogeography Anyway? Platial. Available at: http://platial.typepad.com/news/2006/05/what_is_neogeog.html (Accessed 15 December 2015).

Shneiderman, B. 2008. Copernican challenges face those who suggest that collaboration, not computation are the driving energy for socio-technical systems that characterize Web 2.0. *Science, 319*: 1349–1350.

Sieber, R. 2006. Public participation geographic information systems: A literature review and framework. *Annals of the Association of American Geographers, 96*(3): 491–507.

Sieber, R. 2007. Geoweb for social change. Available at: http://www.ncgia.ucsb.edu/projects/vgi/supp.html (Last accessed 12 September 2012).

Silvertown, J. 2009. A new dawn for citizen science. *Trends in ecology & evolution, 24*(9): 467–471.

Sui, D., & DeLyser, D. 2012. Crossing the qualitative-quantitative chasm I Hybrid geographies, the spatial turn, and volunteered geographic information (VGI). *Progress in Human Geography, 36*(1): 111–124.

SOCIENTIZE. 2013. Green Paper on Citizen Science. Citizen Science for Europe: Towards a Better Society of Empowered Citizens and Enhanced Research. GreenPaper. Available at: ec.europa.eu/newsroom/dae/document.cfm?doc_id=4121 (Last accessed 15 September 2012).

Stefanidis, A., Crooks, A., & Radzikowski, J. 2013. Harvesting Ambient Geospatial Information from Social Media Feeds. *GeoJournal, 78*(2): 319–338.

Tapscott, D., & Williams. A. D. 2006. *Wikinomics: How Mass Collaboration Changes Everything.* New York: Portfolio.

Toffler, A. 1980. The third wave: The classic study of tomorrow. New York, NY: Bantam.

Tomaszewski, B. 2010. Geocollaboration. *Encyclopedia of Geography.* SAGE.

Trumbull, D. J., Bonney, R., Bascom, D., & Cabral, A. 2000. Thinking scientifically during participation in a citizen-science project. *Science education, 84*(2): 265–275.

Tsou, M. H. 2011. Revisiting Web Cartography in the United States: the Rise of User-Centered Design. *Cartography and Geographic Information Science, 38*: 250–257.

Waters, T., & Evans, A. 2003. Tools for web-based GIS mapping of a "fuzzy" vernacular geography. In: *Proceedings of the 7th International Conference on GeoComputation.*

Wikipedia. 2015. Augmented Reality. Available at: https://en.wikipedia.org/wiki/Augmented_reality (Last accessed 16 October 2015).

Wikipedia. 2015b. Collaborative mapping Available at: https://en.wikipedia.org/wiki/Collaborative_mapping [Last accessed 16 October 2015]

Wikipedia. 2015c. Digital Footprint. Available at: https://en.wikipedia.org/wiki/Digital_footprint [Last accessed 16 October 2015]

Wikipedia. 2015d. Enabling technology. Available at: https://en.wikipedia.org/wiki/Enabling_technology [Last accessed 16 October 2015]

Wilson, M.W., & Graham, M. 2013b. Neogeography and Volunteered Geographic Information: A Conversation with Michael Goodchild and Andrew Turner. *Environment and Planning A, 45*(1): 10–18.

Whatmore, S. 2002. *Hybrid geographies: Natures cultures spaces.* Sage.

Zacharias, V., de Melo Borges, J. and Abecker, A. 2012. Anforderungen an eine Technologie für Participatory Sensing Anwendungen. In: Proceedings of the geoinformatik 2012 conference, 28–30 March 2012. Braunschweig, Germany, pp. 311–318.

Zook, M., & Graham, M. 2007. The creative reconstruction of the Internet: Google and the privatization of cyberspace and DigiPlace. *Geoforum*, 38: 1322–1343.

Zook, M., & Poorthuis, A. 2014. Offline Brews and Online Views: Exploring the geography of beer Tweets. In: *The Geography of Beer*. Springer Netherlands, pp. 201–209.

www.ingramcontent.com/pod-product-compliance
Lightning Source LLC
Chambersburg PA
CBHW042312210326
41598CB00042B/7365